T0190098

Dynamic Systems with Time Delays: Stability and Control

Ju H. Park · Tae H. Lee ·
Yajuan Liu · Jun Chen

Dynamic Systems with Time Delays: Stability and Control

Ju H. Park
Department of Electrical Engineering
Yeungnam University
Gyeongsan-si, Korea (Republic of)

Tae H. Lee
Division of Electronic Engineering
Chonbuk National University
Jeonju, Korea (Republic of)

Yajuan Liu
Control and Computer Engineering
North China Electric Power University
Beijing, China

Jun Chen
School of Electrical Engineering
and Automation
Jiangsu Normal University
Xuzhou, China

ISBN 978-981-13-9256-6 ISBN 978-981-13-9254-2 (eBook)
https://doi.org/10.1007/978-981-13-9254-2

This Springer imprint is published by the registered company Springer Nature Singapore Pte Ltd.
The registered company address is: 152 Beach Road, #21-01/04 Gateway East, Singapore 189721, Singapore

To our parents and loved ones.

Preface

Aim of the Book

In the past 200 years of scientific history, research on various topics on dynamic systems has led to tremendous advances. We often get asked the question, "What kind of dynamic systems are most interesting from a control engineering point of view with long scientific history?". We believe that there are many scholars, who answer this question as "**Time-Delay Systems**". The reason for the answer is simply that, because this topic is one of the hottest research areas in control engineering ever as the system has complex and special dynamic characteristics.

With this motivation above, the authors have argued the need to write a new book that will introduce readers to the latest research trends and techniques on the stability and control problems of dynamic systems with time delays.

It should be noted that the common basic mathematical knowledge provided in many books or survey papers on Time-Delay Systems will not be dealt with in order to avoid redundancy as much as possible and such review is also not the main purpose of this book.

The aim of this book is

1. *To provide an introduction to recent research on the various topics on stability and control of dynamical systems and networks with time delays, and*
2. *To present novel techniques on stability analysis and on synthesis frameworks of stabilizing controllers, filters, and state estimators for dynamic systems and networks with time delays.*

Meanwhile, over the past 20 years, the rapid development of IT technology has led to the miniaturization and advancement of the hardware components of the control system including sensors and actuators. Of course, software technology has also become increasingly intelligent. In addition, traditional control systems have become increasingly networked. Large and complex systems that have not been addressed in the past have become interconnected systems called complex networks through wired or wireless communication networks. The flow of signals within the system was also faster in this environment. Although the control system is intelligent, advanced, and miniaturized in every perspective, there is a key element that cannot be ignored in the control system. It is just a time delay which is caused by various factors in the system.

As systems become more sophisticated, the possibility of signal congestion increases as more information is processed. Thus, interest in dynamic systems or networks with time delays has increased in this developing situation.

To show the progress of research on hot topics from various perspectives, in this book, several dynamic system models including time delays in the continuous or discrete-time domain such as Lur'e systems, fuzzy systems, chaotic systems, neural networks, and complex networks, have been considered for stability and control problems.

From a detailed view of the contents of this book, it collects some research works carried out recently by the authors in this field. It covers the development of some mathematical inequalities and lemmas, stability analysis for linear systems in continuous and discrete domain and neural networks, $\mathscr{H}_\infty$ control of neural networks, event-triggered control of linear systems, dynamic output control of complex networks, and reliable sampled-data control for chaotic Lur's systems, $\mathscr{H}_\infty$ filtering of discrete systems, dissipative filtering of fuzzy modeled systems, state estimation of genetic regulatory networks, and, secure communication of chaotic systems.

Here, it is noted that some preliminary results of the material presented in this book have been published in eminent international journals in recent years by the authors.

This book attempts to consolidate the previous results and extend them to have new results on related topics. In this respect, the book is likely to be of use for the wide and heterogeneous group of science and engineering senior students, graduate students, and researchers who are focusing on control theory and corresponding real applications for various dynamic systems as a monograph and a place to look for basic information and recent trend and direction on research in a tutorial style as well as advanced materials developed most recently.

In order to understand fully the contents covered by the book, the background required of the readers is basic and advanced knowledge of linear algebra, calculus, control system theory, Lyapunov stability theory, convex optimization theory, and simulation techniques.

Outline of the Book

This book is organized as follows:

Chapter 1 gives an overall introduction and motivation of the work presented in this book.

Chapter 2 presents a brief review of basic knowledge and the research direction on stability and stabilization of time-delay systems in the last few decades. In particular, this chapter reviews the latest techniques widely used in stability analysis of the systems.

Chapter 3 is devoted to the investigation of integral inequalities. Some of the well-known integral inequalities are firstly recalled such as the Jensen, Wirtinger-based, auxiliary-function-based, and Bessel-Legendre integral inequalities and conventional free matrix based integral inequalities. Then, improved general free-matrix-based inequalities are proposed as a key tool of stability analysis of time-delay systems.

Chapter 4 focuses on the investigation of summation inequalities for analysis of time-delay systems in discrete-time domain.

In other words, we look at the discrete-time domain version of the mathematical inequalities discussed in Chap. 3.

Chapter 5 is concerned with the stability problem for linear continuous- and discrete-time systems with time-varying delays by using the Lyapunov–Krasovskii functional method. Integral and summation inequalities developed in Chaps. 3 and 4 are applied to this problems.

Chapter 6 describes a stability problem of neural networks with a time-varying delay as an extension of the topics dealt with Chaps. 3–5. For both continuous- and discrete-time cases, new L–K functionals are constructed in which the involved vectors are augmented by adding state- and delay-related vectors for analyzing the problem.

Chapter 7 introduces an $\mathscr{H}_\infty$ control problem for neural networks with time-varying delays and external disturbance as the first issue in Part III: Control Problems of Dynamics Systems with Time-Delays.

Chapter 8 is devoted to the topic of the event-triggered $\mathscr{H}_\infty$ control for networked control systems. It is noted that the event-triggered control scheme is an effective method under limited network resources and for saving control cost.

Chapter 9 introduces the design procedure of dynamic output feedback controllers, unlike the static feedback controller which is widely used in control systems. This controller design problem is examined through the synchronization problem of dynamical networks.

Chapter 10 discuss how to control the system in the event of failure of an actuator that is directly related to a real issue of practical control systems. This problem is applied to one interesting chaotic Lur'e system. As in Chap. 9, the synchronization problem is addressed.

Chapter 11 is about to design an $\mathscr{H}_\infty$ filter for guaranteeing stability and performance of discrete nonlinear systems with disturbances and time-vary delays. The filter in case of linear systems as a special case is also investigated.

Chapter 12 concerns the dissipative filter for delayed nonlinear interconnected systems via Takagi–Sugeno fuzzy modelling as a series of filter design problems in Chap. 11.

Chapter 13 considers a design problem of the state estimator for genetic regulatory networks with three types of delays as an application of control problems.

Chapter 14 considers a real practical application example called secure communication of synchronization principle using chaotic signals. Naturally, a dynamic model in which the propagation delay is considered in signal transmission processes is used in this applicable problem.

Gyeongsan, Korea (Republic of) Ju H. Park
Jeonju, Korea (Republic of) Tae H. Lee
Xuzhou, China Jun Chen
Beijing, China Yajuan Liu
March 2019

Acknowledgements

The material presented in this book has been the outcome based on several years of research activity on time-delay systems and their applications to dynamic systems and networks by the authors. The authors have been studying control theories for a long time, and have been carrying out related research with great interest, especially on time-delay systems.

Three authors, Dr. T. H. Lee, Dr. Y. Liu, and Dr. J. Chen, have been working as postdoctoral and visiting researchers in the first author's laboratory named "Advanced Nonlinear Dynamics and Control" in Yeungnam University, Republic of Korea since 2015, and have jointly announced various research results on the control problems of dynamic systems and networks with time delays.

In early 2018, although the authors are well aware that many research papers and books related to time delay systems have already been published to date, the authors have concluded that it is worthwhile to introduce the latest research results related to the time delay systems being performed by the authors' team in a new perspective to many related researchers. This became the direct motivation for writing this professional book.

Most of the content presented in this book is based on the authors' previous research achievements, including more extensive content and entirely new research results.

In fact, such developments are not solely ours, but because we get inspiration and ideas from the work of many friends, seniors, and eminent scholars. Their scientific dedication and especially their influence on our research lead to writing this book in 2018. In this regard, it is a great pleasure to be able to thank many people, who have contributed to creating the material in the text. Without their role, help, and encouragement, this book would not have been possible.

First of all, the authors are grateful to their family and former and current graduate students and postdoctoral members in their research laboratories for long-term understanding and strong support in academic and personal.

The authors would like to express their heartful gratitude to Prof. Shengyuan Xu, Nanjing University of Science and Technology, PR China, Prof. Guang-Hong Yang, Northeastern University, PR China, Prof. Jinde Cao, Southeast University, PR China, Prof. Shouming Zhong, University of Electronic Science and Technology of China, PR China, Prof. Dong Yue, Nanjing University of Post and Telecommunications, PR China, Prof. P. Balasubramaniam, Gandhigram Rural Institute—Deemed University, India, and Prof. Ho-Youl Jung, Yeungnam University, Republic of Korea, who have dedicated authors' research team and members for a long time.

Personally, Ju H. Park gives special thanks to professors in Yeungnam University, Prof. Yong-Wan Park, Prof. Chang-Hyeon Park, Prof. Suk-Gyu Lee, and Prof. Soon-Hak Kwon, for their help and care during staying period of his graduate and postdoctoral members in Yeungnam University.

Also, he would like to give other special thanks to Prof. Qiankun Song, Chongqing Jiaotong University, PR China, Prof. Chuandong Li, Southwest University, PR China, Prof. Xinsong Yang, Chongqing Normal University who have always warmly welcomed him every time he visits their universities as an academic companion and friend over the past few years.

Over the past decade, the research team of the first author, J. H. Park, has been actively participating in various research works on dynamic systems and networks including time-delay systems, and thanking his members for bringing out good academic results, he would like to say that there is a uniform presence in his heart to the members: they are Dr. O. M. Kwon, Dr. S. M. Lee, Dr. Z. G. Wu, Dr. S. Lakshmanan, Dr. B. Song, Dr. H. Zhao, Dr. H. Shen, Dr. X. Liu, Dr. H. B. Zeng, Dr. J. Xia, Dr. G. Arthi, Dr. K. Mathiyalagan, Dr. X. Z. Jin, Dr. X. H. Chang, Dr. Z. Q. Zhang, Dr. B. C. Zheng, Dr. D. Ye, Dr. J. Zhou, Dr. H. Bao, Dr. M. Shen, Dr. X. Song, Dr. T. H. Lee, Dr. L. Hao, Dr. J. Li, Dr. F. Zhou, Dr. Z. Yan, Dr. Y. Liu, Dr. X. Chen, Dr. J. Cheng, Dr. Y. Long, Dr. X. Feng, Dr. Y. Wei, Dr. Y. Kao, Dr. X. Xie, Dr. Z. M. Li, Dr. J. Chen, Dr. X. Yao, Dr. L. Zhou, Dr. X. Xiao, Dr. G. Liu, Dr. H. Zhang. Dr. J. Liu, Dr. Z. Tang, Dr. C. Ge, Dr. S. Ding, Dr. W. Qi, Dr. W. Zhao, Dr. T. Jiao, Dr. Q. Kong, Dr. J. Xiong, Dr. J. Zhao, Dr. L. Wu, Dr. R. Zhang, Dr. D. Zeng, Dr. R. Rakkiyappan, Dr. D. H. Ji, Dr. S. J. Choi, Dr. J. H. Koo, Dr. M. J. Park, and Dr. X. Wang.

Although Ju H. Park cannot mention all names here, he would like to thank his friends and acquaintances for their encouragement to him. He finally dedicates this book to his grandmother, K. S. Kang, who passed away in December 2018.

Tae H. Lee would like to convey his special thanks to his wife Mijin and daughter Hyemin for their solicitude and support for every single second. He is also sincerely obliged to Prof. Hieu M. Trinh and Dr. Van Thanh Huynh at Deakin University, Australia, Prof. Kil To Chong at Chonbuk National University, Prof. Sang-Moon Lee, Oh-Min Kwon, and Myeong-Jin Park for their encouragement and valuable advice on his research.

Jun Chen would like to acknowledge Prof. Shengyuan Xu and Prof. Baoyong Zhang at Nanjing University of Science and Technology, PR China. Heartfelt thanks to his parents, Zhengya Chen and Shuying Chen. Special thanks to his wife, Li Liu, and son, Pengyang Chen. He is also grateful to Jiangsu Normal University and his colleagues.

Yajuan Liu would like to thanks Prof. Sang-Moon Lee and Prof. Ju H. Park during her stay in Korea for 6 years. And she would like to give her thanks to the researchers and students in Prof. Park's lab, especially Dr. Ze Tang's kind help for these years.

Last, but not least, the authors would like to express our gratitude to the Book Editors Ms. Annie Kang, Mr. Smith Chae and Editorial Assistant Ms. Eugene Hong at the Springer Nature Korea and the Book manager, Muruga Prashanth in the division of Scientific Publishing Services at the Springer-Nature for their professional and powerful handling of this book. Without their editorial comments and detail examination, the publication of the book would not have gone so smoothly.

Finally, the book was supported in part by Basic Science Research Programs through the National Research Foundation of Korea (NRF) funded by the Ministry of Education (Grant number NRF-2017R1A2B2004671), in part by the National Research Foundation of Korea (NRF) grant funded by the Ministry of Science and ICT (Grant number NRF-2018R1C1B5036886), and in part by the National Natural Science Foundation of China (NSFC) under Grant 61773186.

About This Book

Dynamic Systems with Time Delays: Stability and Control presents up-to-date research developments and novel methodologies to solve various stability and control problems of dynamic systems with time delays. First, it provides the new introduction of integral and summation inequalities for stability analysis of nominal time-delay systems in continuous and discrete time domain and presents corresponding stability conditions for the nominal system and an applicable nonlinear system. Next, it investigates several control problems for dynamic systems with delays including

- $\mathscr{H}_\infty$ control problem;
- Event-triggered control problem;
- Dynamic output feedback control problem;
- Reliable sampled-data control problem.

Finally, some application topics covering filtering problem, state estimation, and synchronization are considered. It can be a valuable resource and guide for graduate students, researchers, scientists, and engineers in the field of system sciences and control communities.

Contents

About the Authors

Ju H. Park received the Ph.D. degree in Electronics and Electrical Engineering from Pohang University of Science and Technology (POSTECH), Pohang, Republic of Korea, in 1997. From May 1997 to February 2000, he was a Research Associate in Engineering Research Center-Automation Research Center (ERC-ARC), POSTECH. He joined Yeungnam University, Kyongsan, Republic of Korea, in March 2000, where he is currently the Chuma Chair Professor. From 2006 to 2007, he was a Visiting Professor in the Department of Mechanical Engineering, Georgia Institute of Technology. His research interests include robust control and filtering, neural/complex networks, fuzzy systems, multi-agent systems, and chaotic systems. He has published a number of papers in these areas. Professor Park severs as an Editor of *Int. J. Control, Automation and Systems*. He is also a Subject Editor/Associate Editor/Editorial Board member for several international journals, including *IEEE Transactions on Fuzzy Systems, IET Control Theory and Applications, Applied Mathematics and Computation, Journal of the Franklin Institute, Nonlinear Dynamics, Journal of Applied Mathematics and Computing, Cogent Engineering* and so on. He has been a recipient of Highly Cited Researcher Award listed by Clarivate Analytics (formerly, Thomson Reuters) since 2015 and a fellow of the Korean Academy of Science and Technology (KAST).

Tae H. Lee received the B.S., M.S., and Ph.D. degrees in electrical engineering from Yeungnam University, Kyongsan, Republic of Korea, in 2009, 2011, and 2015, respectively. He was a postdoctoral researcher in the same university from 2015 to 2017 and a Alfred Deakin Postdoctoral Research Fellow at Institute for Intelligent Systems Research and Innovation, Deakin University, Australia in 2017. He began with Chonbuk National University, Jeonju-si, Republic of Korea in September 2017, where he is currently an Assistant Professor. His research interests are in the field of complex dynamical networks, sampled-data control systems, chaotic/biological systems, and networked-control systems.

Yajuan Liu received the B.S. degree in Mathematics and Applied Mathematics from Shanxi Normal University, Linfen, PR China, in 2010, M.S. degree in Applied

Mathematics, University of Science and Technology Beijing, Beijing, China, in 2012, and Ph.D. degree at the Division of Electronic Engineering in Daegu University, Deagu, Republic of Korea, in 2015. From 2015 to 2018, she was a postdoctoral research fellow at the Department of Electrical Engineering at Yeungnam University, Kyongsan, Republic of Korea. She is now working as a Lecture at the School of Control and Computer Engineering, North China Electric Power University, Beijing, China. Her research focus is control of dynamic systems including neural networks and complex systems.

Jun Chen received the Ph.D. degree in Control Theory and Control Engineering from Nanjing University of Science and Technology, Nanjing, China, in 2017. From August 2017 to July 2018, he was a postdoctoral researcher at the Department of Electrical Engineering, Yeungnam University, Republic of Korea. He has been with School of Electrical Engineering and Automation, Jiangsu Normal University from August 2004, where he is currently an associate professor. His current research interests include Takagi-Sugeno systems, time-delay systems and neural networks.

Notations

$\mathbb{R}^n$	An n-dimensional Euclidean spaces
$\mathbb{R}^{m\times n}$	The set of $m \times n$ real matrices
$\mathbb{N}$ or $\mathbb{N}^+$	The sets of nonnegative or positive integers
$\mathbb{S}^n$ or $\mathbb{S}^n_+$	The sets of $n \times n$ symmetric or symmetric positive definite matrices
$\mathbb{D}^n_+$	The sets of $n \times n$ positive diagonal matrices
$\Re(\cdot)$	Real part of a complex number
$X > 0$	X is a symmetric positive definite matrix
$X < 0$	X is a symmetric negative definite matrix
$X \geq 0$	X is a symmetric positive semi-definite matrix
$X \leq 0$	X is a symmetric negative semi-definite matrix
I or I_n	The identity matrix with appropriate or $n \times n$ dimensions
0 or $0_{n\times m}$	The zero matrix with appropriate or $n \times m$ dimensions
$*$	The elements below the main diagonal of a symmetric matrix
$\|\cdot\|$	The Euclidean vector norm and the induced matrix norm
$\mathbb{E}\{x\}$	The expectation of the stochastic variable x
$\Pr\{x\}$	The occurrence probability of the event x
$\mathrm{diag}\{\cdots\}$	A block diagonal matrix
$\mathrm{Sym}\{X\}$	$X + X^T$
^{T}	Superscript; transpose transformation
$\forall$	For all
$X_{[f(t)]}$	The elements of the matrix X include the values of $f(t)$
$\binom{k}{l}$	The binomial coefficients given by $\frac{k!}{(k-l)!l!}$
$\mathscr{L}\{\cdots\}$	A linear combination of members
col	Column vector
min	Minimize
$X^\perp$	A basis for the null-space of X
$\mathscr{L}_p[a,b]$	The L_p function space on the interval $[a,b]$

Acronyms

AEDE	Advanced Functional Differential Equation
B–L	Bessel–Legendre
CDN	Complex Dynamical Network
CE	Characteristic Equation
CLS	Chaotic Lur'e System
DNA	Deoxyribo Nucleic Acid
FDE	Functional Differential Equation
FMB	Free-Matrix-Based
GFMB	General Free-Matrix-Based
GRN	Genetic Regulatory Network
L–K	Lyapunov–Krasovskii
LKF	Lyapunov–Krosovskii Functional
LMI	Linear Matrix Inequality
LT	Laplace Transform
MAUB	Maximal Allowable Upper Bound
NCS	Networked Control System
NDE	Neutral Functional Differential Equation
NN	Neural Network
NV	Number of Decision Variable
ODE	Ordinary Differential Equation
PID	Proportional-Integral-Derivative
RCL	Reciprocally Convex Lemma
RDE	Retarded Functional Differential Equation
RNA	Ribo Nucleic Acid
TDS	Time-Delay System
T–S	Tagaki–Sugeno
WII	Wirtinger-based Integral Inequality
ZOH	Zero-Order-Hold

List of Figures

List of Tables

Part I
Introductory Remarks and Preliminaries

Chapter 1
Introduction

Many things that happen around us can be called systems in view of causality. From centuries ago, we began to elucidate the phenomena that take place in these systems around us. There are several factors that have a decisive influence on the dynamic characteristics of the systems. Among such factors, an important element that exists anywhere in the world is the time delay. In the view of science history, the existence of this time lag since the 18th century has become known, and it has become clear that in many systems its existence cannot be absolutely negligible in dynamic properties. Since the industrial revolution of the nineteenth century, as the rapid development of various engineering including mechanical engineering, more accurate system analysis was required. Of course, it has already been shown that the dynamic effects of these time delays are also widespread in biological and sociological phenomena. Since the time delay may become long especially in the hardware and signal processing of the system, there is a big difference from the actual phenomenon when the system analysis is performed with the mathematical model ignoring the time delay. It is also found that when the system naturally has a time delay, the stability and performance of the system gradually deteriorates. We also found that when the system naturally has a time delay, the stability and performance of the system gradually deteriorates as the magnitude of the delay increases.

In this respect, over the last 30 years, tremendous research from a control engineering perspective has been made on the stability and stabilization of time-delay systems. With that effort, theoretical stability analysis, especially for key time delay system models, has reached a nearly perfect stage by discovering notable mathematical interpretation techniques. Many related scientific papers and major books have been published by prominent scholars. As mentioned earlier, due to the scientific importance of this system, a great deal of research has been done to this day, but the field is still of high importance in the areas of control engineering and mathematics. Although there are some well-publicized books on fundamental topics of time delay systems and detailed study of them, the authors feel that there is a need to summarize

J. H. Park et al., *Dynamic Systems with Time Delays: Stability and Control*,
https://doi.org/10.1007/978-981-13-9254-2_1

recent published results and introduce new results to readers in a few specific areas of this research field. This is the direct motivation for writing this book.

This chapter consists of four sections, and it is intended to inform readers about the overall structure of this book, an outline of the contents to be covered, and the contents of each chapter.

1.1 An Overview of This Book

A dynamical system is a system whose behavior changes over time, often in response to external stimulation or forcing. To understand the characteristics of dynamic systems, we first need to do mathematical modeling of these dynamic systems. As is well known, some systems must include time delays in system modeling because of the system's configuration characteristics. In this book, we look at recent hottest control problems for dynamic systems or networks with time delays. To this end, the priority issue is the stability of dynamic systems with time delays. In order to have a thorough understanding of all the systems covered in control engineering or applied mathematics, we need to get a solution of the mathematical model of the dynamic system. However, it is not easy to obtain system solutions from mathematical models. In particular, in a dynamic system with a time delay, it is difficult to obtain a system solution of closed-form because the initial condition is given as a function with intervals, and a recursive method for fining the solution must be used in consideration of the time delay period. Thus, the stability problem of dynamic systems with time delays is of more interest than the stability problems of any other dynamic systems, and numerous studies have been done to date. After reviewing the latest techniques for stability of the time delay system without control inputs and related new results, we will discuss controller design issues to stabilize dynamics systems or networks with time delays. Starting from PID control, to the modern times, innovative theories such as optimal control, adaptive control, and robust control for stabilization of dynamical systems have been developed. In this book, we will introduce and apply control techniques that have been of great interest in the related field, especially over the last few years. Finally, by looking at various application problems such as state estimation, filtering, and synchronization, we would like to conclude the research topics that this book covers.

In relation to this field of dynamic systems with time delays, many technical papers and monographs have been published in the past decades, but there is a need to publish a new book that can organize more recent research achievements and provide direction for future research. This book is intended to satisfy such needs of readers. Now, the following subsections show readers a general overview of a few dynamic systems covered in chapters and their relevance to the content of this book.

1.1.1 Lur'e Systems

In control theory, nonlinear control is one of the most important branches which studies dynamical systems in some nonlinear forms, time variant forms or in both. Among many kinds of nonlinear feedback control systems, the Lur'e system is absolutely a special one,

$$\dot{x}(t) = Ax(t) + Bf(y(t)) \tag{1.1}$$

with the output signal $y(t) = Cx(t)$.

In early 1940s, it was first formulated by a Soviet engineer and applied mathematician, Anatoliy Isakovich Lur'e, who induced a kind of systems which could be viewed as linear systems $\dot{x}(t) = Ax(t)$ coexisted with some nonlinearity feedback terms $Bf(Cx(t))$ as seen in Eq. (1.1). In general, the nonlinear feedback function $f(\cdot)$ is always assumed to satisfy with certain sector conditions $\dot{f}(\cdot) \in [a, b]$. Many classical and common chaotic systems, such as chaotic Chua's circuit, Lorenz system, Chen system, n-double scroll circuits, the Goodwin model, the repressilator, the toggle switch and the swarm model could be come down to Lur'e systems [1]. The most important problem in Lur'e systems is the absolute stability issue. It denotes how to derive some conditions which only contain the transfer matrix and a, b such that $x(t) = 0$ is a globally and asymptotically stable equilibrium of the Lur'e system. Before totally solving this problem, there were two famous conjectures proposed by M. Aizerman in 1949 [2] and R. E. Kalman in 1957 [3]. But, shortly, Aizerman's conjecture and Kalman's conjecture were proved to be wrong by introducing some counterexamples. After that, the Popov criterion [4] and the circle criterion [5] as two theorems which discussed the absolute stability of Lur'e systems have been put forward. However, two famous theorems are well-established with the precondition that the nonlinearity function $f(\cdot)$ must satisfy certain sector conditions. In synchronization and control problem for the Lur'e dynamical networks, many excellent works have been done in [6, 7].

Over the past decade, research on the Lur'e system with time delays has attracted much attention to scholars. For examples, the problem of delay-dependent criteria for absolute stability of uncertain time-delayed Lur'e dynamical systems was investigated by Lee and Park [8]. In 2010, by Ji et al. [9], a delay-dependent synchronization criterion using a delayed feedback proportional-derivative (PD) controller scheme is derived for chaotic systems represented by the Lur'e system with sector and slope restricted nonlinearities. Robust stability criteria for the uncertain neutral Lur'e systems with interval time-varying delays are derived by constructing a set of Lyapunov–Krasovskii functional by Liu et al. [10]. Refer to more related recent works on Lure'e systems with delays [11–15]. Also, more recent works for stabilization and synchronization of Lur'e systems using sampled-data control are conducted in [16, 17].

Chapter 10 of this book deals with a control problem for synchronization of Lur'e systems with time delays.

1.1.2 Neural Networks

Artificial neural networks in machine learning and cognitive science are statistical learning algorithms inspired by biological neural networks such as the brain, especially in the central nervous system of animals. An artificial neural network refers to an entire model that has artificial neurons that form a network of synapses by changing the binding strength of synapses through learning.

The past decades have witnessed a wide range of applications of neural networks in various engineering and research areas, such as image processing, pattern recognition, and optimization problem [18–21]. In particular, in the era of the 4th Industrial Revolution, artificial intelligence (AI) is developing at a tremendous rate based on deep learning technology in conjunction with big data and internet of thing (IoT). The core element of AI is the neural network (NN).

Representative models of artificial neural networks include cellular neural networks, Hopfield neural networks, Cohen-Crossberg neural networks, convolutional neural networks, and bidirectional associative memory neural networks.

Here, let's briefly look at a new model of NNs that is under intensive research of receiving great attention during the last 5 years. To do this, we need to introduce a new physical device first. As the fourth basic circuit element along with resistor (R), inductor (L), and capacitor (C), memristor was originally postulated by Chua in 1971 [22]. The memristor describes the relationship between magnetic flux and electric charge and it has gained much attention when its prototype memristor device was realized by Hewlett-Packard Laboratories [23, 24]. The memristor is a two-terminal element with variable resistance called memristance. The value of memristance depends on the magnitude and polarity of the voltage applied to it and the length of the time that the voltage has been applied. When the voltage is turned off, the memristor remembers its most recent value until it is turned on next time. Therefore, the memristor has a memory feature. Based on the memory feature, the memristor can effectively simulate biological synapses. Due to the special physical properties and wide applications of the memristor, the study of memristive recurrent neural networks (MRNNs) is getting big spotlight [25–29].

On the other hand, usually, a practical NN that can perform some complex tasks has a very large number of neurons and layers, since more neurons and layers bring higher accuracy output results. Inevitably, there exist time delays between the neurons, and they lead to poor performance, or cause undesired dynamic behaviors, or even destroy the stability of the NN. Therefore, prior to the actual application of the developed NN, the problem of analyzing the stability of NNs with time delays should be considered. Over the past 15 years, this topic is an important research issue and has been a subject of extensive research [30–38].

For instance, hottest attention gave rise to the memristor-based neural network with time-varying delays [29, 39]:

$$\dot{x}(t) = -C(x(t))x(t) + A(x(t))f(x(t)) + B(x(t-\tau(t)))f(x(t-\tau(t))) + J, \tag{1.2}$$

where $x(t) = (x_1(t), x_2(t), \ldots, x_n(t))^T \in \mathbb{R}^n$ for $i = 1, 2, \ldots, n$ denotes the state variable associated with the neurons; n denotes the number of neurons; $C(x(t)) = \mathrm{diag}\{c_1(x_1(t)), c_2(x_2(t)), \ldots, c_n(x_n(t))\}$, $i = 1, 2, \ldots, n$;

$$c_i(x_i(t)) = \begin{cases} c_i^*, & |x_i(t)| > T_j, \\ c_i^{**}, & |x_i(t)| < T_j, \end{cases}$$

where $c_i^* > 0,\ c_i^{**} > 0,\ c_i(\pm T_j) = c_i^*$ or c_i^{**}, $T_j > 0$ are the switching jumps; $f(x(t)) = (f_1(x_1(t)), f_2(x_2(t)), \ldots, f_n(x_n(t)))^T$ and $f(x(t-\tau(t))) = (f_1(x_1(t-\tau_1(t))), f_2(x_2(t-\tau_2(t))), \ldots, f_n(x_n(t-\tau_n(t))))^T$ are the neuron activation functions of the neurons at time t and $t-\tau(t)$, $\tau_1(t), \ldots, \tau_n(t)$ are the time-varying delays and satisfy $0 \le \tau_i(t) \le \tau$, τ is a positive constant; $J = (J_1, J_2, \ldots, J_n) \in \mathbb{R}^n$ is a constant external input vector; $A(x(t)) = [a_{ij}(x_j(t))]_{n\times n}$, $B(x(t-\tau(t))) = [b_{ij}(x_j(t-\tau_j(t)))]_{n\times n}$, $i, j = 1, 2, \ldots, n$, are connection memristive weight matrix and the delayed connection memristive weight matrix, respectively:

$$a_{ij}(x_j(t)) = \begin{cases} a_{ij}^*, & |x_j(t)| > T_j, \\ a_{ij}^{**}, & |x_j(t)| < T_j, \end{cases} \quad b_{ij}(x_j(t-\tau_j(t))) = \begin{cases} b_{ij}^*, & |x_j(t-\tau_j(t))| > T_j, \\ b_{ij}^{**}, & |x_j(t-\tau_j(t))| < T_j. \end{cases}$$

$a_{ij}(\pm T_j) = a_{ij}^*$ or a_{ij}^{**}, $b_{ij}(\pm T_j) = b_{ij}^*$ or b_{ij}^{**} for $i,\ j = 1, 2, \ldots, n$, a_{ij}^*, a_{ij}^{**}, b_{ij}^* and b_{ij}^{**} are all constants.

In connection with the topic of stability of NNs, Chap. 6 of this book investigates the stability problem for neural networks with delays, followed by a control problem for the networks in Chap. 7.

1.1.3 T–S Fuzzy Systems

Unlike conventional control systems, the fuzzy control system using fuzzy logic was initially introduced as a model-free control design method based on a representation of the knowledge and the reasoning process of a human operator [40, 41]. Fuzzy logic can capture the continuous nature of human decision processes and as such is a definite improvement over methods based on binary logic. Later, model-based fuzzy control design and identification got much attention to analyze the problems of stability, performance, and robustness [42]. Thus, fuzzy-based models have been recognized as a powerful approach to handling nonlinear systems or even nonanalytic systems.

For the most up-to-date research trends on fuzzy control systems, recent research papers on control problems using fuzzy modeling of several nonlinear dynamic systems can be found [43–45].

Among various fuzzy models, the Takagi–Sugeno (T–S) model has attracted much attention from the control community because T–S fuzzy models provide an efficient method for representing complex nonlinear systems using simple local linear dynamics connected by IF-THEN rules [46, 47]. As a result, the conventional linear system theory can be applied to analysis and synthesis of the class of nonlinear control systems.

The T–S fuzzy model is described by fuzzy *IF-THEN* rules which represent local linear input-output relations of a nonlinear system

$$\begin{aligned} \text{Plant Rule } i : &\text{ if } \sigma_1(k) \text{ is } M_{1i} \text{ and } \ldots \sigma_p(k) \text{ is } M_{pi} \\ &\text{then } x(k+1) = A_i x(k) + B_i w(k) \\ &\qquad y(k) = C_i x(k) + D_i w(k) \end{aligned}$$

where $\sigma_1(k)$, $\sigma_2(k)$, $\ldots$, $\sigma_p(k)$ denote premise variables, M_{di}, $i = 1, 2, \ldots, r$, $d = 1, 2, \ldots, p$ denote the fuzzy sets, and r denotes the number of fuzzy rules.

During last 20 years, a great number of stability analysis and control synthesis results for the class of T–S fuzzy systems in both the continuous-time and discrete-time contexts have been extensively discussed in the literature [48–54]. Of course, studies on fuzzy modeling, fuzzy control, and fuzzy observer for various nonlinear dynamic systems including time delay have been actively conducted [55–61].

In this book, we present a latest result on filtering problem of a T–S fuzzy modeled system with time delays in Chap. 12.

1.1.4 Complex Dynamical Networks/Genetic Regulatory Networks

We can say that everything around us is actually connected to networks. Historically, the study of networks has been mainly the domain of a branch of discrete mathematics known as graph theory. Since its birth in 1736, when the Swiss mathematician Leonhard Euler published the solution to the Königsberg bridge problem, graph theory has witnessed many exciting developments and has provided answers to a series of practical questions [62]. In addition to the developments in mathematical graph theory, the study of networks has seen important achievements in some specialized field from the 1920s, for example, social networks. Since the 1990s, there has been a new field of complex dynamical networks (or, simply, complex networks) in which dynamic systems with state changes evolving in time are interconnected to form a single network. During two decades, extensive empirical studies have shed light on the topology of food webs, electrical power grids, cellular and metabolic networks, the World-Wide Web, epidemic spreading networks, the Internet backbone, the neural network and genetic networks, telephone call graphs, and coauthorship and citation networks of scientists [63].

One of the hot research topics in the study of complex dynamical networks is synchronization. Synchronization processes are ubiquitous in nature and play a very important role in many different contexts such as biology, ecology, climatology, sociology, technology, or even in arts. It is known that synchrony is rooted in human life from the metabolic processes in our cells to the highest cognitive tasks we perform as a group of individuals [64]. Hence, the topic of the synchronization of complex dynamical networks has been extensively studied on account of its important applications in chemical reactions, secure communication, power grids, and biological systems since the investigation of synchronization for the network is meaningful in both basic theory and technological practice [65–69].

Here, let's look at a mathematical model that has time delays in connecting between nodes in complex networks. Consider the complex dynamical networks with interconnection delays consisting of N identical coupled nodes as:

$$\dot{x}_i(t) = \varphi(x_i(t)) + c\sum_{j=1}^{N} G_{ij}\Gamma x_j(t-\tau(t)) + u_i(t), \tag{1.3}$$

where $i = 1, 2, \ldots, N$, $x_i(t) \in \mathbb{R}^n$ denotes the state variable; $u_i(t) \in \mathbb{R}^n$ represents the control input; $\varphi : \mathbb{R}^n \rightarrow \mathbb{R}^n$ is a continuous vector-valued function; $c > 0$ is the coupling strength; $\Gamma = (\gamma_{ij})_{n\times n} \in \mathbb{R}^{n\times n}$ is a matrix describing the inner-coupling of the nodes; $G = (G_{ij})_{N\times N}$ describes the outer coupling matrix of the network, where G_{ij} is defined as: if there exists a connection from the node j to node i $(j \neq i)$, then $G_{ij} > 0$; otherwise, $G_{ij} = 0$, the scalar $\tau(t)$ is a time-varying delay, and the diagonal elements of matrix G are defined by: $G_{ii} = -\sum_{j=1, j\neq i}^{N} G_{ij},\ i = 1, 2, \ldots, N$.

Now, as a special case, let's look at genetic regulatory networks (GRNs), which have been extensively studied as a model of complex networks in the field of biology. Biologically based networks such as neural networks and GRNs have enabled new applications in integrated circuits like neurochips, as well as in biological and biomedical science [70–74]. In living cells, a large amount of genes and proteins are interacting with one another by activation and repression. The mathematical modeling of GRNs can provide insight into the complicated biological and chemical processes linked by gene regulation [75, 76]. Normally, GRNs, which comprise different segments of Deoxyribo Nucleic Acid (DNA), interacts directly/indirectly through gene products and other substances (inputs) in a cell for its gene expression as mRNAs and proteins (outputs). GRN controls the rate and level of informational processing of gene expression within cells through feedback and/or feed-forward regulatory inputs and thereby controls the cellular metabolic and functional behaviors. The increased output signal (gene product) leads to feed-back inhibition of GRN and the excess of input signal (molecules and proteins) governs the feed-forward process of GRN [77, 78].

It is worth noting here that recent GRNs studies are also interested in models with time delays. Let's look at an example.

Genetic regulatory networks with leakage and time-varying delays consisting of n mRNAs and n proteins can be represented as the following functional differential equation model:

$$\begin{cases} \dot{m}_i(t) = -a_i m_i(t-\rho) + k_i(p_1(t-h(t)), \ldots, p_n(t-h(t))), \\ \dot{p}_i(t) = -c_i p_i(t-\sigma) + d_i m_i(t-\tau(t)),\ i = 1, 2, \ldots, n, \end{cases} \tag{1.4}$$

where $m_i(t)$, $p_i(t) \in \mathbb{R}$ are concentrations of mRNA and protein of the ith node, a_i and c_i are positive real numbers that represent the degradation rates of mRNA and protein respectively, d_i is a positive constant representing the ith translation rate from mRNA to protein, ρ and σ are leakage delays, $h(t)$ and $\tau(t)$ are time-varying delays, and $k_i(\cdot)$ represents the feedback regulation of the protein, which contains the information about the connection and topology structure of proteins. For a more detailed discussion of the above model of equations, see the references mentioned above.

In this book, we present the latest results on the synchronization problem of complex dynamical networks in Chap. 9 and the state estimation of genetic regulatory networks in Chap. 13.

1.1.5 Chaotic Systems

Since E.N. Lorenz found the first a chaotic attractor in a simple three-dimensional autonomous systems in 1963, chaos theory in dynamic systems has received great deal of interest during the last four decades. It should be pointed out that chaotic systems have complex behaviors that possess some special features, such as being extremely sensitive to tiny variations of initial conditions, having bounded trajectories in phase space and so on.

Since Mackey and Glass [79] first found chaos in time-delay system in 1977, there has been increasing attention in time-delay chaotic systems [80–83]. One of the most frequent objectives is the stabilization of chaotic behaviors to one of unstable fixed points or unstable periodic orbits embedded within a chaotic attractor. In other words, the goal is to design a stabilizing controller so that the closed-loop system dynamics converges to the fixed point or periodic orbit [84, 85]. In 2005, a novel control method for stabilization of a class of time-delay chaotic systems was proposed by Park and Kwon [86] by using the Lyapunov stability theory and LMI technique.

As another academic frontier in chaos theory, chaos synchronization has been a very hot topic in the nonlinearity community since the concept of the synchronization was introduced by Pecora and Carroll [87], and has attracted much interest of scientists and engineers due to its potential applications in biology, chemistry, engineering, secure communication and some other nonlinear fields. For example, in 2013, Lee, Park, Lee and Kwon [88] investigated the robust synchronization problem via a stochastic sampled-data control for uncertain nonlinear chaotic systems in which the norm-bounded uncertainties enter into the systems in randomly occurring ways and they obey certain Bernoulli distributed white noise sequences. More recent studies on chaos synchronization using different control schemes can be found in the following works: [89–94], and reference therein.

For this topic, Chap. 10 looks at a control problem for chaotic systems of Lur'e type and Chap. 14 presents a new result on secure communication by using a class of chaotic systems with a channel delay.

1.2 Main Features of This Book

This book presents both theoretical development and applications to the real world in the hottest research fields of dynamical control systems with time delays including nonlinear Lur'e systems, neural networks, fuzzy systems, chaotic systems, and complex networks for the topics: stability, stabilization, synchronization, and state estimation via specific techniques—new integral and summation inequalities, new Lyapunov–Krosovskii functionals, an event-triggered Control, a sampled-data control, a dynamic output feedback control, and an $\mathscr{H}_\infty$ theory.

Appropriate features suggested by this book include:

1. A tutorial-style overview of a general class of time-delay systems
2. A brief introduction of dynamical systems and networks such as systems with time delays, Lur'e systems, neural networks, fuzzy systems, complex networks, and genetic regulatory networks, which have recently been studied.
3. A brief overview of recent progress of time-delay systems and some key techniques on integral and summation inequalities.
4. A new development of Lyapunov–Krosovskii functionals (LKF) to get improved stability criteria.
5. A novel design framework for synthesis of stabilizing controllers based on stability and performance objectives of dynamic systems with time delays.
6. Fuzzy modelling and its control to deal with nonlinear dynamic systems.
7. Introduction of state estimator design methods and techniques for a few practical systems.
8. Development of various control schemes to ensure system stability and maximum performance.
9. A new mathematical approach for the interpretation of system stability and stabilization and the development of mathematical tools such as mathematical inequalities and lemmas.
10. Theoretical proof and derivation of algorithms on stabilizing controllers and estimators for guaranteeing control objectives.
11. The applicability of the proposed control techniques to actual real systems via practical examples and corresponding simulations.

1.3 Organization of This Book

The book is composed of an introductory part and 3 technical parts (Stability analysis of Time-Delay Systems, Control Problems of Dynamics Systems with Time-Delays, Applications of Dynamics Systems with Time-Delays).

Part I starts in Chap. 1, ends in Chap. 2, and introduces some basics and recent research directions on stability and stabilization of standard time-delay systems.

Part II starts in Chap. 3 and ends in Chap. 6. First, we briefly review the mathematical inequalities that have been widely used in the stability analysis of time delay systems over the last 20 years, and particularly introduce recent results and expanded new results on integral inequality and summation inequality in Chaps. 3 and 4. Second, Chap. 5 presents new results on stability conditions for time-delay systems using the developed inequalities in the previous chapters. Finally, the stability problem for neural networks with time delays in continuous and discrete-time domain by using the methods and techniques developed in the previous chapters is discussed in Chap. 6 as an extended work of previous chapters in Part II.

Part III deals with control problems of dynamics systems with time delays and starts with a $\mathscr{H}_\infty$ controller design problem for neural networks as an application model of the real world in Chap. 7, followed by another control scheme called event-triggered control for linear networked control systems with delays in Chap. 8. Chapter 9 introduces a design method of the dynamic output feedback control for complex dynamical networks. Part III ends with Chap. 10, which introduces a design method of the sampled-data control for a class of chaotic systems.

Part IV considers special application topics on control and filtering problems of dynamics systems with time delays and uses the methods and techniques presented in the stability and control problems in Parts II and III to look at applications such as filtering problems in Chaps. 11 and 12, state estimation problem in Chap. 13, and synchronization issues in Chap. 14 that are important and practical topics in real-world systems.

Here, we briefly review the outline of the state estimation and synchronization problem in dynamic systems.

The evolution of collective behaviors in large scale systems or networked systems is one of the significant goals. Among the various collective behaviors of such systems, synchronization undoubtedly plays a critical role in achieving the common behaviors among the interacted systems. Synchronization aims to regulate the rhythm of those interacted systems in the network in order to force them to behave uniformly [95].

According to different objectives in the natural world and the artificial society, different types of synchronization of dynamic systems or networks have been discussed such as complete synchronization, lag synchronization, intermittent synchronization, cluster synchronization, impulsive synchronization, anti-synchronization, and quasi-synchronization by using several control schemes [67, 96–107].

On the other hand, the state estimator (or called state observer) is a system that provides an estimate of internal states of a given real dynamics system from measurements of the input and output of the real system. Knowing the state of the system is the basis for dealing with control problems, and it is possible to design a suitable stabilizing controller. Therefore, the state estimation problem is important in both control theory and practical applications because some system states in real systems are not completely available.

To see the recent results on state estimation problems in various dynamic systems and networks, refer to the following results of our team [108–115], and reference therein.

Part I Introductory Remarks and Preliminaries

Chapter 1 Introduction

This chapter briefly introduces the various information on systems and problems covered by this book.

Chapter 2 Basics and Preliminaries of Time-Delay Systems

This chapter first looks at the fundamental facts about the basic model, called the time delay system, before dealing with dynamic systems and networks with time delays. In the last decades of research related to time delay systems, we concentrate on the development process of control theory related research. We introduce the major techniques developed for the stability analysis problem of time delay systems, particularly the mathematical inequalities and Lyapunov functional that are the main tools of the analysis. Finally, the basic facts related to stabilization to guarantee the performance of the time delay system will be discussed. The research trend of this field by listing some themes on control schemes that have recently received the spotlight in this research field is reviewed.

Part II: Stability Analysis of Time-Delay Systems

Chapter 3 Integral Inequalities

This chapter addresses integral inequalities which have been widely used in stability analysis of time delay systems in continuous time domain. It is well known that integral inequalities play an essential role in the stability analysis and corresponding derivation of stability criteria. So, developing an accurate integral inequality is of particular importance. In this chapter, we will focus on the study of integral inequalities. Various single- and multiple-integral inequalities will be presented, including some existing well-known ones such as the Jensen, Wirtinger-based and free-matrix-based inequalities. All of these inequalities can be classified into two types: those without free matrices and those with free matrices. The relationship between the two corresponding inequalities with and without free matrices is discussed. It is worth pointing out that polynomials, especially, orthogonal polynomials, are usually employed to develop integral inequalities. Moreover, more polynomials considered, tighter bounds produced.

Chapter 4 Summation Inequalities

As an extension of Chap. 3 to the discrete case, this chapter is concerned with the study of summation inequalities as they are widely used in the stability analysis of discrete-time systems with time delays. According to whether free matrices are involved or not, summation inequalities are classified into two types: those without free matrices and those with free matrices. Different types of summation inequalities may lead to different ways of estimating summation terms. In this chapter, various summation inequalities are proposed, including some well-known ones such as the Jensen and Wirtinger-based summation inequalities, and several series of summation inequalities with an undetermined parameter N. It is worth pointing out that orthogonal polynomials defined in summation inner spaces play an essential role in achieving summation inequalities.

Chapter 5 Stability Analysis for Linear Systems with Time-Varying Delay
This chapter is concerned with the stability problem for linear systems with a time-varying delay in continuous- and discrete-time domain by using the Lyapunov–Krasovskii (L–K) functional method. The aim is to achieve more relaxed conditions for the linear delayed systems, based on which the maximum allowable upper bounds could be obtained for different low bounds. In this case, the considered linear systems can be guaranteed to be stable with the delay varying in an interval delay as large as possible. As for the Lyapunov–Krosovskii functional method, there are two basic ways to reduce the conservatism: to construct an appropriate Lyapunov–Krosovskii functional and to tightly estimate the integral terms arising in the derivative of the Lyapunov–Krosovskii functional (the continuous-time case) or the summation terms arising in the forward difference of the Lyapunov–Krosovskii functional (the discrete-time case). How to construct a proper Lyapunov–Krosovskii functional is an important issue since an Lyapunov–Krosovskii functional containing more information of the system does not definitely lead to a less conservative stability condition.

Integral/summation inequalities proposed in the previous chapters are utilized and corresponding L–K functionals are deliberately constructed. According to the ways of estimating the single-integral/summation term arising in the derivative or the forward difference of L–K functionals, two types of stability conditions are consequently obtained: those obtained by the combination of integral/summation inequalities without free matrices and the reciprocally convex lemma and those obtained by integral/summation inequalities with free matrices. The relationship between them is closely studied. It is pointed out that the conservatism of the two types of corresponding conditions cannot be compared in theory.

Chapter 6 Stability Analysis for Neural Networks with Time-Varying Delay
This chapter is devoted to the study of the stability of time-delay neural networks in both continuous and discrete contexts. Augmented Lyapunov–Krasovskii (L–K) functionals are deliberately constructed for both continuous and discrete cases, in which state and delay information of neural networks is fully taken into account. During the process of dealing with the time derivative (or the forward difference) of L–K functionals, the integral (or summation) inequalities are employed to estimate integral (or summation) terms. Consequently, more relaxed conditions are derived in the forms of linear matrix inequalities. Several numerical examples are presented to show the effectiveness of the proposed approach.

Part III: Control Problems of Dynamics Systems with Time-Delays

Chapter 7 $\mathscr{H}_\infty$ Control for the Stabilization of Neural Networks with Time-Varying Delay
This chapter is concerned with the $\mathscr{H}_\infty$ control problem for neural networks with time-varying delays and external disturbance. In order to derive less conservative conditions for the existence of a stabilizing controller, the time-varying delay interval divided into two subintervals in which the weighting factor α is introduced to make nonequal subintervals. By Lyapunov stability theory, we derive linear matrix inequalities for designing the controller and show the effectiveness of them by a numerical example.

Chapter 8 Event-Triggered Control for Linear Networked Systems
This chapter studies the event-triggered $\mathscr{H}_\infty$ control problem for networked control systems. The event-triggered mechanism is described by a time-delay model and some latest approach are used to deal with the induced time-delay. Based on this model, criteria for asymptotical stability and control design of event-triggered linear networked control systems are derived by using Lyapunov functional. These criteria are established in the form of linear matrix inequalities. A simulation example is used to demonstrate the effectiveness of the proposed approach.

Chapter 9 Design of Dynamic Controller for the Synchronization of Complex Dynamical Networks with a Coupling Delay
This chapter is about synchronization problem of complex networks. As an advanced control method to solve the problem, the chapter presents ways to the design of a dynamic output feedback controller that can provide better dynamic characteristics compared to a static feedback controller. To this end, two types of controller types, a centralized scheme and a decentralized scheme, are considered.

Chapter 10 Reliable Sampled-Data Control for Synchronization of Chaotic Lur'e Systems with Actuator Failures
This chapter is devoted to the asymptotic synchronization for chaotic Lur'e systems with actuator failures via reliable sampled-data control. Different from some existing results obtained under a full reliability condition, actuator failures are considered, which can reflect more realistic behaviors of the chaotic Lur'e systems. A new Lyapunov–Krasovskii functional is constructed, which can fully capture the information of sawtooth structural sampling pattern. Based on the functional, developed synchronization criteria are obtained, and the desired controllers are designed.

Part IV Applications of Dynamics Systems with Time-Delays

Chapter 11 $\mathscr{H}_\infty$ Filtering for Discrete-Time Nonlinear Systems
This chapter is about the design of $\mathscr{H}_\infty$ filter for a class of nonlinear systems in discrete-time domain. It is well known that $\mathscr{H}_\infty$ control approach is an effective way to deal with uncertainties and disturbances robustly. That's why this control theory is chosen to solve this filtering problem in this chapter. Especially, interval time-varying delays and randomly occurring gain variations are taken into account from a practical point of view in real systems.

Chapter 12 Design of Dissipative Filter for Delayed Nonlinear Interconnected Systems via Takagi–Sugeno Fuzzy Modelling
This chapter considers the problem of dissipative filter design for nonlinear interconnected systems with interval time-varying delays. To deal with the nonlinear systems, Takagi–Sugeno fuzzy modelling is used to express the nonlinear system to a combination of linear systems. Then, a dissipative filter including the property of $\mathscr{H}_\infty$ performance is designed by use of the delay-dependent Lyapunov–Krasovskii functional and a new integral inequality.

Chapter 13 State Estimation of Genetic Regulatory Networks with Leakage, Constant, and Distributed Time-Delays
This chapter is concerned with the design of a state estimator for genetic regulatory networks (GRNs) with leakage and time-varying delays. To approximate the original concentrations of the mRNA and protein, the state estimator is designed using available measurement outputs. Three types of delays are considered for modeling of GRNs, that is, leakage delay, normal constant delay, and distributed delay. By constructing novel Lyapunov functionals and using integral inequalities, delay-dependent criteria for the existence of a state estimator to guarantee that the estimation error dynamics can be globally asymptotically stable are derived. The effectiveness of the proposed methods with several techniques is shown by two numerical examples.

Chapter 14 Secure Communication Based on Synchronization of Uncertain Chaotic Systems with Propagation Delays
This chapter investigates to develop the less conservative secure communication criteria based on chaotic synchronization. By using the sensitivity to initial conditions and system parameters of chaotic systems, the synchronization of chaotic systems widely used for improving security in communication. A communication delay is considered because it is a natural phenomenon in transmitting signals. Firstly, some criteria for designing such a controller which synchronized a receiver chaotic system up to a transmitter chaotic system is derived in terms of linear matrix inequalities by combining extended reciprocal lemma, matrix-refined-function based Lyapunov functional, and extended Wirtinger-based integral inequality. And then, we recover the original message by using the output of the receiver chaotic system which can be checked as a simulation result.

1.4 Background Material

In this section, we introduce some of the mathematical tools that are widely used in the analysis of control systems and are also used in the text of this book.

1.4.1 Linear Matrix Inequalities

Many conditions for system and control theory in dynamic systems can be formulated as convex optimization problems involving linear matrix inequalities (LMIs).

An LMI is any constraint of the form:

$$F(x) := F_0 + \sum_{i=1}^{m} x_i F_i \geq 0,$$

where $x_i \in \mathbb{R}^n$ are the variables and the hermitian matrices $F_i \in \mathbb{R}^{n\times n}$ for $i = 1, 2, \ldots, m$ are given.

For more on LMIs, the reader is referred to [116] or any of the many works on the subject.

In this book, it is noted that some criteria to obtain solution variables to various control, filtering, and application problems are presented in the form of LMIs.

1.4.2 Schur Complement

As a fundamental tool of LMIs theory to transform a Riccati-type inequality into an LMI, the following is frequently used to deal with inequalities in the book.
(**Schur complement** [116]): Given constant symmetric matrices Σ_1, Σ_2, Σ_3 where $\Sigma_1 = \Sigma_1^T$ and $0 < \Sigma_2 = \Sigma_2^T$, then $\Sigma_1 + \Sigma_3^T \Sigma_2^{-1} \Sigma_3 < 0$ if and only if

$$\begin{bmatrix} \Sigma_1 & \Sigma_3^T \\ \Sigma_3 & -\Sigma_2 \end{bmatrix} < 0, \text{ or } \begin{bmatrix} -\Sigma_2 & \Sigma_3 \\ \Sigma_3^T & \Sigma_1 \end{bmatrix} < 0.$$

The following inequality is widely used in control and system theory.
For any real vectors a, b and positive constant ε, it follows that:

$$\pm 2a^T b \le \varepsilon a^T a + \varepsilon^{-1} b^T b.$$

1.4.3 Kronecker Product

The Kronecker product, commonly denoted by $\otimes$, is an operation on two matrices of arbitrary size resulting in a block matrix. Then, the following properties of the Kronecker product are easily established [117]:

1. $(\alpha A) \otimes B = A \otimes (\alpha B)$
2. $(A + B) \otimes C = A \otimes C + B \otimes C$
3. $(A \otimes B)(C \otimes D) = (AC) \otimes (BD)$
4. $(A \otimes B)^T = A^T \otimes B^T$

References

1. Tang Z, Park JH, Feng J (2018) Novel approaches to pin cluster synchronization of complex dynamical networks in Lur'e forms. Commun Nonlinear Sci Numer Simul 57:422–438
2. Aizerman M (1949) On a problem concerning the stability in the large of a dynamical system. Uspekhi Matematicheskikh Nauk 4:187–188

3. Kalman RE (1957) Physical and mathematical mechanisms of instability in nonlinear automatic control systems. Trans ASME 79:553–566
4. Park PG (1997) A revisited Popov criterion for nonlinear Lur'e systems with sector-restrictions. Int J Control 68:461–470
5. Jayawardhana B, Logemann H, Ryan EP (2011) The circle criterion and input-to state stability. IEEE Control Syst Mag 31:32–67
6. Suykens JAK, Curran PF, Chua LO (1999) Robust synthesis for master-slave synchronization of Lur'e systems. IEEE Trans Circuits Syst I Fundam Theory Appl 46:841–850
7. Yalcin ME, Suykens JAK, Vandewalle J (2001) Master-slave synchronization of Lur'e systems with time delay. Int J Bifurc Chaos 11:1707–1722
8. Lee SM, Park JH (2010) Delay-dependent criteria for absolute stability of uncertain time-delayed Lur'e dynamical systems. J Frankl Inst 347:146–153
9. Ji DH, Park JH, Lee SM, Koo JH, Won SC (2010) Synchronization criterion for Lur'e systems via delayed PD controller. J Optim Theory Appl 147:298–317
10. Liu Y, Lee SM, Kwon OM, Park JH (2015) Robust delay-dependent stability criteria for time-varying delayed Lur'e systems of neutral-type. Circuits Syst Signal Process 34:1481–1497
11. Yin C, Zhong S, Chen W (2010) On delay-dependent robust stability of a class of uncertain mixed neutral and Lur'e dynamical systems with interval time-varying delays. J Frankl Inst 347:1623–1642
12. Wang Y, Zhang X, He Y (2012) Improved delay-dependent robust stability criteria for a class of uncertain mixed neutral and Lur'e dynamical systems with interval time-varying delays and sector-bounded nonlinearity. Nonlinear Anal Real World Appl 13:2188–2194
13. Ramakrishnan K, Ray G (2011) An improved delay-dependent stability criterion for a class of Lur'e systems of neutral type. ASME J Dyn Syst Meas Control 134(011008):1–6
14. Duan W, Du B, You J, Zou Y (2015) Improved robust stability criteria for a class of Lur'e systems with interval time-varying delays and sector-bounded nonlinearity. Int J Syst 46:944–954
15. Li T, Qian W, Wang T, Fei S (2014) Further results on delay-dependent absolute and robust stability for time-delay Lur'e system. Int J Robust Nonlinear Control 24:3300–3316
16. Ge C, Wang B, Park JH, Hua C (2018) Improved synchronization criteria of Lur'e systems under sampled-data control. Nonlinear Dyn 94:2827–2839
17. Lee TH, Park JH (2017) Improved sampled-data control for synchronization of chaotic Lur'e systems using two new approaches. Nonlinear Anal Hybrid Syst 24:132–145
18. Cohen MA, Grossberg S (1983) Absolute stability of global pattern formation and parallel memory storage by competitive neural networks. IEEE Trans Syst Man Cybern 5:815–826
19. Chua LO, Yang L (1998) Cellular neural networks: applications. IEEE Trans Circuits Syst 35:1273–1290
20. Haykin S (1998) Neural networks: a comprehensive foundation. Prentice-Hall, Englewood Cliffs
21. Cochocki A, Unbehauen R (1993) Neural networks for optimization and signal processing. Wiley, Hoboken
22. Chua LO (1971) Memristor-the missing circuit element. IEEE Trans Circuit Theory 18:507–519
23. Strukov DB, Snider GS, Stewart DR, Williams RS (2008) The missing memristor found. Nature 453:80–83
24. Tour JM, He T (2008) Electronics: the fourth element. Nature 453:42–43
25. Zhang G, Shen Y (2013) New algebraic criteria for synchronization stability of chaotic memristive neural networks with time-varying delays. IEEE Trans Neural Netw Learn Syst 24:1701–1707
26. Wang G, Shen Y (2014) Exponential synchronization of coupled memristive neural networks with time delays. Neural Comput Appl 24:1421–1430
27. Yang X, Cao J, Yu W (2014) Exponential synchronization of memristive Cohen-Grossberg neural networks with mixed delays. Cogn Neurodyn 8:239–249

28. Zhang G, Hu J, Shen Y (2015) New results on synchronization control of delayed memristive neural networks. Nonlinear Dyn 81:1167–1178
29. Zhang R, Park JH, Zeng D, Liu Y, Zhong S (2018) A new method for exponential synchronization of memristive recurrent neural networks. Inf Sci 466:152–169
30. He Y, Liu GP, Rees D, Wu M (2007) Stability analysis for neural networks with time-varying interval delay. IEEE Trans Neural Netw 18:1850–1854
31. Wang Z, Liu L, Shan QH, Zhang H (2015) Stability criteria for recurrent neural networks with time-varying delay based on secondary delay partitioning method. IEEE Trans Neural Netw Learn Syst 26:2589–2595
32. Zhang CK, He Y, Jiang L, Lin WJ, Wu M (2017) Delay-dependent stability analysis of neural networks with time-varying delay: a generalized free-weighting-matrix approach. Appl Math Comput 294:102–120
33. Lee TH, Park JH, Park MJ, Kwon OM, Jung HY (2015) On stability criteria for neural networks with time-varying delay using Wirtinger-based multiple integral inequality. J Frankl Inst 352:5627–5645
34. Thuan MV, Trinh H, Hien LV (2016) New inequality-based approach to passivity analysis of neural networks with interval time-varying delay. Neurocomputing 194:301–307
35. Zhang XM, Han QL (2009) New Lyapunov-Krasovskii functionals for global asymptotic stability of delayed neural networks. IEEE Trans Neural Netw 20:533–539
36. Zhang XM, Han QL (2011) Global asymptotic stability for a class of generalized neural networks with interval time-varying delays. IEEE Trans Neural Netw 22:1180–1192
37. Zeng HB, Park JH, Zhang CF, Wang W (2015) Stability and dissipativity analysis of static neural networks with interval time-varying delay. J Frankl Inst 352:1284–1295
38. Lee TH, Trinh HM, Park JH (2018) Stability analysis of neural networks with time-varying delay by constructing novel Lyapunov functionals. IEEE Trans Neural Netw Learn Syst 29:4238–4247
39. Bao H, Park JH, Cao J (2015) Matrix measure strategies for exponential synchronization and anti-synchronization of memristor-based neural networks with time-varying delays. Appl Math Comput 270:543–556
40. Zadeh LA (1973) Outline of a new approach to the analysis of complex systems and decision processes. IEEE Trans Syst Man Cybern 1:28–44
41. Mamdani EH (1977) Application of fuzzy logic to approximate reasoning using linguistic systems. Fuzzy Sets Syst 26:1182–1191
42. Sala A, Guerra TM, Babuska R (2005) Perspectives of fuzzy systems and control. Fuzzy Sets Syst 156:432–444
43. Xie X, Yue D, Park JH, Li H (2018) Relaxed fuzzy observer design of discrete-time nonlinear systems via two effective technical measures. IEEE Trans Fuzzy Syst 26:2833–2845
44. Wang B, Zhang D, Cheng J, Park JH (2018) Fuzzy-model-based non-fragile control of switched discrete-time systems. Nonlinear Dyn 93:2461–2471
45. Shen H, Li F, Wu ZG, Park JH, Sreeram V (2018) Fuzzy-model-based non-fragile control for nonlinear singularly perturbed systems with semi-Markov jump parameters. IEEE Trans Fuzzy Syst 26:3428–3439
46. Sugeno M (1985) Industrial applications of fuzzy control. Elsevier, New York
47. Tanaka K, Wang HO (1977) Fuzzy control systems design and analysis: a linear matrix inequality approach. Wiley, New York
48. Tseng CS, Chen BS, Uang HJ (2001) Fuzzy tracking control design for nonlinear dynamic systems via T-S fuzzy model. IEEE Trans Fuzzy Syst 9:381–392
49. Liu X, Zhang Q (2003) Approaches to quadratic stability conditions and $\mathscr{H}_\infty$ control designs for T-S fuzzy systems. IEEE Trans Fuzzy Syst 11:830–839
50. Rhee BJ, Won S (2006) A new fuzzy Lyapunov function approach for a Takagi-Sugeno fuzzy control system design. Fuzzy Sets Syst 157:1211–1228
51. Qiu J, Feng G, Gao H (2013) Static-output-feedback $\mathscr{H}_\infty$ control of continuous-time T-S fuzzy affine systems via piecewise Lyapunov functions. IEEE Trans Fuzzy Syst 21:245–261

52. Lee TH, Lim CP, Nahavandi S, Park JH (2018) Network-based synchronization of T-S Fuzzy chaotic systems with asynchronous samplings. J Frankl Inst 355:5736–5758
53. Liu Y, Guo BZ, Park JH, Lee SM (2018) Event-based reliable dissipative filtering for T-S fuzzy systems with asynchronous constraints. IEEE Trans Actions Fuzzy Syst 26:2089–2098
54. Cheng J, Zhong S, Park JH, Kang W (2017) Sampled-data reliable control for T-S fuzzy semi-Markovian jump system and its application to single link robot arm model. IET Control Theory Appl 11:1904–1912
55. Shen H, Su L, Park JH (2017) Reliable mixed $\mathscr{H}_\infty$/passive control for T-S fuzzy delayed systems based on a semi-Markov jump model approach. Fuzzy Sets Syst 314:79–98
56. Kwon OM, Park MJ, Park JH, Lee SM (2016) Stability and stabilization of T-S fuzzy systems with time-varying delays via augmented Lyapunov-Krasovskii functionals. Inf Sci 372:1–15
57. Zeng HB, Park JH, Xia JW, Xiao SP (2014) Improved delay-dependent stability criteria for T-S fuzzy systems with time-varying delay. Appl Math Comput 235:492–501
58. Su X, Shi P, Wu L, Song YD (2012) A novel approach to filter design for T-S fuzzy discrete-time systems with time-varying delay. IEEE Trans Fuzzy Syst 20:1114–1129
59. Wu L, Su X, Shi P, Qiu J (2011) A new approach to stability analysis and stabilization of discrete-time T-S fuzzy time-varying delay systems. IEEE Trans Syst Man Cybern Part B (Cybernetics) 41:273–286
60. Chen B, Liu XP, Tong SC, Lin C (2008) Observer-based stabilization of T-S fuzzy systems with input delay. IEEE Trans Fuzzy Syst 16:652–663
61. Peng C, Tian YC, Tian E (2008) Improved delay-dependent robust stabilization conditions of uncertain T-S fuzzy systems with time-varying delay. Fuzzy Sets Syst 159:2713–2729
62. Boccaletti S, Latora V, Moreno Y, Chavez M, Hwang DU (2006) Complex networks: structure and dynamics. Phys Rep 424:175–308
63. Strogatz SH (2001) Exploring complex networks. Nature 410:268–276
64. Arenas A, Diaz-Guilera A, Kurths J, Moreno Y, Zhou C (2008) Synchronization in complex networks. Phys Rep 469:93–153
65. Lee TH, Park JH, Ji DH, Kwon OM, Lee SM (2012) Guaranteed cost synchronization of a complex dynamical network via dynamic feedback control. Appl Math Comput 218:6469–6481
66. Liang J, Wang Z, Liu X, Louvieris P (2012) Robust synchronization for 2-D discrete-time coupled dynamical networks. IEEE Trans Neural Netw Learn Syst 26:942–953
67. Yang X, Cao J, Lu J (2012) Stochastic synchronization of complex networks with nonidentical nodes via hybrid adaptive and impulsive control. IEEE Trans Circuits Syst I Regular Pap 59:371–384
68. Wang JL, Wu HN, Huang TW (2015) Passivity-based synchronization of a class of complex dynamical networks with time-varying delay. Automatica 56:105–112
69. Zhang R, Zeng E, Park JH, Liu Y, Zhong S (2018) Non-fragile sampled-data synchronization for delayed complex dynamical networks with randomly occurring controller gain fluctuations. IEEE Trans Syst Man Cybern Syst 48:2271–2281
70. Chaouiya C, Jong HD, Thieffry D (2006) Dynamical modeling of biological regulatory networks. BioSystems 84:77–80
71. Keller AD (1995) Model genetic circuits encoding autoregulatory transactions on cription factors. J Theor Biol 172:169–185
72. Noor A, Serpedin E, Nounou M, Nounou HN (2012) Inferring gene regulatory networks via nonlinear state-space models and exploiting sparsity. IEEE/ACM Trans Comput Biol Bioinf 9:1203–1211
73. Stelling J, Gilles ED (2004) Mathematical modeling of complex regulatory networks. IEEE Trans NanoBiosci 3:172–179
74. Elowitz MB, Leibler S (2000) A synthetic oscillatory network of transcriptional regulators. Nature 403:335–338
75. Lee TH, Lakshmanan S, Park JH, Balasubramaniam P (2013) State estimation for genetic regulatory networks with mode-dependent leakage delays, time-varying delays, and Markovian jumping parameters. IEEE Trans NanoBiosci 12:363–375

76. Lakshmanan S, Park JH, Jung HY, Balasubramaniam P, Lee SM (2013) Design of state estimator for genetic regulatory networks with time-varying delays and randomly occurring uncertainties. Biosystems 111:51–70
77. Alberts B, Bray D, Hopkin K, Johnson AD, Lewis J, Raff M, Roberts K, Walter P (2009) Essential cell biology. Garland Science
78. Lakshmanan S, Rihan FA, Rakkiyappan R, Park JH (2014) Stability analysis of differential genetic regulatory networks model with time-varying delays and Markovian jumping parameters. Nonlinear Anal Hybrid Syst 14:1–15
79. Mackey M, Glass L (1977) Oscillation and chaos in physiological control system. Science 197:287–289
80. Ott E, Grebogi G, Yorke JA (1990) Controlling chaos. Phys Rev Lett 64:1196–1199
81. Lu H, He Z (1996) Chaotic behavior in first-order autonomous continuous-time systems with delay. IEEE Trans Circuits Syst 43:700–702
82. Tian Y, Gao F (1998) Adaptive control of chaotic continuous-time systems with delay. Phys D 117:1–2
83. Chen G, Yu H (1999) On time-delayed feedback control of chaotic systems. IEEE Trans Circuits Syst 46:767–772
84. Guan XP, Chen CL, Peng HP, Fan ZP (2003) Time-delayed feedback control of time-delay chaotic systems. Int J Bifurc Chaos 13:193–205
85. Sun J (2004) Delay-dependent stability criteria for time-delay chaotic systems via time-delay feedback control. Chaos Solitons Fractals 21:143–150
86. Park JH, Kwon OM (2005) LMI optimization approach to stabilization of time-delay chaotic systems. Chaos Solitons Fractals 23:445–450
87. Pecora LM, Carroll TL (1990) Synchronization in chaotic systems. Phys Rev Lett 64:821–825
88. Lee TH, Park JH, Lee SM, Kwon OM (2013) Robust synchronisation of chaotic systems with randomly occurring uncertainties via stochastic sampled-data control. Int J Control 86:107–119
89. Zhang Z, Park JH, Shao H (2015) Adaptive synchronization of uncertain unified chaotic systems via novel feedback controls. Nonlinear Dyn 81:695–706
90. Chen X, Park JH, Cao J, Qiu J (2017) Sliding mode synchronization of multiple chaotic systems with uncertainties and disturbances. Appl Math Comput 308:161–173
91. Liu Y, Park JH, Guo BZ, Shu Y (2018) Further results on stabilization of chaotic systems based on fuzzy memory sampled-data control. IEEE Trans Fuzzy Syst 26:1040–1045
92. Lee TH, Park JH (2018) New methods of fuzzy sampled-data control for stabilization of chaotic systems. IEEE Trans Syst Man Cybern Syst 48:2026–2034
93. Ge C, Wang H, Liu Y, Park JH (2018) Stabilization of chaotic systems under variable sampling and state quantized controller. Fuzzy Sets Syst 344:129–144
94. Chen X, Cao J, Park JH, Huang T, Qiu J (2018) Finite-time multi-switching synchronization behavior for multiple chaotic systems with network transmission mode. J Frankl Inst 355:2892–2911
95. Tang Z, Park JH, Shen H (2018) Finite-time cluster synchronization of Lur'e networks: a nonsmooth approach. IEEE Trans Syst Man Cybern Syst 48:1213–1224
96. Chen J, Lu JA, Wu X, Zheng WX (2009) Generalized synchronization of complex dynamical networks via impulsive control. Chaos 19:043119
97. Chen WH, Luo S, Zheng WX (2016) Impulsive synchronization of reaction-diffusion neural networks with mixed delays and its application to image encryption. IEEE Trans Neural Netw Learn Syst 27:2696–2710
98. Chen WH, Lu X, Zheng WX (2015) Impulsive stabilization and impulsive synchronization of discrete-time delayed neural networks. IEEE Trans Neural Netw Learn Syst 26:734–748
99. Tang Y, Qian F, Gao H, Kurths J (2014) Synchronization in complex networks and its application: a survey of recent advances and challenges. Ann Rev Control 38:184–198
100. Zhao M, Zhang HG, Wang ZL, Liang HJ (2014) Observer-based lag synchronization between two different complex networks. Commun Nonlinear Sci Numer Simul 19:2048–2059

101. Zhao M, Zhang HG, Wang ZL, Liang HJ (2015) Synchronization of complex networks via aperiodically intermittent pinning control. IEEE Trans Autom Control 60:3316–3321
102. Cao YT, Wen SP, Chen MZQ, Huang TW, Zeng ZG (2016) New results on anti-synchronization of switched neural networks with time-varying delays and lag signals. Neural Netw 81:52–58
103. Park JH (2007) Adaptive controller design for modified projective synchronization of Genesio-Tesi chaotic system with uncertain parameters. Chaos Solitons Fractals 34:1154–1159
104. Park JH, Ji DH, Won SC, Lee SM (2008) $\mathscr{H}_\infty$ synchronization of time-delayed chaotic systems. Appl Math Comput 204:170–177
105. Ji DH, Park JH, Yoo WJ, Won SC, Lee SM (2010) Synchronization criterion for Lur'e type complex dynamical networks with time-varying delay. Phys Lett A 374:1218–1227
106. Tang Z, Park JH, Feng J (2018) Impulsive effects on quasi-synchronization of neural networks with parameter mismatches and time-varying delay. IEEE Trans Neural Netw Learn Syst 29:908–919
107. Tang Z, Park JH, Lee TH (2016) Topology and parameters recognition of uncertain complex networks via nonidentical adaptive synchronization. Nonlinear Dyn 85:2171–2181
108. Rakkiyappan R, Maheswari K, Velmurugan G, Park JH (2018) Event-triggered $\mathscr{H}_\infty$ state estimation for semi-Markov jumping discrete-time neural networks with quantization. Neural Netw 105:236–248
109. Chen X, Cao J, Park JH, Qiu J (2017) Stability analysis and estimation of domain of attraction for the endemic equilibrium of an SEIQ epidemic model. Nonlinear Dyn 87:975–985
110. Song XM, Park JH, Yan X (2017) Linear estimation for measurement-delay systems with periodic coefficients and multiplicative noise. IEEE Trans Autom Control 62:4124–4130
111. Song XM, Park JH (2017) Linear optimal estimation for discrete-time measurement-delay systems with multi-channel multiplicative noise. IEEE Trans Circuits Syst II Express Lett 64:156–160
112. Song XM, Duan Z, Park JH (2016) Linear optimal estimation for discrete-time systems with measurement-delay and packet dropping. Appl Math Comput 284:115–124
113. Shen H, Zhu Y, Zhang L, Park JH (2017) Extended dissipative state estimation for Markov jump neural networks with unreliable links. IEEE Trans Neural Netw Learn Syst 28:346–358
114. Rakkiyappan R, Sakthivel N, Park JH, Kwon OM (2013) Sampled-data state estimation for Markovian jumping fuzzy cellular neural networks with mode-dependent probabilistic time-varying delays. Appl Math Comput 221:741–769
115. Lee TH, Park JH, Kwon OM, Lee SM (2013) Stochastic sampled-data control for state estimation of time-varying delayed neural networks. Neural Netw 46:99–108
116. Boyd S, El Ghaoui L, Feron E, Balakrishnan V (1994) Linear matrix inequalities in system and control theory. SIAM Press
117. Graham A (1982) Kronecker products and matrix calculus: with applications. Wiley, New York

Chapter 2
Basics and Preliminaries of Time-Delay Systems

2.1 Time-Delay Systems

The aim of this section is to review and present the basic theory for time-delay systems. The contents include the basics on mathematical models of time-delay systems and related theories.

In the mathematical modeling of physical systems, the simplest method is that future dynamic behavior depends only on the current state and not on the history of the past. Furthermore, changes in the current state are instantaneous. This assumption leads to a mathematical model of the dynamic system as a series of ordinary differential equations (ODEs).

A general form of ODEs in state-space for such modelling of nonlinear dynamic systems is

$$\dot{x}(t) = f(t, x(t), u(t)) \tag{2.1}$$

where $x(t) \in \mathbb{R}^n$ is the n-dimensional state vector and $u(t)$ is the control input.

It is noted that the value of the state vector $x(t),\ t_0 \leq t < \infty$, for any initial time t_0, can be found once the initial condition

$$x(t_0) = x_0 \tag{2.2}$$

is given.

Various research problems including control problems of many dynamic systems have been studied for a long time based on this system model. However, sometimes physical processes are faced with an inevitable time delay in the response, processing, and transmission of signals. Such kind of dynamic systems is called time-delay systems (shortly, TDS), hereditary systems, time-lag systems, differential-difference equations, or systems with memory, aftereffect or dead-time, which represent a class of functional differential equations (FDEs) which are infinite-dimensional systems with a selective memory of discrete point or pointwise, as opposed to ODEs.

J. H. Park et al., *Dynamic Systems with Time Delays: Stability and Control*,
https://doi.org/10.1007/978-981-13-9254-2_2

In systems with these time delays, if we use a mathematical model that ignores the time delay, the dynamic behavior of the actual system will result in a significantly different one from the response of the mathematical model. Also, the existence of time delay has a considerable effect on the stability and performance in the time response of the dynamic system. Actually, time delays are ubiquitous in nature and human made systems including various mechanics, physics, biology, economy, AIDS epidemics, population dynamic models, large-scale systems, automatic control systems, neural networks, chaotic systems, and so on. For this reason, tremendous research has been carried out by mathematicians and engineers on dynamic systems with time delays over the past 60 years, and this research topic is still one of the areas of greatest interest.

A simplest form of nonlinear time-delay systems is

$$\dot{x}(t) = f(t, x(t), x(t-h), u(t)) \tag{2.3}$$

where the constant $h > 0$ is the time-delay.

One of the areas of greatest interest and effort among the various research themes is to find criteria that guarantee the stability of the time-delay systems. It is noteworthy that long-term studies on the stability of some specific time-delay systems have shown that the maximum allowable bound of the time-delay of the developed criteria is close to analytical values of the delay guaranteeing stability.

In order to understand the basic knowledge of time-delay systems, knowledge on functional differential equations and Lyapunov function (or functional) theory including Lyapunov–Razumikhin functional approach which are fundamental tools in the theory of the stability analysis are required. The Refs. [1–5] are recommended for readers as good guidebooks to get such mathematical knowledge.

It is well known that the first functional differential equations were considered by great mathematicians including L. Euler, P. S. Laplace, S. M. Poisson, J. L. Lagrange, and so on in the 18th century.

From early 20th century, numerous practical problems have been modeled by FDEs. These include viscoelasticity problem in 1909 and predator-prey model in population dynamics in 1928–1931 by Volterra, mathematical biology problems in 1934 by Kostyzin, and ship stabilization problem in 1942 by Minorsky [6]. Also, in addition to mathematical analysis, intensive engineering approaches to time-delay systems began in earnest at the same time. Some systematic figures and mathematical models of practical engineering systems with time delays such as chemical/nuclear reactor systems, microwave oscillator, metal cutting systems, transmission line, predator-prey models, and continuous rolling mill, are detailed in the above references.

The purpose of this book is to introduce the latest results, analyzing methods, and trends in research on stability and control problems in dynamic systems and networks with time delays, so an introduction to the basic knowledge of functional differential equations and Lyapunov function theory is omitted. Also, the great number of monographs, survey papers, and special issues from prominent journals have been published after 1990 in the field of TDSs. Readers are encouraged to refer to the following studies in [7–13] in order to obtain a general understanding of this area.

Here, let us look at a brief classification of FDEs:

- Retarded functional differential equations (RDEs)
  - Differential equations with lumped delays
    Differential equations with fixed point delays
    Differential equations with noncommensurate delays
    Differential equations with commensurate delays
  - Differential equations with distributed delays
- Neutral functional differential equations (NDEs)
- Advanced functional differential equations (AFDE).

Among above classifications, in the next section, we will briefly review the basics and research topics of RDEs and NDEs.

An equation of the advanced type may represent a system in which the rate of change of a quantity depends on its present and future values of the quantity and of the input signal, for instance, $\dot{x}(t-\tau) = f(t, x(t), x(t-\tau))$. In reality, differential equations with delay in advanced arguments occur in many applied problems. For example, issues related to various government economic policies are susceptible to anticipated future economic and social and international circumstances. Mathematical modeling of these economic problems naturally leads to AFDEs. However, since this book does not cover studies on AFDEs, please refer to some mathematical monography dealing with general knowledge about FDEs.

2.2 Models of Time-Delay Systems

2.2.1 Basic Retarded Differential Equations

The simplest fundamental FDE is the linear first-order delay differential equation, also referred to as retarded differential equations (RDEs), which is given by

$$\dot{x}(t) = ax(t) + bx(t-\tau) + f(t), \ t \geq 0 \tag{2.4}$$

where a and b are constants, $f(t)$ is some continuous function, and the variable $x(t)$ is a scalar which represents signal or information in the system.

From the literature, let's look at some real examples of retarded differential equations. According to Gorelik [14], the equation

$$\ddot{x}(t) + 2rx(t) + \omega^2 x(t) + 2qx(t-1) = \varepsilon x^3(t-1) \tag{2.5}$$

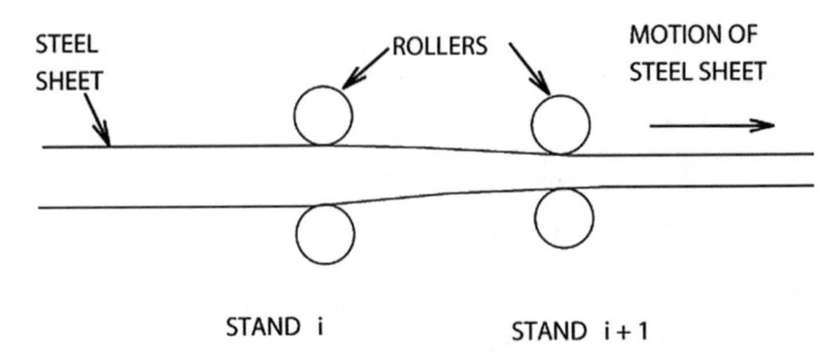

Fig. 2.1 Cold rolling mill

describes the vacuum-tube oscillator and takes into account the time lag caused by the finiteness of the time that the electrons require to pass from the cathode to the anode of a vacuum tube [3].

Let $N(t)$ denote the number of individuals in an isolated population at time t. The life-span of every individual is assume to be a fixed constant L. Then, a well-known single species growth mode developed by Hutchinson [15] is

$$\dot{N}(t) = [\alpha - \gamma N(t - L)]N(t), \ \alpha, \gamma > 0. \tag{2.6}$$

For the detail dynamic characteristics of the Eq. (2.6) according to the interval of αL, refer to the book [3].

As another one, let us see a cold rolling mill process as a real physical model. Delays in the system are due to high-speed transport of steel strips between stands. Consider the process of producing a sheet of steel of uniform thickness by rolling tempered steel through a feeder to a multi-stand mill. The sheet thickness is controlled by adjusting the spacing between the rollers according to measurements (by **X**-ray diffraction) of thickness made on the sheets leaving the roller. A cold rolling mill is depicted in Fig. 2.1. This measurement, however, cannot be made as soon as the sheet leaves the roller. It has to be made further down the line where the thickness has reached an equilibrium value due to cooling. Thus, there is a delay between the time a sheet leaves the rollers at the ith stand and the time it reaches the rollers at the $(i + 1)$st stand at which corrections are applied to the spacing between the rollers. For details of modeling of the system, see the book [16].

Now, let us look at two mathematical models of retarded differential equations to see how the time delays in dynamic systems affect the stability and response of the system.

The first one is a simple equation without forcing input described by

$$\dot{x}(t) = ax(t) + bx(t - h) \tag{2.7}$$

where the system parameters are selected as $a = -2$, $b = 1.5$, $x(0) = 2$, and $x(t) = 0, -h < t < 0$.

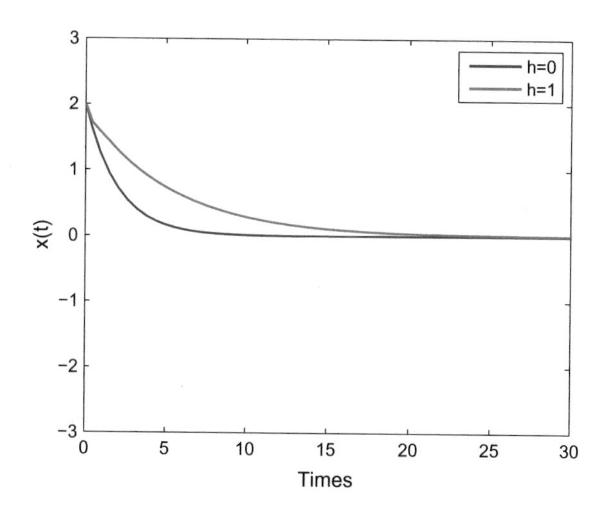

Fig. 2.2 A scalar time-delay system

Figure 2.2 shows that the time response of $x(t)$ in case of $h = 0$ and $h = 1$. As you can see in the figure, the response of the system is slowed down by the 1s delay. It can be easily predicted that the larger the value of the time delay, the slower the response will be.

The other example as shown in Fig. 2.3 is a standard 2nd-order closed-loop system with a delay in feedback loop. It is noted that many engineering systems can be represented by such the 2nd-order system. Two parameters, the damping ratio ζ and the natural frequency ω_n, are selected as 0.5 and 2, respectively, in a simulation.

Figure 2.4 shows the response of the output $y(t)$ to the time delay T value when a step function is applied as the reference input $r(t)$. When $T = 0$ (in case of no delay), the output is well tracked to the target value 1 after 3 s. However, as can be seen in the figure, the larger the time delay, the worse the output response is. In case of $T = 1$, the system is unstable as the output of the system diverges.

In this example, as the value of the time delay increases, it demonstrates that the stability of the system can be broken.

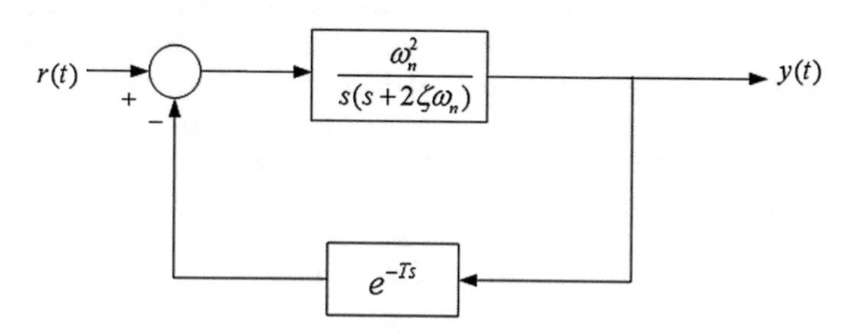

Fig. 2.3 Feedback systems with delay

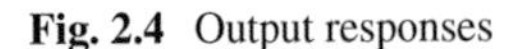

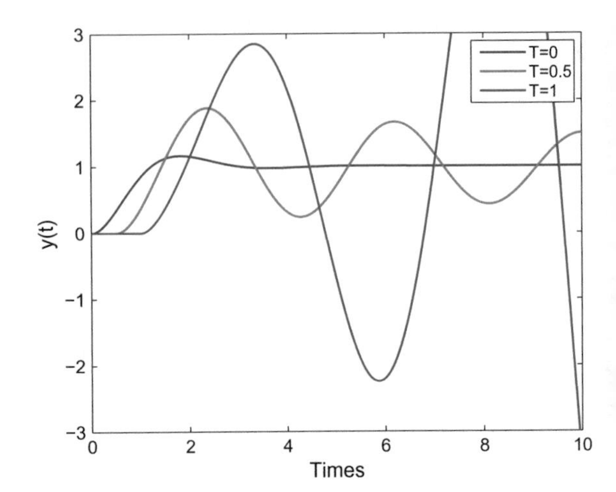

Fig. 2.4 Output responses

Numerous studies on several types of retarded differential equations have been made by mathematicians and scientists through various analyzing methods since the 1930s. The most basic mathematical study of this equations or systems is about the existence and uniqueness of the solution. Next, studies have been made on the conditions under which the solution of the system oscillates. But, above all, the hottest and important research on the equations or systems was about determining stability and instability and finding corresponding conditions.

Before closing this subsection, we briefly introduce time-delay systems in discrete-time domain which are described by delay-difference equations.

In some modern dynamic systems with digital components including digital computers, the information in the systems flows in communication networks. The fundamental character of the digital computer with the necessary input/output hardware to implement most real systems is that it takes to compute answers at discrete steps. Because of the finite information processing speed or the finite switching speed in system components, the influence on time delay naturally exists in such discrete-time systems. For these reasons, the study on stability analysis for discrete-time systems with time delays (or called discrete-delay systems) has been widely investigated, readers are referred to [17–28] and the references therein. It should be noted that a linear discrete time system with a constant delay can be converted to a new linear discrete system without a time delay by introducing a state augmentation method and a new state variable. However, the above conversion does not apply in the following cases with a time-varying delay of integer type:

$$
\begin{aligned}
x(k+1) &= Ax(k) + A_d x(k - h(k)) \\
x(k) &= \phi(k), \ k = -h_M, -h_M + 1, \ldots, 0
\end{aligned}
\tag{2.8}
$$

where k is the discrete time, $x(k) \in \mathbb{R}^n$ is the state vector, $\phi(k)$ is a known given initial condition sequence, and the delay $h(k)$ is interval time-varying and satisfies

$$0 < h_m \leq h(k) \leq h_M$$

where h_m and h_M are any positive integers.

In Chaps. 4 and 5, when dealing with the stability problem of time-delay systems in discrete-time domain, we will see this kind of systems again.

2.2.2 *Retarded Differential Equations of n-Dimension*

With the development of IT technology since the 1980s, various control problems on engineering systems have been actively researched. Since then, the system under study has been essentially multi-input multi-output (MIMO).

Hence, stability analysis for the following type of retarded differential equations of n-dimension was of great interest:

$$\dot{x}(t) = Ax(t) + \sum_{i=1}^{n} A_i x(t - h_i) \tag{2.9}$$

where $h_i > 0$ are the constant time-delays and $n > 0$ is an integer number.

For convenience, the equation is simply called the time-delay systems in engineering fields. The issue of finding the maximum possible time-delay limit to ensure system stability became the hot topic.

In general, there are two ways to interpret the stability of system (2.9). The first is to obtain a characteristic equation (CE) expressed in frequency space using a Laplace transform (LT). It is well know that a RDE is asymptotically stable if and only if all the roots $(s_i, i = 1, 2, \ldots, \infty)$ of the CE have negative real parts in s-plane. However, as is also well known, CE schemes can only be applied to systems with constant time delays. The second method is to use the Lyapunov stability analysis using the most widely used Lyapunov function or functional.

By taking the LT in (2.9), the corresponding CE is

$$\Delta(s) = \det\left(sI_n - A - \sum_{i=1}^{n} A_i e^{-sh_i}\right). \tag{2.10}$$

The Eq. (2.10) has infinite number of roots by the term e^{-sh_i} which leads the system as an infinite-dimensional one.

In order to determine the stability of time-delay systems from such CEs, it is necessary to investigate where all the roots of the CE are located on the s-plane. Generally, simple form conditions guaranteeing that all roots of the CE exist in the left half s-plane can be obtained by using some mathematical tools including matrix norm or spectral radius. The limitation of the CE approach is that it cannot be applied to dynamic systems with time-varying delays or nonlinear perturbations.

We will look more closely at the other sections on how to find stability conditions using the Lyapunov stability theory.

A main mathematical model for studies on time-delay systems since 1980 is as follows:

$$\begin{cases} \dot{x}(t) = Ax(t) + A_d x(t - h(t)), \quad t \geq 0 \\ x(t) = \phi(t), \quad t \in [-h_M, 0] \end{cases} \tag{2.11}$$

where $x(t) \in \mathbb{R}^n$ is the state vector; A and A_d are constant system matrices; $\phi(t)$ is an initial condition function; the time-varying delay $h(t)$ is assumed to satisfy the following constraints:

$$0 \leq h(t) \leq h_M, \quad d_m \leq \dot{h}(t) \leq d_M < 1, \tag{2.12}$$

where h_M, d_m and d_M are constant scalars.

Here, it is noted that studies that exclude the constraint for $\dot{h}(t) < 1$ are also being carried out.

2.2.3 *Neutral Differential Equations/Systems*

The study of neutral differential equations (NDEs) (or called neutral differential systems, neutral time-delay systems, neutral systems) is of great interest but are more complicated and an increased mathematical complexity than RDEs. That is, NDEs also are delay systems, but involve the same highest derivation order for some components of $x(t)$ at both time t and past time(s) $t^- < t$.

A general expression of NDEs can be described by

$$\begin{aligned} &\dot{x}(t) = f(t, x_t, \dot{x}_t), \ t \geq t_0 \\ &\dot{x}(t) = \dot{x}(t + \theta), \ x_t = x(t + \theta), \ -\infty < \theta \leq 0. \end{aligned} \tag{2.13}$$

Neutral systems are a generalized class of TDSs, wherein time delay is present in the highest derivative of the states of the system [1, 2]. They have significant importance due to their wide range of real world applications in population dynamical modelling [3], lossless transmission lines in power systems [12], tele-operation systems [29], robot systems in contact with the environment [30], hot and cold rolling and milling machines [16], and so on [31]. Because of its wider application including the subjects mentioned above, the problem of the stability of delay-differential neutral system has received considerable attention by many researchers in the last two decades. In fact, stability of these systems proves to be a more complex issue compared with the case of RDEs.

A study on the stability problem of n-dimensional neutral time-delay systems with a constant time-delay since the 1990s has received much attention from researchers.

Consider neutral time-delay systems described by:

$$\begin{aligned} \dot{x}(t) - C\dot{x}(t-d) &= Ax(t) + A_d x(t-d), \\ x(t) &= \phi(t), \quad t \in [-d, 0], \end{aligned} \tag{2.14}$$

where $x(t) \in \mathbb{R}^n$ is the state vector, $d > 0$ is the constant time delay, $\phi(t)$ is a compatible vector valued initial function, and A, A_d and C are known real constant matrices with appropriate dimensions.

Without loss of generality, it is assumed that $\rho(C) < 1$ where $\rho(\cdot)$ is the spectral radius. The assumption guarantees that the differential equation $\mathscr{D}x_t = x(t) - Cx(t-d) = 0$ is asymptotically stable for all d.

Unlike the case of RDEs, it should be noted that a strong stability condition is needed for the characteristic root location of an NDE. In other words, it requires that the real parts of all the roots should be located in $\Re(s_i) \leq -\alpha$ for a positive α. This means that it is possible that a NDE is not asymptotically stable although all $\Re(s_i) < 0$ but $\lim_{i\to\infty} \Re(s_i) = 0$ [1, 6].

To the best of authors' knowledge, the first work without seeking eigenvalues on simple stability criteria for the n-dimensional linear neutral systems $\dot{x}(t) = Ax(t) + B(x-h) + C\dot{x}(t-h)$ with a constant delay was given by Li [32] in 1988. Then, an extended work on the same topic was conducted by Hu and Hu [33] in 1996. It is noted that their conditions [32, 33] are given in terms of matrix measure and matrix norms. Verriest and Niculescu [34] in 1997 focused delay-independent stability of linear neutral systems by using a Riccati equation approach and formulated the Riccati equation type condition as a linear matrix inequality (LMI). Also, Park and Won [35] in 1999 tried another approach to asymptotic stability of the neutral delay-differential systems by using the characteristic equation $\big(\Delta(s) = \det[sI - A - (B + Cs)\exp(-hs)] = 0)\big)$ of the system and got a stability condition expressed by a spectral radius. As a different approach, Chen [36] in 1995 studied a method in frequency domain to compute maximal delay intervals for which both retarded and neutral delay systems expressed in state-space forms maintain stability. His method is based on computing the generalized eigenvalues of certain frequency-dependent matrices.

In 1997, Hui and Hu [37] was firstly studied the stability problem for n-dimensional neutral systems with multiple delays and obtained a stability condition in terms of matrix measure and matrix norms. In 1999, Park and Won [38] further studied asymptotic stability of neutral systems with constant multiple delays:

$$\dot{x}(t) = Ax(t) + \sum_{i=1}^{m} \big[B_i x(t-h_i) + C_i \dot{x}(t-h_i)\big] \tag{2.15}$$

by using Lyapunov method and present some sufficient conditions for the stability of systems in terms of LMIs.

For the stability analysis of the system (2.14), several analyzing methods have been attempted. One popular way to use mathematical operators has been developed and widely applied. Let us consider an operator $\mathscr{D}(x_t) : \mathscr{C}_{n,h} \to \mathbb{R}^n$ as

$$\mathscr{D}(x_t) = x(t) + \int_{t-h}^{t} Gx(s)ds - Cx(t-h), \tag{2.16}$$

where $x_t = x(t+s), s \in [-h, 0]$ and $G \in \mathbb{R}^{n\times n}$ is a constant matrix which can be chosen to make the system asymptotically stable.

In order to apply this operator to the stability problem of the system (2.14), the stability of the operator (2.16) itself must first be ensured. In this regard, the following lemma was developed and utilized in literature.

Lemma 2.1 ([39]) *For given positive scalar h, the operator $\mathscr{D}(x_t)$ is stable if there exist a positive definite matrix Γ and positive scalars β_1 and β_2 such that*

$$\beta_1 + \beta_2 < 1, \quad \begin{bmatrix} C^T\Gamma C - \beta_1\Gamma & hC^T\Gamma G \\ * & h^2G^T\Gamma G - \beta_2\Gamma_1 \end{bmatrix} < 0. \tag{2.17}$$

For more basic information and further research topics about asymptotic or robust stability on time-delay systems of neutral type, readers can see a lot of results from the following references: the publications before 2005 [40–51], and further publications between 2005 and 2010 [52–65]. The results of many subsequent studies on NDEs since 2010 can be readily available to readers through academic search sites including Google Scholar, so we skip introducing them in this subsection.

2.3 Stability of Time-Delay Systems

In stability problems of TDSs, to determine the maximal range of delay so that the system remains stable is crucial objective of studying the problem. The maximal range is also called the delay margin, which is defined by the critical value of delay at which a system becomes unstable. Therefore, all the efforts to derive less conservative criteria guaranteeing the stability of TDSs is made for several decades. The maximal allowable upper bound (MAUB) of time-delay is the most core one of the important indexes to evaluate conservatism of stability criteria in the systems. Hence, many researchers have tried to develop such new and improved conditions which ensure the stability for MAUB of time-delay as large as possible [66].

To achieve this goal, the two most fundamental key points are: First, most of the recent stability analysis is based on the Lyapunov stability theory. In this way, it is most important to define which Lyapunov–Krasovskii (L–K) functional for a given time delay system. As for the L–K functional method, to construct an appropriate L–K functional is obviously essential in reducing the conservatism of stability criteria. The superiority of the stability conditions derived from the selection of L–K functional will vary considerably. The second is to use some mathematical tools properly in deriving the stability of given TDSs. This is basically an important step in the stability analysis of any dynamic systems, but especially in the case of TDSs. In this subsection, we review some of the notable results of these two key points for TDSs.

2.3.1 Mathematical Lemmas and Inequalities

Since the early 1990s, the stability study of TDSs has shown explosive interest in a new perspective, it has fundamentally presented two types of stability conditions as a result of many studies. One is 'delay-independent criteria' that guarantee the stability of the system independently of the value of time-delay, the other is 'delay-dependent criteria' that depend on the value of the time-delay while making fully use of information on the length of the delay. In general, latter is less conservative than delay-independent ones.

For given TDSs, the most commonly used method from late 90s to early 2000s to obtain such delay-dependent criteria was to take 'model transformations' of the original systems [67].

The first type is a first-order transformation, but it is known that the transformed system is not equivalent to the original one since additional eigenvalues are introduced into the transformed system [68]. The second type is "a neutral transformation" and it is also known that this transformation is not equivalent to the original system [69].

On the other hand, in the stability analysis of dynamic systems, it is often used to take mathematical norms or expectation on certain terms or to apply mathematical inequalities to separate nonlinear terms (called cross-term bounding technique) in which some variables are mixed.

A well known cross-term bounding technique developed by introducing a free matrix, M, to obtain a less conservative inequality $-2a^T b \leq (a + Mb)^T X (a + Mb) + b^T X b + 2b^T M b$ where $a, b \in \mathbb{R}^n$ by Park [70] in 1999. This inequality has been widely used at the time of its discovery, since it is less restrictive than the $-2a^T b \leq a^T X a + b^T X^{-1} b$ inequality with $X > 0$ that was commonly used. Later, the inequality [70] has been extended by Moon et al. [71] to a more general one.

These two inequalities were used in the model transformation of TDSs with the Leibniz–Newton formula on the term $x(t-h)$, and another model transformation by Fridman and Shaked [72] was extended by using a descriptor model transformation [42].

Through the early 2000s, the techniques mentioned above continue to evolve in a more advanced direction.

From the mid-2000s, several mathematical equations, inequalities, and lemmas have been developed for use with Lyapunov–Krasovskii functionals (L–K functionals) in dealing with the stability analysis of TDSs. The most notable mathematical techniques in the middle of proof stage of stability analysis are as follows:

- Adding zero equalities
- Introducing free weighting matrices
- Partitioning delay intervals
- Defining augmented vectors and corresponding augmented Lyapunov functions
- Reduction approach for Jensen's inequality
- Reciprocally convex approach.

Adding zero equalities: Adding zero equalities is one of the simplest methods to increase the degree of freedom of stability conditions in the proof of stability analysis. The Leibniz–Newton formula

$$x(t) - \int_{t-h(t)}^{t} \dot{x}(s)ds - x(t - h(t)) = 0 \tag{2.18}$$

is employed to obtain a delay-dependent stability condition in TDSs.

Multiplication of the Eq. (2.18) by other terms does not affect the equality of the whole equation. To further improve the efficiency of this technique, use 'free weighting matrices' together.

Specially, the terms on the left side of the equation

$$2[x^T(t)Y + x^T(t - h(t))T] \times \left[x(t) - \int_{t-h(t)}^{t} \dot{x}(s)ds - x(t - h(t))\right] = 0 \tag{2.19}$$

are added to the derivative of certain Lyapunov functional [66]. In this Eq. (2.19), the free weighting matrices Y and T indicate the relationship between the terms in the Leibniz–Newton formula.

Delay-partitioning method: Another popular technique in reducing the conservatism of stability criteria is delay-partitioning method. Since Gu [73] firstly proposed this method, it is well recognized that the delay-partitioning approach can increase the feasible region of stability criteria owing to the fact that this method can obtain more tighter upper bounds obtained by calculation of the time-derivative of L–K functional, which leads to less conservative results. In this regard, Han [65] in 2009 proposed a novel method for construction of a discrete delay decomposition approach which uniformly divides the discrete delay interval into multiple segments and chooses proper functionals with different weighted matrices corresponding to different segments for deriving stability criteria of linear time-delay systems of both retarded and neutral types. Kwon, Park, and Lee [74] proposed the delay partitioning method which divides the delay interval into two ones for uncertain dynamic systems and different free weighting utilization technique at each intervals. This proposed scheme has also been extended for application to neutral delay systems [75]. With another expansion of the delay-partitioning approach, a kind of interior division method of the delay variation by introducing virtual delay-central points was proposed to obtain maximum delay bounds for systems with asymmetric bounds on delay derivative in [76].

Augmented vectors and corresponding augmented Lyapunov functional: In general, when using the Lyapunov function (or functional) for stability analysis of a dynamic system, the components of the Lyapunov function are expressed as a function of the basic state vector $x(t)$. However, in the stability analysis of TDSs, other state vector information is required in addition to the basic state vector $x(t)$ in order to derive a more improved condition equation. For this reason, another approach to reduce the conservatism of stability criteria is to utilize integral terms of states as another state vectors. For instance, $\int_{t-h_L}^{t} x(s)ds$, $\int_{t-h(t)}^{t-h_L} x(s)ds$, $\int_{t-h_U}^{t-h(t)} x(s)ds$, $\int_{t-h_L}^{t}\int_{s}^{t} x(u)duds$, and $\int_{t-h_U}^{t-h_L}\int_{s}^{t-h_L} x(u)duds$ can be possible choices where h_L and h_U are the lower and upper bound of time-varying delay $h(t)$.

By augmenting these vectors, thus, more information from the system can be utilized, which can increase the feasible region of stability criteria [77]. For instance, a simple augmented vector for analyzing a TDS can be

$$z(t) = \begin{bmatrix} \int_{t-h(t)}^{t} x(s)ds \\ x(t) \\ \dot{x}(t) \end{bmatrix} \tag{2.20}$$

which was used in [78]. Another example [79] of augmented vectors is

$$\alpha(t) = col\left\{x(t),\ x(t-h_L),\ x(t-h_U),\ \int_{t-h_L}^{t} x(s)ds,\ \int_{t-h_U}^{t-h_L} x(s)ds,\right.$$
$$\left.\int_{t-h_L}^{t}\int_{s}^{t} x(u)duds,\ \int_{t-h_U}^{t-h_L}\int_{s}^{t-h_L} x(u)duds\right\}. \tag{2.21}$$

The augmented vector given in Eq. (2.21) is much more complex than one given in (2.20). More complex forms of augmented vectors can be easily seen in the literature.

Meanwhile, when suggesting Lyapunov functional for TDSs including NDE type, it is generally composed of several terms. Looking at an example of a simple Lyapunov functional, the following function consists of three terms:

$$V = \int_{t-\tau}^{t} (s-t+\tau)x^T(s)R_1x(s)ds + \int_{t-\tau}^{t} x^T(s)R_2x(s)ds$$
$$+\left(x(t) + A_1\int_{t-\tau}^{t} x(s)ds\right)^T P\left(x(t) + A_1\int_{t-\tau}^{t} x(s)ds\right)^T \tag{2.22}$$

which was used in [80] for the stability analysis of the following system $\dot{x}(t) = Ax(t) + A_1x(t-\tau) + f(t, x(t)) + f_1(t, x(t-\tau))$ in which $f(\cdot)$ and $f_1(\cdot)$ are nonlinear perturbations with bounded norms.

Gradually, it is found that better results in the stability analysis can be derived if augmented vectors are used instead of using the state vectors individually as in Eq. (2.22).

Using the augmented vectors described above, augmented Lyapunov functional gradually began to be applied to the stability analysis of TDSs, and the proposed Lyapunov functional was composed of several terms. For instance, He et al. [77] in 2005 proposed an augmented Lyapunov functional to investigate the asymptotic stability of neutral systems. Parlakci [81] developed an augmented Lyapunov function to treat the cross-terms of variables for uncertain time-varying state-delayed systems in 2006. Also, Suplin et al. [82] in 2006 proposed delay-dependent stability conditions for time-delay systems based on the augmented Lyapunov–Krasovskii functional and Finsler's lemma. Afterwards, some triple-integral terms in Lyapunov–Krasovskii functional were proposed to reduce the conservatism of stability criteria in [58, 83]. Since then, various problems [84–87] have been tacked by utilizing triple-integral terms of Lyapunov–Krasovskii functional. Also, Kim [88] constructed the quadrable-integral terms in Lyapunov–Krasovskii functional, and then new delay-dependent stability criteria for linear systems with interval time-varying delays have been presented.

Reduction approach for Jensen's inequality: On the other hand, there is an integral term $\int_a^b x^T(s)Rx(s)ds$ that is always encountered in the proof of the stability analysis of TDSs in conjunction with the cross-term bounding technique. However, since this integral term can not be generally used to obtain stability conditions of linear type, an inequality in place of this integral term must be used. The following Jensen's inequality is an essential inequality:

$$\int_a^b x^T(s)Rx(s)ds \geq \frac{1}{b-a}\int_a^b x^T(s)ds\ R \int_a^b x(s)ds, \tag{2.23}$$

where $b > a$ are positive constants and R is a positive definite matrix.

The relationship in Eq. (2.23) naturally implies that there is a gap between values of two integrals, and it has been stimulated to the researchers to find a new inequality that can reduce this gap whenever possible. So far, extended and improved versions of Jensen's inequality have published in many research papers. Here, some of them are introduced.

Lemma 2.2 ([89]) *For a matrix $Z \in \mathbb{S}_+^n$ and a scalar $d > 0$, if there exists a vector function $x(s) : [0, d] \to \mathbb{R}^n$ such that the following integrations are well defined, then*

$$-g_l(t) \leqslant -f_l^T(t)Zf_l(t), \tag{2.24}$$

where l is non-negative integer and

$$g_l(t) = \frac{d^{l+1}}{(l+1)!} \int\limits_{t-d}^{t} \int\limits_{v_l}^{t} \cdots \int\limits_{v_1}^{t} x^T(s) Z x(s) \mathrm{d}s \mathrm{d}v_1 \cdots \mathrm{d}v_l$$

$$f_l(t) = \int\limits_{t-d}^{t} \int\limits_{v_l}^{t} \cdots \int\limits_{v_1}^{t} x(s) \mathrm{d}s \mathrm{d}v_1 \cdots \mathrm{d}v_l.$$

Remark 2.1 When $l = 0$, that is, $g_0(t) = d\int_{t-d}^{t} x^T(s)Zx(s)\mathrm{d}s$ and $f_0(t) = \int_{t-d}^{t} x(s)\mathrm{d}s$, inequality (2.24) reduces to the well-known Jensen inequality [90]. When $l = 1$, inequality (2.24) reduces to the inequality of [58]. Thus, the existing inequalities can be regarded as the special cases of our inequality [89].

In order to reduce the conservativeness of the celebrated Jensen's inequality [91], a Wirtinger-based integral inequality (WII) was developed by Seuret and Gouaisbaut [92] in 2013 to find more tighter lower bound of a single integral form of a quadratic term.

Lemma 2.3 ([92]) *For a matrix $M \in \mathbb{S}_+^n$, the following inequality holds for all continuously differentiable function φ in $[a, b] \to \mathbb{R}^n$:*

$$\begin{aligned} &(b-a)\int_a^b \varphi^T(s)M\varphi(s)ds \\ &\geq \left(\int_a^b \varphi(s)ds\right)^T M \left(\int_a^b \varphi(s)ds\right) + 3\Theta^T M \Theta, \end{aligned} \tag{2.25}$$

where $\Theta = \int_a^b \varphi(s)ds - \frac{2}{b-a}\int_a^b\int_a^s \varphi(u)duds$.

Later, some parts in Seuret and Gouaisbaut [92] was commented by Zheng et al. [93].

Lemma 2.4 (Extended Wirtinger Integral Inequality [91]) *Let $x(t) \in W_n[a, b)$ and $x(a) = 0$. Then for a matrix $R \in \mathbb{S}_+^n$ the following inequality holds:*

$$\int_a^b x^T(s)Rx(s)ds \leq \frac{4(b-a)^2}{\pi^2} \int_a^b \dot{x}^T(s)R\dot{x}(s)ds.$$

Remark 2.2 It is shown in [94, 95] that the inequality (2.25) can be alternatively written as

$$\int_a^b \varphi^T(s)M\varphi(s)ds \geq \frac{2}{b-a}\chi^T \begin{bmatrix} 2M & \frac{-3}{b-a}M \\ * & \frac{6}{(b-a)^2}M \end{bmatrix} \chi \tag{2.26}$$

where $\chi^T = \left[\int_a^b \varphi^T(s)ds \quad \int_a^b\int_u^b \varphi^T(s)dsdu\right]$.

Moreover, the Wirtinger-based single integral inequality has been very extended to double-integral inequalities in [94, 95].

Here, it is noted that the WII cannot be applied directly in finding lower bound of $\int_{t-h}^{t}\int_{s}^{t} x^T(u)Mx(u)duds$ $(M > 0)$ which can be obtained by calculating the triple integral form of Lyapunov–Krasovskii functional $\int_{t-h}^{t}\int_{s}^{t}\int_{u}^{t} x^T(v)Mx(v)dvduds$. It should be pointed out that only Jensen's inequality has been used to find an lower bound of $\int_{t-h}^{t}\int_{s}^{t} x^T(u)Mx(u)duds$ [94].

By the observation mentioned above, a double integral form of the WII, called as Wirtinger-based double integral inequality), was introduced in the following lemma.

Lemma 2.5 ([94]) *For a given matrix $M \in \mathbb{S}_+^n$, given scalars a and b satisfying $a < b$, the following inequality holds for all continuously differentiable function x in $[a, b] \to \mathbb{R}^n$:*

$$\begin{aligned} &\frac{(b-a)^2}{2}\int_a^b\int_s^b x^T(u)Mx(u)duds \\ &\geq \left(\int_a^b\int_s^b x(u)duds\right)^T M\left(\int_a^b\int_s^b x(u)duds\right) \\ &\quad +2\Theta_d^T M\Theta_d, \end{aligned} \tag{2.27}$$

where $\Theta_d = -\int_a^b\int_s^b x(u)duds + \frac{3}{b-a}\int_a^b\int_s^b\int_u^b x(v)dvduds$.

The following lemma is also widely used for getting equivalent form of given inequalities.

Lemma 2.6 (Finsler's lemma [96]) *Let $\zeta \in \mathbb{R}^n$, $\Phi \in \mathbb{S}^n$, and $B \in \mathbb{R}^{m\times n}$ such that* $\mathrm{rank}(B) < n$. *The following statements are equivalent:*
(i) $\zeta^T\Phi\zeta < 0$, $B\zeta = 0$, $\zeta \neq 0$,
(ii) ${B^{\perp}}^T\Phi B^{\perp} < 0$.

In 2015 by Zeng et al. [97], a new free-matrix-based integral inequality that includes the Wirtinger inequality as a special case was developed by the fact that the free-weighting matrices combined with integral-inequality techniques can be used to derive improved delay-dependent criteria for the stability analysis of time varying delay systems because they avoid both the use of a model transformation and the technique of bounding cross terms.

Lemma 2.7 ([97]) *Let x be a differentiable function: $[\alpha, \beta] \to \mathbb{R}^n$. For matrices $R \in \mathbb{S}_+^n$ and $Z_1, Z_3 \in \mathbb{S}^{3n}$, $Z_2 \in \mathbb{R}^{3n\times 3n}$, and $N_1, N_2 \in \mathbb{R}^{3n\times n}$ satisfying*

$$\begin{bmatrix} Z_1 & Z_2 & N_1 \\ * & Z_3 & N_2 \\ * & * & R \end{bmatrix} \geq 0,$$

the following inequality holds:

$$-\int_{\alpha}^{\beta} \dot{x}^T(s)R\dot{x}(s)ds \leq \varpi_1^T(\alpha,\beta)\Psi_1\varpi_1(\alpha,\beta),$$

where

$$\varpi_1(\alpha,\beta) = \left[x^T(\beta),\ x^T(\alpha),\ \tfrac{1}{\beta-\alpha}\int_\alpha^\beta x^T(s)ds\right]^T,$$

$$\Psi_1 = (\beta-\alpha)\left(Z_1 + \frac{1}{3}Z_3\right) + Sym\Big\{N_1[I,\ \ -I,\ \ 0] + N_2[-I,\ \ -I,\ \ 2I]\Big\}.$$

In 2017 by Lee and Park [98], another new inequality which is the modified version of free-matrix-based integral inequality was developed.

Lemma 2.8 ([98]) *Let $x \in \mathbb{R}^n$ be a continuous function and admits a continuous derivative differentiable function in $[\alpha,\ \beta]$. For matrices $R \in \mathbb{S}^n$, $Z_1 \in \mathbb{S}^{2n}$, $Z_2 \in \mathbb{R}^{2n\times n}$ satisfying*

$$\begin{bmatrix} Z_1 & Z_2 \\ * & R \end{bmatrix} \geq 0,$$

the following inequality holds:

$$-\int_{\alpha}^{\beta} x^T(s)Rx(s)ds \leq \varpi_2^T(x,\alpha,\beta)\Psi_2\varpi_2(x,\alpha,\beta),$$

where

$$\varpi_2(x,\alpha,\beta) = \left[\int_\alpha^\beta x^T(s)ds,\ \tfrac{1}{\beta-\alpha}\int_\alpha^\beta\int_v^\beta x^T(s)dsdv\right]^T,$$

$$\Psi_2 = \frac{\beta-\alpha}{3}Z_1 + Sym\Big\{Z_2[-I,\ \ 2I]\Big\}.$$

Note that, for the results of [97, 98], the proposed bounding inequality for $-\int_\alpha^\beta \dot{x}^T(s) R\dot{x}(s)ds$ is considered in [97] while the bounding inequality for $-\int_\alpha^\beta x^T(s)Rx(s)ds$ is considered in [98].

Additionally, based on the Legendre polynomials and Bessel inequality, a novel integral inequality, called the Bessel–Legendre inequality (B–L inequality), was developed by Seuret and Gouaisbaut [99] in 2015. However, the Bessel–Legendre inequality is not suitable for applying stability analysis to time-varying delay case. In order to address the time-varying case, a new class of integral inequalities (called Auxiliary function-based single integral inequalities and Auxiliary function-based double integral inequalities) was proposed in [100] via auxiliary functions.

In 2016 by Chen et al. [101], a theoretical framework called *"Third-order B–L integral inequality"* for integral inequalities was established, in which the existing integral inequalities can be almost combined into the two general inequalities defined in the upper and lower forms, respectively.

Lemma 2.9 ([101]) *For scalars a and b satisfying $a \leq b$ and a matrix $R \in \mathbb{S}_+^n$, the following inequality*

$$\int_a^b \dot{x}^T(s)R\dot{x}(s)ds \geq \frac{1}{b_a}\sum_{k=0}^{3}(2k+1)\vartheta_k^T R\vartheta_k \tag{2.28}$$

holds, where

$$\begin{aligned}
\vartheta_0 &= x(b) - x(a),\\
\vartheta_1 &= x(b) + x(a) - \frac{2}{b_a}\Theta_{0^1},\\
\vartheta_2 &= x(b) - x(a) - \frac{6}{b_a}\Theta_{0^1} + \frac{12}{(b_a)^2}\Theta_{0^2},\\
\vartheta_3 &= x(b) + x(a) - \frac{12}{b_a}\Theta_{0^1} + \frac{60}{(b_a)^2}\Theta_{0^2} - \frac{120}{(b_a)^3}\Theta_{0^3}.
\end{aligned}$$

In 2017, Zhang et al. [102], an improved reciprocally convex inequality including some existing ones as its special cases is derived. Compared with an extended one recently reported in [103] over the original reciprocally convex inequality by Park et al. [104], the improved reciprocally convex inequality can provide a maximum lower bound with less slack matrix variables for some reciprocally convex combinations.

Lemma 2.10 ([102]) *For a scalar $\alpha \in (0, 1)$, matrices $R_1, R_2 \in \mathbb{S}_+^n$ and $Y_1, Y_2 \in \mathbb{R}^{n\times n}$, the following matrix inequality*

$$\begin{bmatrix} \frac{1}{\alpha}R_1 & 0 \\ * & \frac{1}{1-\alpha}R_2 \end{bmatrix} \geq \begin{bmatrix} R_1 + (1-\alpha)X_1 & Y(\alpha) \\ * & R_2 + \alpha X_2 \end{bmatrix} \tag{2.29}$$

holds, where $X_1 = R_1 - Y_1R_2^{-1}Y_1^T$, $X_2 = R_2 - Y_2^T R_1^{-1}Y_2$ and $Y(\alpha) = \alpha Y_1 + (1-\alpha)Y_2$.

Based on the improved inequality (2.29), an augmented Lyapunov–Krasovskii functional is devised in order to use a second-order B–L inequality, and then a improved stability criterion is derived. In the stage of stability proof, the following lemma was used.

Lemma 2.11 ([105]) *For a given quadratic function $f(s) = a_2s^2 + a_1s + a_0$, where $a_2, a_1, a_0 \in \mathbb{R}$, one has $f(s) < 0$ for $s \in [0, h]$ if the following inequalities hold: $f(0) < 0$; $f(h) < 0$; $-a_2h^2 + f(0) < 0$.*

Reciprocally convex approach: A popular technique for integral inequality is the reciprocally convex approach. In 2009, Shao [106] has achieved an improved work of reducing the conservativeness of Jensen inequality lemma by the basic idea that is to approximate the integral terms of quadratic quantities into a convex combination of quadratic terms of the integral quantities. In 2011, Park, Ko, and Jeong [107] suggests a lower bound lemma for such a linear combination of positive functions with inverses of convex parameters as the coefficients to decrease the number of decision variables:

Lemma 2.12 ([107]) *Let $f_1, f_2, \ldots, f_N : \mathscr{R}^m \rightarrow \mathscr{R}$ have positive values in an open subsets D of $\mathscr{R}^m$. Then, the reciprocally convex combination of f_i over D satisfies*

$$\min_{\{\alpha_i|\alpha_i>0,\sum_i \alpha_i=1\}} \sum_i \frac{1}{\alpha_i} f_i(t) = \sum_i f_i(t) + \max_{g_{i,j}(t)} \sum_{i\neq j} g_{ij}(t) \tag{2.30}$$

subject to

$$\left\{g_{ij}(t) : \mathscr{R}^m \rightarrow R,\, g_{ij}(t) = g_{ji}(t), \begin{bmatrix} f_i(t) & g_{ij}(t) \\ g_{ji}(t) & f_j(t) \end{bmatrix} \geq 0\right\}. \tag{2.31}$$

Later, this lemma is extended to second-order case [108].

In addition to some of the mathematical integration inequalities introduced in this subsection, many modified integral inequalities have been used for stability analysis of TDSs, so readers are encouraged to look through various literature reviews to gain additional understanding.

2.3.2 *New Lyapunov–Krosovskii Functionals*

In order to obtain the improved conditions for the stability of time delay systems, it is indispensable to find and utilize new Lyapunov–Krosovskii functions (or functionals) in addition to the various inequalities introduced in the previous subsection.

So, in this subsection, we introduce some recently introduced Lyapunov functions which are applied to TDSs.

In general, for Lyapunov analysis of TDSs including neural cases, there are two ways to improve conditions for the stability of TDSs. As described in previous subsection, the first one is to develop new inequalities to estimate tighter lower bound of $J_R(\dot{x}, a, b) = \int_a^b \dot{x}^T(s)R\dot{x}(s)ds$. As well known, Jensen's inequality had widely used for the purpose in the last decade. Later, some remarkable inequalities including the Wirtinger-based integral inequality or summation inequalities (in case of discrete TDSs) are developed as explained in the last subsection. The other way is to construct suitable Lyapunov functionals. In this view point, an augmented vector has been commonly used to be Lyapunov functional. This approach employs more state information in the augmented vector like $\eta^T(t)P\eta(t)$.

Now, let us consider the following TDS of time-varying delay:

$$\begin{cases} \dot{x}(t) = Ax(t) + Bx(t - h(t)), \\ x(t) = \phi(t), \quad t \in [-h, 0] \end{cases} \tag{2.32}$$

where $x(t) \in \mathbb{R}^n$ is the state vector, $h(t)$ is the time-varying delay satisfying $0 \leq h(t) \leq h$ and $|\dot{h}(t)| \leq \mu < 1, \phi(t)$ is a compatible vector valued initial function, and A, B are known real constant matrices with appropriate dimensions.

Since the various Lyapunov functional proposals associated with the system (2.32) have been published in numerous papers over the past two decades, only a few recently proposed Lyapunov Functionals are presented here.

In 2017 by Lee and Park [109], a new function which consisted of two quadratic functions with a special structural matrix called the *Matrix-refined-function* (MRF) is established to be a Lyapunov functional candidate. It is proved that the proposed Lyapunov functional plays key role to decrease the conservatism of derived conditions.

Lemma 2.13 ([109]) *For the system (2.32) with given a scalar h, matrices $X_4, Y_4 \in \mathbb{S}_+^{3n}$, $X_1, Y_1 \in \mathbb{S}^{3n}$, $X_6, Y_6 \in \mathbb{S}^n$, $X_2, Y_2 \in \mathbb{R}^{3n \times 3n}$, $X_3, X_5, Y_3, Y_5 \in \mathbb{R}^{3n \times n}$ satisfying the following LMIs*

$$X = \begin{bmatrix} X_1 & X_2 & X_3 \\ * & X_4 & X_5 \\ * & * & X_6 \end{bmatrix} > 0, \quad Y = \begin{bmatrix} Y_1 & Y_2 & Y_3 \\ * & Y_4 & Y_5 \\ * & * & Y_6 \end{bmatrix} > 0, \tag{2.33}$$

then the following function can be Lyapunov functional candidate:

$$\begin{aligned} V_L(t) = {} & \eta_1^T(t)\mathscr{X}_{[h(t)]}\eta_1(t) + \int_{t-h(t)}^{t} \dot{x}^T(s)X_6\dot{x}(s)ds \\ & + \eta_2^T(t)\mathscr{Y}_{[h(t)]}\eta_2(t) + \int_{t-h}^{t-h(t)} \dot{x}^T(s)Y_6\dot{x}(s)ds, \end{aligned} \tag{2.34}$$

where

$$\begin{aligned} \eta_1(t) &= \left[x^T(t),\ x^T(t-h(t)),\ \textstyle\int_{t-h(t)}^{t} x^T(s)ds \right]^T, \\ \eta_2(t) &= \left[x^T(t-h(t)),\ x^T(t-h),\ \textstyle\int_{t-h}^{t-h(t)} x^T(s)ds \right]^T, \\ \mathscr{X}_{[h(t)]} &= h(t)\mathscr{X}_1 + \mathscr{X}_2, \quad \mathscr{Y}_{[h(t)]} = (h - h(t))\mathscr{Y}_1 + \mathscr{Y}_2, \\ \mathscr{X}_1 &= X_1 + \frac{h^2}{3}X_4 - Sym\Big\{[X_5,\ \ X_5,\ \ 0]\Big\}, \\ \mathscr{X}_2 &= Sym\Big\{[X_3,\ \ -X_3,\ \ 2X_5]\Big\}, \end{aligned}$$

$$\mathscr{Y}_1 = Y_1 + \frac{h^2}{3} Y_4 - Sym\Big\{[Y_5, \ \ Y_5, \ \ 0]\Big\},$$
$$\mathscr{Y}_2 = Sym\Big\{[Y_3, \ \ -Y_3, \ \ 2Y_5]\Big\}.$$

Based on this lemma, three new theorems was developed by applying the novel Lyapunov functional named MRF (2.34) to three existing works [92, 97, 110] and then the superiority of established theorems was shown as three numerical simulations. That is, it can be concluded that the MRF is effective to improve stability of the TDSs (2.32). For detail, see the paper [109].

In 2017, another novel L–K functional was proposed by Zhang et al. [102] as below:

$$V_Z(t, x_t, \dot{x}_t) = V_1(t, x_t) + h_M \int_{t-h_M}^{t} \int_{\theta}^{t} \dot{x}(s)^T R \dot{x}(s) ds d\theta \tag{2.35}$$

where

$$\begin{aligned}
V_1(t, x_t) &= \int_{t-h(t)}^{t} \tilde{\eta}_1(t, s)^T Q_1 \tilde{\eta}_1(t, s) ds + \int_{t-h_M}^{t-h(t)} \tilde{\eta}_2(t, s)^T Q_2 \tilde{\eta}_2(t, s) ds, \\
\eta_0(t) &= col\{x(t), x(t - h(t)), x(t - h_M)\}, \qquad (2.36)\\
\tilde{\eta}_1(t, s) &= col\left\{\dot{x}(s), x(s), \eta_0(t), \int_{t-h(t)}^{s} x(\theta) d\theta\right\}, \\
\tilde{\eta}_2(t, s) &= col\left\{\dot{x}(s), x(s), \eta_0(t), \int_{t-h_M}^{s} x(\theta) d\theta\right\}.
\end{aligned}$$

Compared with the L–K functional proposed in [92], a quadratic functional was removed in $V_Z(t, x_t, \dot{x}_t)$ given in (2.35) and meanwhile, the integrand vectors $\tilde{\eta}_1(t, s)$ and $\tilde{\eta}_2(t, s)$ were deliberately augmented by adding some state-related vectors to cooperate with the use of the second-order B–L integral inequality. Based on $V_Z(t, x_t, \dot{x}_t)$, a new stability condition was obtained in [102] that is less conservative than some previous ones. However, there still exists some room to improve. The obtained stability conditions could be further relaxed by constructing new L–K functionals.

In this regard, consider the following new L–K functional proposed by Chen et al. [111] in 2018:

$$V(t, x_t, \dot{x}_t) = V_0(t, x_t) + V_1(t, x_t) + V_2(t, \dot{x}_t) \tag{2.37}$$

where

$$V_0(t, x_t) = \chi^T(t)P\chi(t),$$
$$V_1(t, x_t) = \int_{t-h_t}^{t} \eta_1^T(t, s)Q_1\eta_1(t, s)ds + \int_{t-h_M}^{t-h_t} \eta_2^T(t, s)Q_2\eta_2(t, s)ds,$$
$$V_2(t, \dot{x}_t) = h_M \int_{t-h_M}^{t} \int_{\theta}^{t} \dot{x}(s)^T R\dot{x}(s)dsd\theta,$$

where

$$h_t = h(t), \quad h_{Mt} = h_M - h_t, \quad x_t(s) = x(t+s),$$
$$\nu_i(t) = \int_{t-h_t}^{t} \left(\frac{t-s}{h_t}\right)^i x(s)ds, \quad i = 0, 1, 2,$$
$$\mu_j(t) = \int_{t-h_M}^{t-h_t} \left(\frac{t-h_t-s}{h_{Mt}}\right)^j x(s)ds, \quad j = 0, 1, 2,$$
$$\chi(t) = col\{\eta_0(t), \nu_0(t), \nu_1(t), \nu_2(t), \mu_0(t), \mu_1(t), \mu_2(t)\},$$
$$\eta_0(t) = col\{x(t), x(t-h(t)), x(t-h_M)\},$$
$$\eta_1(t, s) = col\left\{\eta_0(t), \dot{x}(s), x(s), \int_{t-h_t}^{s} x(u)du, \int_{s}^{t} x(u)du\right\},$$
$$\eta_2(t, s) = col\left\{\eta_0(t), \dot{x}(s), x(s), \int_{t-h_M}^{s} x(u)du, \int_{s}^{t-h_t} x(u)du\right\}.$$

Compared with $V_Z(t, x_t, \dot{x}_t)$ in (2.35), the quadratic functional $V_0(t, x_t)$ is added into this new L–K functional $V(t, x_t, \dot{x}_t)$ (2.37). Obviously, the crossing information of the terms such as the instant state $x(t)$, the delayed states $x_t(-h_t)$ and $x_t(-h_M)$, the integral terms of the state $\nu_i(t)$ and $\mu_i(t)$, $i \in \{0, 1, 2\}$, is taken into account. Specially, in order to coordinate with the third-order B–L integral inequality (2.28), more integral terms $\nu_2(t)$ and $\mu_2(t)$ are included in $\chi(t)$.

On the other hand, the two vectors $\eta_1(t, s)$ and $\eta_2(t, s)$ in the new L–K functional (2.37) are further augmented by adding $\int_s^t x(s)ds$ and $\int_s^{t-h_t} x(u)du$, respectively. From the view of integral regions, the added term $\int_s^t x(u)du$ can be seen as a complement to the term $\int_{t-h_t}^{s} x(u)du$ since the combination of the two integral regions is just the whole integral region of $\int_{t-h_t}^{t} (\cdot)du$. So, when the term $\int_s^t x(u)du$ is added, the coupling relationship between the two terms $\int_{t-h_t}^{s} x(u)du$ and $\int_s^t x(u)du$ will be fully considered through the positive-definite matrix Q_1. In the same way, the term $\int_s^{t-h_t} x(u)du$ is for the coupling relationship between the two terms $\int_{t-h_M}^{s} x(u)du$ and $\int_s^{t-h_t} x(u)du$. For stability condition by use of the function $V(t, x_t, \dot{x}_t)$ (2.37) and corresponding numerical results, refer to the paper [111].

2.4 Control Problems of TDSs

2.4.1 PID Control

For several decades, the control problem for stabilization of time-delay systems has received considerable attention and has become one of the most interesting topics in control theory. This is due to both theoretical interest as well as the need to develop tools for practical system analysis and design, since delay phenomena are frequently encountered in mechanics, physics, applied mathematics, biology, economics, social networks, and engineering systems, and are a source of instability and poor performance.

As the fist control method for systems with pure time delay, the Smith predictor [112] was invented by O. J. M. Smith in 1957 as a type of predictive controller.

As well known, the most typical classic control method widely used in many dynamic systems is the proportional-integral-derivative (PID) controller described by in the s-domain as

$$U(s) = \mathrm{P} + \mathrm{I} + \mathrm{D} = K_p + \frac{K_i}{s} + K_d s, \tag{2.38}$$

where K_p, K_i, and K_d are control gains for proportional, integral, derivative part of the controller, respectively.

Its corresponding form in time-domain is

$$u(t) = K_p e(t) + K_i \int e(t)dt + K_d \frac{de(t)}{dt}, \tag{2.39}$$

when an error signal $e(t)$ is used as an input of the controller.

The key to PID control is how to tune P, I, and D gain, respectively, for best performance of concerned systems. In particular, the topic on PID gain tuning methods including trial-and-error tuning, Ziegler–Nichols method, Cohen-Coon parameters, relay tuning method, and various auto tuning methods, has been extensively studied in the control of dynamic systems, including time delay systems of low-order in chemical engineering [113]. Refer to the following papers for examples of applying PID control to TDSs [114–119].

2.4.2 Feedback Control

For a long time, all control methods for stabilization problems of TDSs, including Bellman equation and optimal control, were then applied to control the TDSs one by one. Many studies from 1990 to 2000 devoted to the control of TDSs are well documented in Table 3 of the paper [9] by J. P. Richard in 2003. The control schemes in the table include Smith predictor and its generalization, robust control, $\mathscr{H}_\infty$ control, model matching problem, self-adjusting, deadbeat control, sliding mode control, optimal control, disturbance decoupling problem, and feedback linearization for linear or nonlinear TDSs.

Since 2000, the stabilization problem of TDSs in earnest has been studied extensively with the attention of the researchers along with stability problems [120–122]. In this subsection, we briefly introduce controller design problems for stabilizing TDSs over the past decade.

One of the simplest linear systems subject to only an input delay or measurement delay is

$$\dot{x}(t) = Ax(t) + Bu(t-h) \tag{2.40}$$

where A and B are multi-dimensional system matrices and h is a constant time-delay. This simple model can describe a number of phenomena commonly present in real processes including transport of signal or information.

A more general model for TDSs with control input can be

$$\dot{x}(t) = Ax(t) + A_1x(t-h) + Bu(t) \tag{2.41}$$

with a system matrix A_1 in delayed state.

In real practice, dynamics systems almost present some uncertainties because it is very difficult to obtain an exact mathematical model due to environmental noise, uncertain or slowly varying parameters, and so on. Therefore, considerable amounts of efforts have been done to the resolution of robust stabilization of uncertain time-delay systems. Such an uncertain TDS model is as follow:

$$\dot{x}(t) = (A + \Delta A)x(t) + (A_1 + \Delta A_1)x(t-h) + (B + \Delta B)u(t) \tag{2.42}$$

where $x(s) = \phi(s),\, s \in [-h, 0]$ is a given continuous vector valued initial function, the parameter uncertainties ΔA, ΔA_1, and ΔB are assumed to be in the form

$$\begin{aligned} \Delta A &= D_1F_1(t)E_1, \\ \Delta A_1 &= D_2F_2(t)E_2, \\ \Delta B &= D_3F_3(t)E_3, \end{aligned} \tag{2.43}$$

in which D_i, E_i $(i = 1, 2, 3)$ are known real constant matrices of appropriate dimensions, and $F_i(t) \in \mathbb{R}^{k_i \times l_i}$ are unknown matrices satisfying

$$F_i^T(t)F_i(t) \leq I, \ (i = 1, 2, 3).$$

For various types of TDSs with control inputs including the models given in (2.40)–(2.42), controller design problems have been extensively researched. To see various approaches and analysis for the problem of TDSs, see the following Refs. [72, 123–131] which were published before 2010.

A simple and standard form of the control input $u(t)$ is the state-feedback control as

$$u(t) = -Kx(t). \tag{2.44}$$

Since 2000s, the most common method of finding the control gain K for stabilization of TDSs was to combine the Lyapunov stability theory with the LMI framework. At this time, the control gain K can be obtained from the terms of the LMI decision variables by solving the proposed LMI stabilization conditions. From a number of papers, you can easily find controller design examples using this approach.

2.4.3 Special Topics on Stabilization of TDSs

In the previous subsection, we briefly introduced the historical background and research trend since 1990 on the control of TDSs. This subsection simply reviews some interesting control problems for TDSs. In addition to the traditional state feedback controller design problems, we can list the issues that have received much attention as follows.

- Robust $\mathscr{H}_\infty$ control
- Observer and observer-based control
- Guaranteed cost control
- Non-fragile control
- Dynamic output feedback control
- Sampled-data control.

Now, let's take a look at the control problems mentioned above in turn.

Robust $\mathscr{H}_\infty$ Control: In the real world, there are several factors which cause instability and poor performance of dynamical systems such as disturbances and uncertainties in addition to time delays. Of these, disturbances and uncertainties are frequently considered and dealt in practical control problems. Since external disturbances or noises are considered as one of the main source leading to undesirable behaviors, therefore, it is necessary to reduce the effects of noises or disturbances to a certain acceptable value. In this regard, it is well known that one of the best control methods to cope with disturbances or perturbations in dynamical systems is 'Robust $\mathscr{H}_\infty$ control'. Hence, studies in this area on TDSs have been very broad. The $\mathscr{H}_\infty$ problem is to minimize the effects of the external disturbances. So, the goal of this problem is to design an $\mathscr{H}_\infty$ controller to robustly stabilize the closed-loop systems while guaranteeing a prescribed level of disturbance attenuation γ in the $\mathscr{H}_\infty$ sense for TDS systems with external disturbances.

Consider the following uncertain systems with a time-delay:

$$\begin{aligned} \dot{x}(t) &= [A + \Delta A(t)]x(t) + [A_1 + \Delta A_1]x(t-h) + [B + \Delta B]u(t) + B_w w(t) \\ z(t) &= Cx(t) + Du(t) \end{aligned} \tag{2.45}$$

where $w(t) \in \mathbb{R}^p$ is the disturbance input which belongs to $\mathscr{L}_2$ and $z(t)$ is the controlled output.

Then, we are interested in designing a memoryless state-feedback controller $u(t) = Kx(t)$ for all admissible uncertainties and any constant time-delay h satisfying $0 \leq h \leq \bar{h}$ such that

(i) the closed-loop system is stable;
(ii) the closed-loop system guarantees, under zero initial condition,

$$\|z(t)\|_2 < \gamma \|w(t)\|_2$$

for all nonzero $w(t)$ and some prescribed constant γ.

This is the Robust $\mathscr{H}_\infty$ control problem.

Observer and observer-based control: In many real-world systems, the standard designed state feedback control fails to guarantee the required performance when some state variables of the system are not measurable by several hard environments the system has. In such situations, designing an observer to estimate state variables and corresponding observer-based controller will be helpful to reconstruct the system states and achieve the required feedback to control the system. That's why state estimation is one of the major topics in modern control theory with several applications. The most standard and popular device of the state estimation, Luenberger observer of full order has been well studied for several decades.

From this point of view, research on the implementation of observer and observer-based controller for TDSs has been actively conducted. In the following literature, the related research contents can be reviewed [132–140] and reference therein.

Here is an example of observer-based control studied in [133]. Consider a class of linear neutral differential system of the form:

$$\begin{aligned} \dot{x}(t) &= A_0 x(t) + A_1 x(t-h) + A_2 \dot{x}(t-h) + Bu(t), \\ y(t) &= Cx(t) \end{aligned} \tag{2.46}$$

where $y(t)$ is the output of the system.

For this system (2.46), by assuming that the pairs (A_0, B) and (A_0, C) are completely controllable and observable, respectively, a Luenberger type observer of full order is constructed as follows:

$$\begin{aligned} \dot{\tilde{x}}(t) &= A_0 \tilde{x}(t) + A_1 \tilde{x}(t-h) + Bu(t) + L(y(t) - C\tilde{x}(t)), \\ \tilde{x}(t_0 + \theta) &= 0, \quad \forall\, \theta \in [-h, 0] \end{aligned} \tag{2.47}$$

where L is the observer gain. Then, the observer-based control with a control gain K for the system is

$$u(t) = -K\tilde{x}(t). \tag{2.48}$$

More recently, a generalized class of state observer called functional observer is an interesting subject for research (see [141, 142]). In this class of observers, one or multiple functions of the states of the system are estimated in lieu of the full set of the states. This feature can be helpful in reducing the order of an observer, as well as mitigating the observability/detectability requirements to less restrictive conditions [142, 143]. By Reza et al. [144] in 2018, a novel practical delay-dependent sliding mode functional observer design algorithm for linear time-invariant systems with unknown time-varying state delays has been proposed.

Guaranteed cost control: When controlling a real plant, it is also desirable to design a control system which is not only stable but also guarantees an adequate level of performance. Chang and Peng [145] has first introduced one way for this problem that is called guaranteed cost control approach. The approach has the advantage of providing an upper bound on a given linear quadratic cost function.

Since 1999, the Guaranteed cost control techniques have been applied to TDSs including neutral time-delay systems as a control method for performance optimization. A guaranteed cost control problem via memoryless state feedback controllers was studied for a class of linear time-delay systems with norm-bounded time-varying parametric uncertainty ($\dot{x}(t) = [A + \Delta A]x(t) + [A_1 + \Delta A_1]x(t - d) + [B + \Delta B] u(t)$) and a given quadratic cost function by Yu and Chu [146].

The cost function associated with the system is generally

$$J = \int_0^\infty \left[x^T(t)Qx(t) + u^T(t)Ru(t)\right] dt \tag{2.49}$$

where Q and R are given positive-definite symmetric matrices.

The objective of this control method is to develop a procedure to designing a memoryless state feedback guaranteed cost control law $u^*(t) = -Kx(t)$ for the concerned uncertain time-delay system such that the closed-loop system is stable and the closed-loop value of the cost function (2.49) satisfies $J \leq J^*$ which J^* is said to be a guaranteed cost.

More related works on the guaranteed cost control for various TDSs has been investigated in the literature [17, 147–156].

Non-fragile control: It is generally known that feedback systems designed for robustness with regard to plant parameters, may require very accurate controllers. In late 1990, it is shown that relatively small perturbations in controller parameters could even destabilize the closed-loop system [157, 158]. Therefore, it is necessary that any controller should be able to tolerate some level of controller gain variations. This raises a new issue in control society: how to design a controller for a given plant with uncertainty such that the controller is non-fragile with regard to its gain variations.

In this approach, although the standard state-feedback controller $u(t) = -Kx(t)$ is designed for TDSs, the actual controller implemented by variation of control parameters is

$$u(t) = -[K + \Delta K]x(t) \tag{2.50}$$

where ΔK represents the multiplicative gain perturbations of the form

$$\Delta K = H\Phi(t)EK \tag{2.51}$$

with H and E being known constant matrices, and uncertain parameter matrix $\Phi(t)$ satisfying $\Phi^T(t)\Phi(t) \leq I$.

Of course, the expression of gain variation different from Eq. (2.51) can be easily seen in the literature. For this topic, there have been many studies to tackle the nonfragile controller design problem for TDSs in [159–165].

Dynamic output feedback control: In practical control applications and realizations, due to the fact that the full state information is difficult to obtain while the system output signals are easy to access, therefore, the output feedback control strategy is very important when the system states are not available. In terms of designing output feedback controller, either a static output feedback case or a dynamic output feedback case is considered. In contrast to dynamic output feedback, static output feedback is more straightforward and reliable in practice in some sense. However, in some situation, there is a strong need to construct a dynamic output feedback controller instead of a static output feedback controller in order to obtain better performance and refined dynamical behavior of the state response.

Consider the following dynamic output feedback controller:

$$\begin{aligned} \dot{\zeta}(t) &= A_c\zeta(t) + B_c y(t), \\ u(t) &= C_c\zeta(t) + D_c y(t), \ \ \zeta(0) = 0 \end{aligned} \tag{2.52}$$

where $\zeta(t) \in \mathbb{R}^n$ is the state vector of the dynamic controller, and A_c, B_c, C_c, and D_c are gain matrices of appropriate dimensions.

In the design of the state feedback controller, only the control gain K can be obtained. However, since the implementation of the dynamic output feedback controller requires four control gains (A_c, B_c, C_c, D_c), the difficulty is much greater than the state feedback case.

In 2004 and 2005, the design problem for a dynamic output feedback controller for asymptotic stabilization of a class of neutral delay differential systems is investigated by Park [166, 167], respectively, and the conditions on existence of the dynamic controller are derived in terms of LMIs. Further results are also found in [168, 169].

Sampled-data control: Rapid growing of communication and digital technology makes a meteoric rise of sampled-data control scheme because of many advantages such as easy installation, high reliability, maintenance with low cost, and efficiency. The sampled-data control method only uses sampled information of the system at its sampling instants to the controller during certain sampling period. So, it has a special character like hybrid systems owing to the coexistence of both continuous and discontinuous signals in a system. Sampling period is the most important factor which effects conservatism, efficiency, and performance of the sampled-data control system. For example, a large sampling period will relax operating conditions such as limited communication capacity and bandwidth. Therefore, to suggest certain criteria that achieve the control aims of the sampled-data controller for a sampling period as large as possible is a significant issue [170]. In fact, the sampled-data control method was developed in the 1970s in relation to the study of digital control systems, but it is attracting more attention again in connection with the rapid development of recent IT technology and the research in the field of networked control systems since 2010.

For this control scheme, the control input $u(t) = (u_1, u_2, \ldots, u_m)^T \in \mathbb{R}^m$ follows

$$u(t) = Kx(t_k), \quad t_k \le t < t_{k+1}, \quad k = 1, 2, \ldots$$

where $K \in \mathbb{R}^{m \times n}$ is the control gain matrix and t_k $(k = 1, 2, \ldots)$ is the sequence of sampling instants.

Variable sampling intervals satisfying

$$\begin{aligned} t_{k+1} - t_k &= h_k, \\ h_k &\in [h_L, \ h_U], \end{aligned}$$

can be considered for this control method, in which h_L and h_U are known positive constant and $0 < h_L \le h_U$ [171].

According to above viewpoint, many researchers have tried to design sampled-data controllers for controlling dynamic systems or networks with time-delays under lager sampling periods. For example, refer the publication for some applications to fuzzy systems [172–174], chaotic systems [175–177], multi-agent systems [178], and neural networks [179–181].

2.5 Conclusion

In this chapter, we first looked briefly at the basic nature of the time delay system. Next, we looked at the development of research related to the stability analysis of this time delay system over the last few decades. In particular, several remarkable techniques that are indispensable in stability analysis of the system have been reviewed. Finally, we revisited the basic facts and several schemes about controller design for stabilization of time delay systems.

References

1. Hale JK (1977) Theory of functional differential equations. Springer, New York
2. Hale JK, Verduyn Lunel SM (1993) Introduction to functional differential equations. Springer, New York
3. Kolmanovskii VB, Nosov VR (1986) Stability of functional differential equations. Academic, London
4. Kolmanovskii V, Myshkis A (1992) Applied theory of functional differential equations. Kluwer Academic Publishers, Dordrecht

5. Kharitonov VL (2013) Time-delay systems: Lyapunov functionals and matrices. Birkhauser, Basel
6. Gu K, Niculescu SI (2003) Survey on recent results in the stability and control of time-delay systems. J Dyn Syst Meas Control 125:158–165
7. Kharitonov V (1998) Robust stability analysis of time delay systems: a survey. In: Proceedings of fourth IFAC conference on system structure and control, pp 1–12
8. Kolmanovskii V, Niculescu SI, Gu K (1999) Delay effects on stability: a survey. In: Proceeding of 38th IEEE conference on decision and control, pp 1993–1998
9. Richard JP (2003) Time-delay systems: an overview of some recent advances and open problems. Automatica 39:1667–1694
10. Fridman E, Shaked U (2003) Delay systems. Int J Robust Nonlinear Control 13:791–937
11. Niculescu SI, Richard JP (2002) Analysis and design of delay and propagation systems. IMA J Math Control Inf 19:1–227
12. Niculescu SI (2001) Delay effects on stability. Lecture notes in control and information sciences, vol 269
13. Niculescu SI, Verriest EI, Dugard L, Dion JM (1997) Stability and robust stability of time-delay systems: a guided tour. Lecture notes in control and information sciences, vol 228
14. Gorelik G (1939) To the theory of feedback with delay. J Tech Phys 9:450–454
15. Hutchinson GE (1948) Circular causal systems in ecology. Ann N Y Acad Sci 50:221–246
16. Malek-Zaverei M, Jamshidi M (1987) Time-delay systems: analysis, optimization and applications. North-Holland, Amsterdam
17. Fridman E, Shaked U (2005) Stability and guaranteed cost control of uncertain discrete delay systems. Int J Control 78:235–246
18. Gao H, Chen T (2007) New results on stability of discrete-time systems with time-varying state delay. IEEE Trans Autom Control 52:328–334
19. Zhang B, Xu S, Zou Y (2008) Improved stability criterion and its applications in delayed controller design for discrete-time systems. Automatica 44:2963–2967
20. He Y, Wu M, Liu GP, She JH (2008) Output feedback stabilization for a discrete-time system with a time-varying delay. IEEE Trans Autom Control 53:2372–2377
21. Yue D, Tian E, Zhang Y (2009) A piecewise analysis method to stability analysis of linear continuous/discrete systems with time-varying delay. Int J Robust Nonlinear Control 19:1493–1518
22. Meng X, Lam J, Du B, Gao H (2010) A delay-partitioning approach to the stability analysis of discrete-time systems. Automatica 46:610–614
23. Shao H, Han QL (2011) New stability criteria for linear discrete-time systems with interval-like time-varying delays. IEEE Trans Autom Control 56:619–625
24. Ramakrishnan K, Ray G (2013) Robust stability criteria for a class of uncertain discrete-time systems with time-varying delay. Appl Math Model 37:1468–1479
25. Peng C (2012) Improved delay-dependent stabilisation criteria for discrete systems with a new finite sum inequality. IET Control Theory Appl 6:448–453
26. Lee SM, Kwon OM, Park JH (2012) Regional asymptotic stability analysis for discrete-time delayed systems with saturation nonlinearity. Nonlinear Dyn 67:885–892
27. Kwon OM, Park MJ, Park JH, Lee SM, Cha EJ (2015) Improved delay-partitioning approach to robust stability analysis for discrete-time systems with time-varying delays and randomly occurring parameter uncertainties. Optim Control Appl Methods 36:496–511
28. Seuret A, Gouaisbaut F, Fridman E (2015) Stability of discrete-time systems with time-varying delays via a novel summation inequality. IEEE Trans Autom Control 60:2740–2745
29. Castanos F, Estrada E, Mondie S, Ramirez A (2015) Passivity-based PI control of first-order systems with I/O communication delays: a complete sigma-stability analysis. arXiv:150701146
30. Lakshmanan S, Senthilkumar T, Balasubramaniam P (2011) Improved results on robust stability of neutral systems with mixed time-varying delays and nonlinear perturbations. Appl Math Model 35:5355–5368

31. Mohajerpoor R, Lakshmanan S, Abdi H, Rakkiyappan R, Nahavandi S, Park JH (2017) Improved delay-dependent stability criteria for neutral systems with mixed interval time-varying delays and nonlinear disturbances. J Frankl Inst 354:1169–1194
32. Li LM (1988) Stability of linear neutral delay-differential systems. Bull Aust Math Soc 38:339–344
33. Hu GD, Hu GD (1996) Some simple stability criteria of neutral delay-differential systems. Appl Math Comput 80:257–271
34. Verriest EI, Niculescu SI (1997) Delay-independent stability of linear neutral systems: a Riccati equation approach. In: Proceedings of 1997 European control conference, pp 3632–3636
35. Park JH, Won S (1999) A note on stability of neutral delay-differential systems. J Frankl Inst 336:543–548
36. Chen J (1995) On computing the maximal delay intervals for stability of linear delay systems. IEEE Trans Autom Control 40:1087–1093
37. Hui GD, Hu GD (1997) Simple criteria for stability of neutral systems with multiple delays. Int J Syst Sci 28:1325–1328
38. Park JH, Won S (1999) Asymptotic stability of neutral systems with multiple delays. J Optim Theory Appl 103:183–200
39. Yue D, Won S, Kwon OM (2003) Delay dependent stability of neutral systems with time delay: an LMI approach. IEE Proc Control Theory Appl 150:23–27
40. Park JH, Won S (2000) Stability analysis for neutral delay-differential systems. J Frankl Inst 337:1–9
41. Park JH, Won S (2001) A note on stability analysis of neutral systems with multiple time-delays. Int J Syst Sci 32:409–412
42. Fridman E (2001) New Lyapunov-Krasovskii functionals for stability of linear retarded and neutral type systems. Syst Control Lett 43:309–319
43. Chen JD, Lien CH, Fan KK (2001) Criteria for asymptotic stability of a class of neutral systems via a LMI approach. IEE Proc Control Theory Appl 148:442–447
44. Park JH (2001) A new delay-dependent criterion for neutral systems with multiple delays. J Comput Appl Math 136:177–184
45. Park JH (2002) Stability criterion for neutral differential systems with mixed multiple time-varying delay arguments. Math Comput Simul 59:401–412
46. Han QL (2002) Robust stability of uncertain delay-differential systems of neutral type. Automatica 38:719–723
47. Park JH (2003) Robust guaranteed cost control for uncertain linear differential systems of neutral type. Appl Math Comput 140:523–535
48. Han QL (2004) On robust stability of neutral systems with time-varying discrete delay and norm-bounded uncertainty. Automatica 40:1087–1092
49. He Y, Wu M, She JH, Liu GP (2004) Delay-dependent robust stability criteria for uncertain neutral systems with mixed delays. Syst Control Lett 51:57–65
50. Wu M, He Y, She JH (2004) New delay-dependent stability criteria and stabilizing method for neutral systems. IEEE Trans Autom Control 49:2266–2271
51. Park JH (2005) LMI optimization approach to asymptotic stability of certain neutral delay differential equation. Appl Math Comput 160:355–361
52. Xu S, Lam J, Zou Y (2005) Further results on delay-dependent robust stability conditions of uncertain neutral systems. Int J Robust Nonlinear Control 15:233–246
53. Kharitonov V, Mondie S, Collado J (2005) Exponential estimates for neutral time-delay systems: an LMI approach. IEEE Trans Autom Control 50:666–670
54. Lien CH (2005) Delay-dependent stability criteria for uncertain neutral systems with multiple time-varying delays via LMI approach. IEE Proc Control Theory Appl 152:707–714
55. Liu XG, Wu M, Martin R, Tang ML (2007) Delay-dependent stability analysis for uncertain neutral systems with time-varying delays. Math Comput Simul 75:15–27
56. Chen WH, Zheng WX (2007) Delay-dependent robust stabilization for uncertain neutral systems with distributed delays. Automatica 43:95–104

57. Zhang J, Shi P, Qiu J (2008) Robust stability criteria for uncertain neutral system with time delay and nonlinear uncertainties. Chaos Solitons Fractals 38:160–167
58. Sun J, Liu GP, Chen J (2009) Delay-dependent stability and stabilization of neutral time-delay systems. Int J Robust Nonlinear Control 19:1364–1375
59. Rakkiyappan R, Balasubramaniam P (2008) LMI conditions for global asymptotic stability results for neutral-type neural networks with distributed time delays. Appl Math Comput 204:317–324
60. Kwon OM, Park JH, Lee SM (2008) On stability criteria for uncertain delay-differential systems of neutral type with time-varying delays. Appl Math Comput 197:864–873
61. Kwon OM, Park JH (2008) On improved delay-dependent stability criterion of certain neutral differential equations. Appl Math Comput 199:385–391
62. Li M, Liu L (2009) A delay-dependent stability criterion for linear neutral delay systems. J Frankl Inst 346:33–37
63. Kwon OM, Park JH, Lee SM (2009) Augmented Lyapunov functional approach to stability of uncertain neutral systems with time-varying delays. Appl Math Comput 207:202–212
64. Wu M, He Y, She JH (2010) Stability analysis and robust control of time-delay systems. Springer, Beijing
65. Han QL (2009) A discrete delay decomposition approach to stability of linear retarded and neutral systems. Automatica 45:517–524
66. Wu M, He Y, She JH, Liu GP (2004) Delay-dependent stability criteria for robust stability of time-varying delay systems. Automatica 40:1435–1439
67. Fridman E, Shaked U (2003) Delay-dependent stability and $\mathscr{H}_\infty$ control: constant and time-varying delays. Int J Control 76:48–60
68. Gu K, Niculescu SI (2000) Additional dynamics in transformed time delay systems. IEEE Trans Autom Control 45:572–575
69. Gu K, Niculescu SI (2001) Further remarks on additional dynamics in various model transformations of linear delay systems. IEEE Trans Autom Control 46:497–500
70. Park PG (1999) A delay-dependent stability criterion for systems with uncertain linear state-delayed systems. IEEE Trans Autom Control 35:876–877
71. Moon YS, Park PG, Kwon WH, Lee YS (2001) Delay-dependent robust stabilization of uncertain state-delayed systems. Int J Control 74:1447–1455
72. Fridman E, Shaked U (2002) An improved stabilization method for linear time-delay systems. IEEE Trans Autom Control 47:1931–1937
73. Gu K (1999) Discretized Lyapunov functional for uncertain systems with multiple time-delay. Int J Control 72:1436–1445
74. Kwon OM, Park JH, Lee SM (2010) An improved delay-dependent criterion for asymptotic stability of uncertain dynamic systems with time-varying delays. J Optim Theory Appl 145:343–353
75. Kwon OM, Park MJ, Lee SM, Park JH, Cha EJ (2012) New delay-partitioning approaches to stability criteria for uncertain neutral systems with time-varying delays. J Frankl Inst 349:2799–2823
76. Ko JW, Park PG (2011) Delay-dependent stability criteria for systems with asymmetric bounds on delay derivative. J Frankl Inst 348:2674–2688
77. He Y, Wang QG, Lin C, Wu M (2005) Augmented Lyapunov functional and delay-dependent stability criteria for neutral systems. Int J Robust Nonlinear Control 15:923–933
78. Park MJ, Kwon OM, Park JH, Lee SM (2011) A new augmented Lyapunov-Krasovskii functional approach for stability of linear systems with time-varying delays. Appl Math Comput 217:7197–7209
79. Kwon OM, Park MJ, Park JH, Lee SM, Cha EJ (2013) Analysis on robust $\mathscr{H}_\infty$ performance and stability for linear systems with interval time-varying state delays via some new augmented Lyapunov-Krasovskii functional. Appl Math Comput 224:108–122
80. Park JH, Kwon OM (2005) Novel stability criterion of time delay systems with nonlinear uncertainties. Appl Math Lett 18:683–688

81. Parlakci MNA (2006) Robust stability of uncertain time-varying state-delayed systems. IEE Proc Control Theory Appl 153:469–477
82. Suplin V, Fridman E, Shaked U (2006) $\mathscr{H}_\infty$ control of linear uncertain time-delay systems-a projection approach. IEEE Trans Autom Control 51:680–685
83. Ariba Y, Gouaisbaut F (2007) Delay-dependent stability analysis of linear systems with time-varying delay. In: Proceedings of the 46th IEEE conference on decision and control, pp 2053–2058
84. Kwon OM, Lee SM, Park JH (2011) On the reachable set bounding of uncertain dynamic systems with time-varying delays and disturbances. Inf Sci 181:3735–3748
85. Sun J, Liu GP, Chen J, Rees D (2010) Improved delay-range-dependent stability criteria for linear systems with time-varying delays. Automatica 46:466–470
86. Sakthivel R, Mathiyalagan K, Marshal Anthoni S (2012) Robust stability and control for uncertain neutral time delay systems. Int J Control 85:373–383
87. Lee WI, Lee SY, Park PG (2014) Improved criteria on robust stability and $\mathscr{H}_\infty$ performance for linear systems with interval time-varying delays via new triple integral functionals. Appl Math Comput 243:570–577
88. Kim JH (2011) Note on stability of linear systems with time-varying delay. Automatica 47:2118–2121
89. Fang M, Park JH (2013) A multiple integral approach to stability of neutral time-delay systems. Appl Math Comput 224:714–718
90. Gu K, Kharitonov VK, Chen J (2003) Stability of time-delay systems. Birkhauser, Boston
91. Liu K, Suplin V, Fridman E (2011) Stability of linear systems with general sawtooth delay. IMA J Math Control Inf 27:419–436
92. Seuret A, Gouaisbaut F (2013) Wirtinger-based integral inequality: application to time-delay systems. Automatica 49:2860–2866
93. Zheng M, Li K, Fei M (2014) Comments on "Wirtinger-based integral inequality: application to time-delay systems [Automatica 49 (2013) 2860–2866]". Automatica 50:300–301
94. Park MJ, Kwon OM, Park JH, Lee SM, Cha EJ (2015) Stability of time-delay systems via Wirtinger-based double integral inequality. Automatica 55:204–208
95. Mohajerpoor R, Lakshmanan S, Abdi H, Rakkiappan R, Nahavandi S, Shi P (2018) New delay-range-dependent stability criteria for interval-time-varying delay systems via Wirtinger-based inequalities. Int J Robust Nonlinear Control 28:661–677
96. Skelton RE, Iwasaki T, Grigoradis KM (1997) A unified algebraic approach to linear control design. Taylor & Francis
97. Zeng HB, He Y, Wu M, She J (2015) Free-matrix-based integral inequality for stability analysis of systems with time-varying delay. IEEE Trans Autom Control 60:2768–2772
98. Lee TH, Park JH, Xu S (2017) Relaxed conditions for stability of time-varying delay systems. Automatica 75:11–15
99. Seuret A, Gouaisbaut F (2015) Hierarchy of LMI conditions for the stability analysis of time-delay systems. Syst Control Lett 81:1–7
100. Park PG, Lee WI, Lee SY (2015) Auxiliary function-based integral inequalities for quadratic functions and their applications to time-delay systems. J Frankl Inst 352:1378–1396
101. Chen J, Xu S, Chen W, Zhang B, Ma Q, Zou Y (2016) Two general integral inequalities and their applications to stability analysis for systems with time-varying delay. Int J Robust Nonlinear Control 26:4088–4103
102. Zhang XM, Han QL, Seuret A, Gouaisbaut F (2017) An improved reciprocally convex inequality and an augmented Lyapunov-Krasovskii functional for stability of linear systems with time-varying delay. Automatica 84:221–226
103. Seuret A, Gouaisbaut F (2016) Delay-dependent reciprocally convex combination lemma. Rapport LAAS n16006
104. Park PG, Ko J, Jeong J (2011) Reciprocally convex approach to stability of systems with time-varying delays. Automatica 47:235–238
105. Kim JH (2016) Further improvement of Jensen inequality and application to stability of time-delayed systems. Automatica 64:121–125

106. Shao H (2009) New delay-dependent stability criteria for systems with interval delay. Automatica 45:744–749
107. Park PG, Ko JW, Jeong CK (2011) Reciprocally convex approach to stability of systems with time-varying delays. Automatica 47:235–238
108. Lee WI, Park PG (2014) Second-order reciprocally convex approach to stability of systems with interval time-varying delays. Appl Math Comput 229:245–253
109. Lee TH, Park JH (2017) A novel Lyapunov functional for stability of time-varying delay systems via matrix-refined-function. Automatica 80:239–242
110. Kwon OM, Park MJ, Park JH, Lee SM, Cha EJ (2014) Improved results on stability of linear systems with time-varying delays via Wirtinger-based integral inequality. J Frankl Inst 351:5386–5398
111. Chen J, Park JH, Xu S (2018) Stability analysis of continuous-time systems with time-varying delay using new Lyapunov-Krasovskii functionals. J Frankl Inst 355:5957–5967
112. Smith OJM (1957) Posicast control of damped oscillatory systems. Proc IRE 45:1249–1255
113. Zhong QC (2006) Robust control of time-delay systems. Springer, London
114. Leva A, Maffezzoni C, Scattolini R (1994) Self-tuning PI-PID regulators for stable systems with varying delay. Automatica 30:1171–1183
115. Shafiei Z, Shenton AT (1994) Tuning of PID-type controllers for stable and unstable systems with time delay. Automatica 30:1609–1615
116. Poulin E, Pomerleau A (1996) PID tuning for integrating and unstable processes. IEE Proc Control Theory Appl 143:429–435
117. Astrom KJ, Hagglund T (2001) The future of PID control. Control Eng Pract 9:1163–1175
118. Ge M, Chiu MS, Wang QG (2002) Robust PID controller design via LMI approach. J Process Control 12:3–13
119. Silva GJ, Datta A, Bhattacharyya SP (2002) New results on the synthesis of PID controllers. IEEE Trans Autom Control 47:241–252
120. Park JH (2001) Robust stabilization for dynamic systems with multiple time-varying delays and nonlinear uncertainties. J Optim Theory Appl 108:155–174
121. Park JH, Jung HY (2003) On the exponential stability of a class of nonlinear systems including delayed perturbations. J Comput Appl Math 159:467–471
122. Park JH (2004) On dynamic output feedback guaranteed cost control of uncertain discrete-delay systems: an LMI optimization approach. J Optim Theory Appl 121:147–162
123. Fridman E, Shaked U (2003) Delay dependent stability and $\mathscr{H}_\infty$ control: constant and time-varying delays. Int J Control 76:48–60
124. Mahmoud MS, Ismail A (2005) New results on delay-dependent control of time-delay systems. IEEE Trans Autom Control 50:95–100
125. Zhang XM, Wu M, She JH, He Y (2005) Delay-dependent stabilization of linear systems with time-varying state and input delays. Automatica 41:1405–1412
126. Michiels W, Van Assche V, Niculescu SI (2005) Stabilization of time-delay systems with a controlled time-varying delay and applications. IEEE Trans Autom Control 50:493–504
127. Parlakli MNA (2006) Improved robust stability criteria and design of robust stabilizing controller for uncertain linear time-delay systems. Int J Robust Nonlinear Control 16:599–636
128. Zhong R, Yang Z (2006) Delay-dependent robust control of descriptor systems with time delay. Asian J Control 8:36–44
129. Lin Z, Fang H (2007) On asymptotic stabilizability of linear systems with delayed input. IEEE Trans Autom Control 52:998–1013
130. Xu S, Lam J (2008) A survey of linear matrix inequality techniques in stability analysis of delay systems. Int J Syst Sci 39:1095–1113
131. Hien LV, Phat VN (2009) Exponential stability and stabilization of a class of uncertain linear time-delay systems. J Frankl Inst 346:611–625
132. Choi HH, Chung MJ (1996) Observer-based $\mathscr{H}_\infty$ controller design for state delayed linear systems. Automatica 32:1073–1075
133. Park JH (2004) On the design of observer-based controller of linear neutral delay-differential systems. Appl Math Comput 150:195–202

134. Kwon OM, Park JH, Lee SM, Won SC (2006) LMI optimization approach to observer-based controller design of uncertain time-delay systems via delayed feedback. J Optim Theory Appl 128:103–117
135. Hua C, Li F, Guan X (2006) Observer-based adaptive control for uncertain time-delay systems. Inf Sci 176:201–214
136. Chen JD (2007) Robust output observer-based control of neutral uncertain systems with discrete and distributed time delays: LMI optimization approach. Chaos Solitons Fractals 34:1254–1264
137. Karimi HR (2008) Observer-based mixed $\mathscr{H}_2/\mathscr{H}_\infty$ control design for linear systems with time-varying delays: an LMI approach. Int J Control Autom Syst 6:1–14
138. Majeed R, Ahmad S, Rehan M (2015) Delay-range-dependent observer-based control of nonlinear systems under input and output time-delays. Appl Math Comput 262:145–159
139. Zhou J, Park JH, Ma Q (2016) Non-fragile observer-based $\mathscr{H}_\infty$ control for stochastic time-delay systems. Appl Math Comput 291:69–83
140. Liu Q, Zhou B (2017) Extended observer based feedback control of linear systems with both state and input delays. J Frankl Inst 354:8232–8255
141. Eskandari N, Dumont GA, Wang ZJ (2017) An observer/predictor-based model of the user for attaining situation awareness. IEEE Trans Hum Mach Syst 46:279–290
142. Mohajerpoor R, Abdi H, Nahavandi S (2016) A new algorithm to design minimal multi-functional observers for linear systems. Asian J Control 18:842–857
143. Fernando TL, Trinh HM, Jennings L (2010) Functional observability and the design of minimum order linear functional observers. IEEE Trans Autom Control 55:1268–1273
144. Mohajerpoor R, Lakshmanan S, Abdi H, Nahavandi S, Park JH (2018) Delay-dependent functional observer design for linear systems with unknown time-varying state delays. IEEE Trans Cybern 48:2036–2048
145. Chang SSL, Peng TKC (1972) Adaptive guaranteed cost control of systems with uncertain parameters. IEEE Trans Autom Control 17:474–483
146. Yu L, Chu J (1999) An LMI approach to guaranteed cost control of linear uncertain time-delay systems. Automatica 35:1155–1159
147. Esfahani SH, Petersen IR (2000) An LMI approach to output feedback guaranteed cost control for uncertain time-delay systems. Int J Robust Nonlinear Control 10:157–174
148. Yu L, Gao F (2001) Optimal guaranteed cost control of discrete-time uncertain systems with both state and input delays. J Frankl Inst 338:101–110
149. Chen WH, Guan ZH, Lu X (2003) Delay-dependent guaranteed cost control for uncertain discrete-time systems with delay. IEE Proc Control Theory Appl 150:412–416
150. Shi P, Boukas EK, Shi Y, Agarwal RK (2003) Optimal guaranteed cost control of uncertain discrete time-delay systems. J Comput Appl Math 157:435–451
151. Park JH, Jung HY (2004) On the design of nonfragile guaranteed cost controller for a class of uncertain dynamic systems with state delays. Appl Math Comput 150:245–257
152. Chen WH, Guan ZH, Lu X (2004) Delay-dependent output feedback guaranteed cost control for uncertain time-delay systems. Automatica 40:1263–1268
153. Park JH (2004) Delay-dependent guaranteed cost stabilization criterion for neutral-delay-differential systems: matrix inequality approach. Comput Math Appl 47:1507–1515
154. Park JH (2005) Delay-dependent criterion for guaranteed cost control of neutral delay systems. J Optim Theory Appl 124:491–502
155. Lien CH (2007) Delay-dependent and delay-independent guaranteed cost control for uncertain neutral systems with time-varying delays via LMI approach. Chaos Solitons Fractals 33:1017–1027
156. Fernando TL, Phat VN, Trinh HM (2013) Output feedback guaranteed cost control of uncertain linear discrete systems with interval time-varying delays. Appl Math Model 37:1580–1589
157. Keel L, Bhattacharyya S (1997) Robust, fragile, or optimal. IEEE Trans Autom Control 42:1098–1105
158. Dorato P (1998) Non-fragile controller design: an overview. In: Proceedings of American control conference, pp 2829–2831

159. Xu S, Lam J, Wang J, Yang GH (2004) Non-fragile positive real control for uncertain linear neutral delay systems. Syst Control Lett 52:59–74
160. Park JH (2004) Robust non-fragile control for uncertain discrete-delay large-scale systems with a class of controller gain variations. Appl Math Comput 149:147–164
161. Yue D, Lam J (2005) Non-fragile guaranteed cost control for uncertain descriptor systems with time-varying state and input delays. Optim Control Appl Methods 26:85–105
162. Xie N, Tang GY (2006) Delay-dependent nonfragile guaranteed cost control for nonlinear time-delay systems. Nonlinear Anal Theory Methods Appl 64:2084–2097
163. Lien CH, Yu KW (2007) Nonfragile control for uncertain neutral systems with time-varying delays via the LMI optimization approach. IEEE Trans Syst Man Cybern Part B (Cybernetics) 37:493–499
164. Chen JD, Yang CD, Lien CH, Horng JH (2008) New delay-dependent non-fragile $\mathscr{H}_\infty$ observer-based control for continuous time-delay systems. Inf Sci 178:4699–4706
165. Liu L, Han Z, Li W (2010) $\mathscr{H}_\infty$ non-fragile observer-based sliding mode control for uncertain time-delay systems. J Frankl Inst 347:567–576
166. Park JH (2004) Design of dynamic output feedback controller for a class of neutral systems with discrete and distributed delay. IEE Proc Control Theory Appl 151:610–614
167. Park JH (2005) Convex optimization approach to dynamic output feedback control for delay differential systems of neutral type. J Optim Theory Appl 127:411–423
168. Li L, Jia Y (2009) Non-fragile dynamic output feedback control for linear systems with time-varying delay. IET Control Theory Appl 3:995–1005
169. Thuan MV, Phat VN, Trinh HM (2012) Dynamic output feedback guaranteed cost control for linear systems with interval time-varying delays in states and outputs. Appl Math Comput 218:10697–10707
170. Lee TH, Park JH (2018) New methods of fuzzy sampled-data control for stabilization of chaotic systems. IEEE Trans Syst Man Cybern Syst 48:2026–2034
171. Lee TH, Park JH (2017) Stability analysis of sampled-data systems via free-matrix-based time-dependent discontinuous Lyapunov approach. IEEE Trans Autom Control 62:3653–3657
172. Wen S, Huang T, Yu X, Chen MZQ, Zeng Z (2016) Aperiodic sampled-data sliding-mode control of fuzzy systems with communication delays via the event-triggered method. IEEE Transactions on Fuzzy Systems 24:1048–1057
173. Wang B, Cheng J, Al-Barakati A, Fardoun HM (2017) A mismatched membership function approach to sampled-data stabilization for T-S fuzzy systems with time-varying delayed signals. Signal Process 140:161–170
174. Ge C, Shi Y, Park JH, Hua C (2019) Robust $\mathscr{H}_\infty$ stabilization for T-S fuzzy systems with time-varying delays and memory sampled-data control. Appl Math Comput 346:500–512
175. Wu ZG, Shi P, Su H, Chu J (2013) Sampled-data synchronization of chaotic Lur'e systems with time delays. IEEE Trans Neural Netw Learn Syst 24:410–421
176. Hua C, Ge C, Guan X (2015) Synchronization of chaotic Lur'e systems with time delays using sampled-data control. IEEE Trans Neural Netw Learn Syst 26:1214–1221
177. Shi K, Liu X, Zhu H, Zhong S, Liu Y, Yin C (2016) Novel integral inequality approach on master-slave synchronization of chaotic delayed Lur'e systems with sampled-data feedback control. Nonlinear Dyn 83:1259–1274
178. Wen G, Duan Z, Yu W, Chen G (2013) Consensus of multi-agent systems with nonlinear dynamics and sampled-data information: a delayed-input approach. Int J Robust Nonlinear Control 23:602–619
179. Wu ZG, Shi P, Su H, Chu J (2013) Stochastic synchronization of Markovian jump neural networks with time-varying delay using sampled data. IEEE Trans Cybern 43:1796–1806
180. Lee TH, Park JH, Kwon OM, Lee SM (2013) Stochastic sampled-data control for state estimation of time-varying delayed neural networks. Neural Netw 46:99–108
181. Rakkiyappan R, Sakthivel N, Cao J (2015) Stochastic sampled-data control for synchronization of complex dynamical networks with control packet loss and additive time-varying delays. Neural Netw 66:46–63

Part II
Stability Analysis of Time-Delay Systems

Chapter 3
Integral Inequalities

3.1 Introduction

Integral inequalities [1] have been widely used in the stability analysis of time-delay systems since they can directly estimate integral terms arising in the derivative of Lyapunov–Krasovskii (L–K) functionals. In essence, the role of an integral inequality is to transform the integral of a product of vectors to the product of integrals of vectors. As a result, tractable stability conditions could be obtained in the form of linear matrix inequalities (LMIs) [2].

Generally speaking, according to whether free matrices are involved or not, integral inequalities [3–6] can be classified into two categories: integral inequalities without free matrices and integral inequalities with free matrices.

For the first category, a number of well-known integral inequalities are reported in the literature [2, 7–15] such as the Jensen inequality [2, 16], Wirtinger-based integral inequality [11, 17], and auxiliary-function-based integral inequality [9]. It is worth noting that the above-mentioned three integral inequalities are developed via three different methods, that is, respectively, the Schur complement, the Wirtinger inequality and the auxiliary-function-based method.

For the second category, there are also various inequalities reported in the literature [18–23], such as the conventional free-matrix-based (FMB) integral inequality [18, 19, 21], the general free-matrix-based (GFMB) inequality [23] and the improved GFMB integral inequality [24]. The improved GFMB inequality provides more freedom to estimate integral terms than the GFMB one while the GFMB inequality provides more freedom than the conventional FMB one. It is noted that the two different categories of integral inequalities lead to two different ways to estimate a single-integral term with a time-varying delay. One way is the combination of an integral inequality without free matrices with the reciprocally convex lemma [25–27]. Another way is just the application of an integral inequality with free matrices.

In this chapter, we are devoted to the investigation of integral inequalities. The remainder of this chapter is divided into three parts.

J. H. Park et al., *Dynamic Systems with Time Delays: Stability and Control*,
https://doi.org/10.1007/978-981-13-9254-2_3

The first part is concerned with integral inequalities without free matrices. Some of existing well-known integral inequalities are firstly recalled such as the Jensen, Wirtinger-based, auxiliary-function-based, and Bessel–Legendre ones. Then, based on orthogonal polynomials defined in integral inner spaces, two series of general integral inequalities without free matrices are developed, respectively, in the upper- and lower-triangular forms. Finally, various single- and multiple-integral inequalities without free matrices are obtained.

The second part is concerned with integral inequalities with free matrices. The conventional FMB integral inequalities are firstly recalled. Then, based on orthogonal polynomials defined in a single-integral inner space, one series of single-integral inequalities with free matrices are developed. Finally, the general FMB (GFMB) and improved GFMB inequalities are, respectively, proposed. It is worth noting that the idea for developing the GFMB and improved GFMB inequalities can be directly extended to the multiple-integral case.

In the third part, the relationships between the two categories of integral inequalities are discussed. It is pointed out that the two corresponding integral inequalities with and without free matrices actually produce the same tight bounds of integral terms but lead to different conservative stability conditions although the same L–K functional is employed.

3.2 Integral Inequalities Without Free Matrices

3.2.1 Some Well-Known Integral Inequalities

3.2.1.1 Jensen Inequalities

Lemma 3.1 (The Jensen single-integral inequality) *For a matrix $R \in \mathbb{S}_+^n$ and a vector function $x(s) : [a, b] \to \mathbb{R}^n$ such that the integrals concerned are well defined, the following inequality holds:*

$$\int_a^b x^T(s)Rx(s)ds \geq \frac{1}{b-a}\left(\int_a^b x(s)ds\right)^T R\left(\int_a^b x(s)ds\right). \tag{3.1}$$

The inequality (3.1) is proved by the Schur complement [2]. With $x(s)$ being replaced by $\dot{x}(s)$, we have the following corollary.

Corollary 3.1 (The Jensen single-integral inequality) *For a matrix $R \in \mathbb{S}_+^n$ and a differentiable vector function $x(s) : [a, b] \to \mathbb{R}^n$, the following inequality holds:*

$$\int_a^b \dot{x}^T(s)R\dot{x}(s)ds \geq \frac{1}{b-a}\left(x(b) - x(a)\right)^T R\left(x(b) - x(a)\right). \tag{3.2}$$

When the double-integral functional $\int_{t-h}^{t}\int_{u}^{t}\dot{x}^T(s)R\dot{x}(s)dsdu$ is included in L–K functionals, the Jensen inequality (3.2) is often used to estimate the single-integral term $\int_{t-h}^{t}\dot{x}(s)^T R\dot{x}(s)ds$ arising in the derivative of L–K functionals. On the other hand, if double-integral terms are countered in the derivative of L–K functionals, the following Jensen double-integral inequalities are needed.

Lemma 3.2 (The Jensen double-integral inequalities) *For a matrix $R \in \mathbb{S}_+^n$ and a vector function $x(t) : [a, b] \to \mathbb{R}^n$ such that the integrals concerned are well defined, the following inequalities hold:*

$$\int_a^b\int_u^b x^T(s)Rx(s)dsdu \geq 2\Phi_{u1}^T R\Phi_{u1}, \tag{3.3}$$

$$\int_a^b\int_a^u x^T(s)Rx(s)dsdu \geq 2\Phi_{l1}^T R\Phi_{l1} \tag{3.4}$$

where

$$\Phi_{u1} = \frac{1}{b-a}\int_a^b\int_u^b x(s)dsdu, \quad \Phi_{l1} = \frac{1}{b-a}\int_a^b\int_a^u x(s)dsdu.$$

The two double-integral inequalities can also be proved by the Schur complement [2]. With $x(s)$ being replaced by $\dot{x}(s)$, we have the following corollary.

Corollary 3.2 (The Jensen double-integral inequalities) *For a matrix $R \in \mathbb{S}_+^n$ and a differentiable vector function $x(t) : [a, b] \to \mathbb{R}^n$, the following inequalities hold:*

$$\int_a^b\int_u^b \dot{x}^T(s)R\dot{x}(s)dsdu \geq 2\tilde{\Phi}_{u1}^T R\tilde{\Phi}_{u1}, \tag{3.5}$$

$$\int_a^b\int_a^u \dot{x}^T(s)R\dot{x}(s)dsdu \geq 2\tilde{\Phi}_{l1}^T R\tilde{\Phi}_{l1} \tag{3.6}$$

where

$$\tilde{\Phi}_{u1} = x(b) - \frac{1}{b-a}\int_a^b x(s)ds, \quad \tilde{\Phi}_{l1} = x(a) - \frac{1}{b-a}\int_a^b x(s)ds.$$

Remark 3.1 Unlike single-integral inequalities, multiple-integral inequalities are in two different forms according to integral regions: in the upper-triangular form and in the lower-triangular form. For example, the double-integral inequality (3.5) is in the upper-triangular form while (3.6) is in the lower-triangular form. Both of them can be used to estimate double-integral terms arising in the derivative of L–K functionals.

3.2.1.2 The Wirtinger-Based Integral Inequality

Lemma 3.3 (The Wirtinger-based integral inequality) *For a matrix $R \in \mathbb{S}_+^n$ and a vector function $x(t) : [a, b] \to \mathbb{R}^n$ such that the integrals concerned are well defined, the following inequality holds:*

$$\int_a^b x^T(s)Rx(s)ds \geq \frac{1}{b-a}\Theta_0^T R\Theta_0 + \frac{3}{b-a}\Theta_1^T R\Theta_1 \tag{3.7}$$

where

$$\Theta_0 = \int_a^b x(s)ds, \quad \Theta_1 = \int_a^b x(s)ds - \frac{2}{b-a}\int_a^b \int_u^b x(s)dsdu.$$

The Wirtinger-based integral inequality (3.7) proved via the Wirtinger inequality [11] produces a tighter bound than what the Jensen inequality (3.1) does since the second summand $\frac{3}{b-a}\Theta_1^T R\Theta_1$ is always non-negative. It is worth noting that the information of the double integral of the state, $\int_a^b \int_u^b x(s)dsdu$, is taken into account in (3.7). With $x(s)$ being replaced by $\dot{x}(s)$, we have the following corollary.

Corollary 3.3 (The Wirtinger-based integral inequality) *For a matrix $R \in \mathbb{S}_+^n$ and a differentiable vector function $x(t) : [a, b] \to \mathbb{R}^n$, the following inequality holds:*

$$\int_a^b \dot{x}^T(s)R\dot{x}(s)ds \geq \frac{1}{b-a}\tilde{\Theta}_0^T R\tilde{\Theta}_0 + \frac{3}{b-a}\tilde{\Theta}_1^T R\tilde{\Theta}_1 \tag{3.8}$$

where

$$\tilde{\Theta}_0 = x(b) - x(a), \quad \tilde{\Theta}_1 = x(b) + x(a) - \frac{2}{b-a}\int_a^b x(s)ds.$$

Due to higher accuracy, the Wirtinger-based integral inequality (3.8) is finding a wider application in the stability analysis of time-delay systems than the Jensen one (3.2).

Remark 3.2 Suppose $x(t)$ is a constant vector. Then, the left-hand side of (3.8) is equal to zero. In this case, every summand on the right-hand side must be equal to zero so that the inequality (3.8) holds. It is found that both $\tilde{\Theta}_0$ and $\tilde{\Theta}_1$ are zero vectors, which verifies the above conclusion. In fact, the above conclusion is suitable for any other integral inequalities, which can be used to check the correctness of integral inequalities.

3.2.1.3 The Auxiliary-Function-Based Inequality

Lemma 3.4 (The auxiliary-function-based single-integral inequality) *For a matrix $R \in \mathbb{S}_+^n$ and an integrable vector function $x(t) : [a, b] \to \mathbb{R}^n$, the following inequality holds:*

$$\int_a^b x^T(s)Rx(s)ds \geq \frac{1}{b-a}\Theta_0^T R\Theta_0 + \frac{3}{b-a}\Theta_1^T R\Theta_1 + \frac{5}{b-a}\Theta_2^T R\Theta_2 \quad (3.9)$$

where Θ_0 and Θ_1 are defined in Lemma 3.3, and

$$\Theta_2 = \int_a^b x(s)ds - \frac{6}{b-a}\int_a^b\int_s^b x(r)drds + \frac{12}{(b-a)^2}\int_a^b\int_u^b\int_s^b x(r)drdsdu.$$

With the help of the two functions $p_1(s) = (s-a) - \frac{(b-a)}{2}$ and $p_2(s) = (s-a)^2 - (b-a)(s-a) + \frac{(b-a)^2}{6}$, the auxiliary-function-based integral inequality (3.9) is developed [9]. Compared with the Wirtinger-based inequality (3.7), the information of the triple integral of the state, $\int_a^b\int_u^b\int_s^b x(r)drdsdu$, is taken into account. So, the auxiliary-function-based integral inequality (3.9) produces a tighter bound than what the inequality (3.7) does.

Lemma 3.5 (The auxiliary-function-based double-integral inequalities) *For a matrix $R \in \mathbb{S}_+^n$ and an integrable vector function $x(s) : [a, b] \to \mathbb{R}^n$, the following inequalities hold:*

$$\int_a^b\int_u^b x^T(s)Rx(s)dsdu \geq 2\Phi_{u1}^T R\Phi_{u1} + 4\Phi_{u2}^T R\Phi_{u2}, \quad (3.10)$$

$$\int_a^b\int_a^u x^T(s)Rx(s)dsdu \geq 2\Phi_{l1}^T R\Phi_{l1} + 4\Phi_{l2}^T R\Phi_{l2} \quad (3.11)$$

where

$$\begin{aligned}
\Phi_{u1} &= \frac{1}{b-a}\int_a^b\int_u^b x(s)dsdu,\\
\Phi_{u2} &= \frac{2}{b-a}\int_a^b\int_u^b x(s)dsdu - \frac{6}{(b-a)^2}\int_a^b\int_u^b\int_r^b x(s)dsdrdu,\\
\Phi_{l1} &= \int_a^b x(s)ds - \frac{1}{(b-a)}\int_a^b\int_u^b x(s)dsdu,\\
\Phi_{l2} &= \int_a^b x(s)ds - \frac{4}{(b-a)}\int_a^b\int_u^b x(s)dsdu + \frac{6}{(b-a)^2}\int_a^b\int_u^b\int_r^b x(s)dsdrdu.
\end{aligned}$$

With the help of the auxiliary functions $p(s) = (b-s) - \frac{(b-a)}{3}$ and $p(s) = (s-a) - \frac{(b-a)}{3}$, the auxiliary-function-based double-integral inequalities (3.10) and (3.11) are, respectively, obtained in the upper- and lower-triangular forms. With $x(t)$ being replaced by $\dot{x}(t)$ in (3.9), (3.10) and (3.11), we have the following corollary.

Corollary 3.4 *For a matrix $R \in \mathbb{S}_+^n$ and a differentiable vector function $x(s) : [a, b] \to \mathbb{R}^n$, the following inequalities hold:*

$$\int_a^b \dot{x}^T(s)R\dot{x}(s)ds \geq \frac{1}{b-a}\tilde{\Theta}_0^T R\tilde{\Theta}_0 + \frac{3}{b-a}\tilde{\Theta}_1^T R\tilde{\Theta}_1 + \frac{5}{b-a}\tilde{\Theta}_2^T R\tilde{\Theta}_2, \quad (3.12)$$

$$\int_a^b \int_u^b \dot{x}^T(s)R\dot{x}(s)dsdu \geq 2\tilde{\Phi}_{u1}^T R\tilde{\Phi}_{u1} + 4\tilde{\Phi}_{u2}^T R\tilde{\Phi}_{u2}, \quad (3.13)$$

$$\int_a^b \int_a^u \dot{x}^T(s)R\dot{x}(s)dsdu \geq 2\tilde{\Phi}_{l1}^T R\tilde{\Phi}_{l1} + 4\tilde{\Phi}_{l2}^T R\tilde{\Phi}_{l2} \quad (3.14)$$

where $\tilde{\Theta}_0$ and $\tilde{\Theta}_1$ are defined in Corollary 3.3, $\tilde{\Phi}_{u1}$ and $\tilde{\Phi}_{l1}$ are defined in Corollary 3.2, and

$$\tilde{\Theta}_2 = x(b) - x(a) + \frac{6}{b-a}\int_a^b x(s)ds - \frac{12}{(b-a)^2}\int_a^b \int_u^b x(s)dsdu,$$

$$\tilde{\Phi}_{u2} = x(b) + \frac{2}{b-a}\int_a^b x(s)ds - \frac{6}{(b-a)^2}\int_a^b \int_u^b x(s)dsdu,$$

$$\tilde{\Phi}_{l2} = x(a) - \frac{4}{b-a}\int_a^b x(s)ds + \frac{6}{(b-a)^2}\int_a^b \int_u^b x(s)dsdu.$$

Remark 3.3 According to the above discussions, it is found that with the information of the double integral of the state considered, the Wirtinger-based integral inequality (3.7) is more accurate than the Jensen inequality (3.1). Similarly, with the information of the triple integral of the state considered, the auxiliary-function-based inequality (3.9) is more accurate than (3.7). Then, a question naturally arises: if more integral information of the state is considered, could a more accurate integral inequality be developed? On the other hand, the integral inequalities discussed above are all the single- or double-integral ones which can be used to estimate only the single- or double-integral terms. Then, another question naturally arises: how can we estimate multiple-integral terms if they arise in the derivative of L–K functionals? As for the first question, the Bessel–Legendre inequality gives the answer. As for the second question, two series of general integral inequalities give the answer.

3.2.1.4 The Bessel–Legendre Inequality

Let $L_k(s)$, $k \in \mathbb{N}$, be Legendre polynomials that are defined as follows [13]:

$$L_k(s) = (-1)^k \sum_{l=0}^{k} p_l^k \left(\frac{s-a}{b-a}\right)^l, \quad k \in \mathbb{N}, \quad s \in [a, b], \quad (3.15)$$

where $p_l^k = (-1)^l \binom{k}{l}\binom{k+l}{l}$. The set of Legendre polynomials forms an orthogonal sequence with respect to the inner product:

$$\langle L_l, L_k \rangle = \int_a^b L_l(s) L_k(s) ds, \quad l, k \in \mathbb{N}.$$

Legendre polynomials defined in (3.15) have the following properties for $l, k \in \mathbb{N}$:

Orthogonality:

$$\langle L_l, L_k \rangle \begin{cases} = 0, & k \neq l, \\ = \frac{b-a}{2k+1}, & k = l. \end{cases}$$

Boundary conditions:

$$L_k(a) = (-1)^k, \quad L_k(b) = 1.$$

Differentiation:

$$\dot{L}_k(s) = \begin{cases} 0, & k = 0, \\ \sum_{i=0}^{k-1} \frac{(2i+1)}{h}\left(1 - (-1)^{k+i}\right) L_i(s), & k \geq 1. \end{cases}$$

The property of differentiation shows that the derivative of any Legendre polynomial can be linearly represented by Legendre polynomials of low degree.

Lemma 3.6 (The Bessel–Legendre inequality) *For a matrix $R \in \mathbb{S}_+^n$ and an integrable vector function $x(s) : [a, b] \to \mathbb{R}^n$, the following inequality holds for any $N \in \mathbb{N}$:*

$$\int_a^b x^T(s) R x(s) ds \geq \frac{1}{b-a} \sum_{k=0}^{N} (2k+1) \Omega_k^T R \Omega_k \tag{3.16}$$

where

$$\Omega_k = \int_a^b L_k(s) x(s) ds.$$

With the help of Legendre polynomials, the Bessel–Legendre inequality (3.16) is developed [13]. As the value of N increases, the Bessel–Legendre inequality (3.16) becomes more accurate. Especially, by setting $N = 0$, $N = 1$ and $N = 2$, the Jensen (3.1), Wirtinger-based (3.7) and auxiliary-function-based integral inequalities (3.12) are, respectively, recovered.

3.2.2 Two Series of General Integral Inequalities

For any integrable function $f(s)$, the two notations are defined:

$$\mathscr{I}^{U}_{m(a,b)}(f(s)) = \int_{a}^{b}\int_{s_1}^{b}\cdots\int_{s_{m-1}}^{b} f(s_m)ds_m\cdots ds_2ds_1,$$
$$\mathscr{I}^{L}_{m(a,b)}(f(s)) = \int_{a}^{b}\int_{a}^{s_1}\cdots\int_{a}^{s_{m-1}} f(s_m)ds_m\cdots ds_2ds_1.$$

When no confusion is possible, the subscript (a, b) may be omitted.

Lemma 3.7 *For given integers $N \in \mathbb{N}$, $m \in \mathbb{N}^{+}$, a matrix $R \in \mathbb{S}^{n}_{+}$, and an integrable vector function $x(s) : [a, b] \to \mathbb{R}^{n}$, the following integral inequalities hold for any polynomial functions $p_k(s)$ of degree k:*

$$\begin{aligned}&\mathscr{I}^{U}_{m}(x^T(s)Rx(s))\\ &\geq \sum_{k=0}^{N}\left(\mathscr{I}^{U}_{m}(p_k(s)^2)\right)^{-1}\Big(\mathscr{I}^{U}_{m}(p_k(s)x(s))\Big)^{T} R\Big(\mathscr{I}^{U}_{m}(p_k(s)x(s))\Big),\end{aligned} \tag{3.17}$$

$$\begin{aligned}&\mathscr{I}^{L}_{m}(x^T(s)Rx(s))\\ &\geq \sum_{k=0}^{N}\left(\mathscr{I}^{L}_{m}(p_k(s)^2)\right)^{-1}\Big(\mathscr{I}^{L}_{m}(p_k(s)x(s))\Big)^{T} R\Big(\mathscr{I}^{L}_{m}(p_k(s)x(s))\Big),\end{aligned} \tag{3.18}$$

satisfying

$$\mathscr{I}^{U}_{m}(p_k(s)p_l(s))\begin{cases} = 0, & k \neq l,\\ \neq 0, & k = l,\end{cases} \tag{3.19}$$

$$\mathscr{I}^{L}_{m}(p_k(s)p_l(s))\begin{cases} = 0, & k \neq l,\\ \neq 0, & k = l,\end{cases} \tag{3.20}$$

for (3.17) and (3.18), respectively.

Proof Let

$$F(s) = \begin{bmatrix} x(s) & p_0(s)I & p_1(s)I & \cdots & p_N(s)I \end{bmatrix}.$$

Then, it follows that

$$F^T(s)RF(s) = \begin{bmatrix} x^T(s)Rx(s) & p_0(s)x^T(s)R & p_1(s)x^T(s)R & \cdots & p_N(s)x^T(s)R \\ * & p_0^2(s)R & p_0(s)p_1(s)R & \cdots & p_0(s)p_N(s)R \\ * & * & p_1^2(s)R & \cdots & p_1(s)p_N(s)R \\ * & * & * & \ddots & \vdots \\ * & * & * & * & p_N^2(s)R \end{bmatrix}.$$

Let $\Phi = \mathscr{I}_m^U(F^T(s)RF(s))$. Note that $\Phi \geq 0$ owing to $R > 0$. From (3.19) we get

$$\Phi = \begin{bmatrix} \mathscr{I}_m^U(x^T(s)Rx(s)) & \mathscr{I}_m^U(p_0(s)x^T(s)R) & \mathscr{I}_m^U(p_1(s)x^T(s)R) & \cdots & \mathscr{I}_m^U(p_N(s)x^T(s)R) \\ * & \mathscr{I}_m^U(p_0^2(s)R) & 0 & \cdots & 0 \\ * & * & \mathscr{I}_m^U(p_1^2(s)R) & \cdots & 0 \\ * & * & * & \ddots & \vdots \\ * & * & * & * & \mathscr{I}_m^U(p_N^2(s)R) \end{bmatrix}.$$

So, applying the Schur complement to $\Phi \geq 0$ directly leads to the inequality (3.17). In a similar way, the inequality (3.18) can be obtained. This completes the proof. ■

Remark 3.4 In Lemma 3.7, based on orthogonal polynomials defined in integral inner spaces, the two series of integral inequalities (3.17) and (3.18) are, respectively, obtained in the upper- and lower-triangular forms. The two series of general integral inequalities actually establish a basic framework. Almost all of existing integral inequalities without free matrices can be recovered from them, no matter the single- or multiple-integral inequalities, no matter in the upper- or lower-triangular forms. For example, by setting $m = 2$ and $N = 1$, the auxiliary-function-based double-integral inequalities (3.10) and (3.11) are, respectively, recovered from (3.17) and (3.18). By setting $m = 1$ and choosing Legendre polynomials as the orthogonal ones, the Bessel–Legendre (B–L) inequality (3.16) is recovered from (3.17). In addition, it is worth noting that the proof of Lemma 3.7 is very simple and straightforward, in which an intermediate matrix is defined and the Schur complement is used only once.

Remark 3.5 In Lemma 3.7, the set of polynomials $p_k(s)$, $k \in \mathbb{N}$, forms an orthogonal base with respect to the integral inner space $\langle p_k, p_l \rangle = \mathscr{I}_m^U(p_k(s)p_l(s))$ or $\langle p_k, p_l \rangle = \mathscr{I}_m^L(p_k(s)p_l(s))$. When $m = 1$, Legendre polynomials are exactly the orthogonal ones defined in the single-integral inner space while when $m \geq 2$, the orthogonal polynomials defined in the multiple-integral inner space can be systematically obtained by the Schmidt orthogonal method with an initially chosen base such as $\{1, (s-a), \ldots, (s-a)^N\}$.

Just like Legendre polynomials, ordinary orthogonal polynomials $p_k(s)$ defined in the integral inner space also have the following three properties: orthogonality, boundary conditions and differentiation. Especially, for the property of differentiation, it follows:

$$\dot{p}_k(s) = \begin{cases} 0, & k = 0, \\ \sum_{i=0}^{k-1} g_i^k p_i(s), & k > 0, \end{cases} \tag{3.21}$$

where g_i^k, $i \in \{0, 1, \ldots, k-1\}$, are constant scalars.

3.2.2.1 Properties of Integral Terms

It is worth pointing out that it is not convenient to directly apply the two series of general integral inequalities to stability analysis of time-delay systems since the orthogonal polynomials are explicitly involved in the bounds. So, how to remove the involved orthogonal polynomials becomes an important topic. In what follows, the properties of the integral terms $\mathscr{I}_m^U(p_k(s)x(s))$ and $\mathscr{I}_m^L(p_k(s)x(s))$ are investigated so that they can be linearly represented by a sequence of integral terms only with respect to the state $x(s)$.

We always suppose $p_0(s) = 1$ in any integral inner space without loss of generality. For convenience, we define the following notations, $k \in \mathbb{N}$, $m \in \mathbb{N}^+$:

$$\Omega_k(\alpha) = \int_\alpha^b p_k(s)x(s)ds, \quad \Omega_{k,0^m}(\alpha) = \int_\alpha^b p_k(s)\Omega_{0^m}(s)ds,$$

$$\Omega_{0^m,k}(\alpha) = \int_\alpha^b \int_{u_1}^b \cdots \int_{u_{m-1}}^b \Omega_k(u_m)du_m \cdots du_2 du_1,$$

$$\Theta_k(\beta) = \int_a^\beta p_k(s)x(s)ds, \quad \Theta_{k,0^m}(\beta) = \int_a^\beta p_k(s)\Theta_{0^m}(s)ds,$$

$$\Theta_{0^m,k}(\beta) = \int_a^\beta \int_a^{u_1} \cdots \int_a^{u_{m-1}} \Theta_k(u_m)du_m \cdots du_2 du_1,$$

where 0^m denotes the abbreviation of $\underbrace{0, \ldots, 0}_{m}$. Especially, we define $\Omega_k = \Omega_k(a)$, $\Omega_{k,0^m} = \Omega_{k,0^m}(a)$, $\Omega_{0^m,k} = \Omega_{0^m,k}(a)$, $\Theta_k = \Theta_k(b)$, $\Theta_{k,0^m} = \Theta_{k,0^m}(b)$, $\Theta_{0^m,k} = \Theta_{0^m,k}(b)$, $\Omega_0 = \Omega_{0^1}$ and $\Theta_0 = \Theta_{0^1}$ for simplicity. According to the above definitions, some instances are listed for clarity:

$$\Omega_0 = \int_a^b x(s)ds, \quad \Omega_{0^2} = \int_a^b \int_u^b x(s)dsdu,$$

$$\Omega_2 = \int_a^b p_2(s)x(s)ds, \quad \Omega_{2,0^3} = \int_a^b p_2(s)\Omega_{0^3}(s)ds,$$

$$\Omega_{0^2,2} = \int_a^b \int_u^b \Omega_2(s)dsdu, \quad \Theta_0 = \int_a^b x(s)ds,$$

$$\Theta_2 = \int_a^b p_2(s)x(s)ds, \quad \Theta_{3,0^2} = \int_a^b p_3(s)\Theta_{0^2}(s)ds.$$

Additionally, it is also seen from the definitions of notations that

$$\Omega_k = \mathscr{I}_1^U(p_k(s)x(s)), \quad \Omega_{k,0^m} = \mathscr{I}_1^U(p_k(s)\Omega_{0^m}(s)),$$
$$\Omega_{0^m,k} = \mathscr{I}_m^U(\Omega_k(s)) = \mathscr{I}_{m+1}^U(p_k(s)x(s)).$$

Lemma 3.8 *For polynomials $p_k(s)$ of degree k, $k \in \mathbb{N}$, satisfying (3.21), the following equations hold, $m \in \mathbb{N}^+$:*

$$\Omega_k = p_k(a)\Omega_0 + \sum_{i=0}^{k-1} g_i^k \Omega_{i,0}, \tag{3.22}$$

$$\Omega_{k,0^m} = p_k(a)\Omega_{0^{m+1}} + \sum_{i=0}^{k-1} g_i^k \Omega_{i,0^{m+1}}, \tag{3.23}$$

$$\Omega_k = \mathscr{L}\{\Omega_0, \Omega_{0^2}, \ldots, \Omega_{0^{k+1}}\}, \tag{3.24}$$

$$\Omega_{k,0^m} = \mathscr{L}\{\Omega_{0^{m+1}}, \Omega_{0^{m+2}}, \ldots, \Omega_{0^{m+k+1}}\}. \tag{3.25}$$

Proof By applying an integration by parts, we have

$$\begin{aligned}
\Omega_k &= \int_a^b p_k(u) d \int_b^u x(s)ds \\
&= p_k(a)\Omega_0 + \int_a^b \dot{p}_k(u) \int_u^b x(s)dsdu \\
&= p_k(a)\Omega_0 + \int_a^b \left(\sum_{i=0}^{k-1} g_i^k p_i(u)\right) \Omega_0(u)du \\
&= p_k(a)\Omega_0 + \sum_{i=0}^{k-1} g_i^k \Omega_{i,0}.
\end{aligned}$$

With similar lines, the Eq. (3.23) can be obtained. Additionally, with (3.22) or (3.23) applied repeatedly, the Eqs. (3.24) and (3.25) can be obtained. This completes the proof. ■

As for the lower-triangular form, similar results can be obtained.

Lemma 3.9 *For polynomials $p_k(s)$ of degree k, $k \in \mathbb{N}$, satisfying (3.21), the following equations hold, $m \in \mathbb{N}^+$:*

$$\Theta_k = p_k(b)\Theta_0 - \sum_{i=0}^{k-1} g_i^k \Theta_{i,0},$$

$$\Theta_{k,0^m} = p_k(b)\Theta_{0^{m+1}} - \sum_{i=0}^{k-1} g_i^k \Theta_{i,0^{m+1}},$$

$$\begin{aligned}\Theta_k &= \mathscr{L}\{\Theta_0, \Theta_{0^2}, \ldots, \Theta_{0^{k+1}}\},\\ \Theta_{k,0^m} &= \mathscr{L}\{\Theta_{0^{m+1}}, \Theta_{0^{m+2}}, \ldots, \Theta_{0^{m+k+1}}\}.\end{aligned}$$

Remark 3.6 From Lemmas 3.8 and 3.9, it can be seen that the integral terms $\mathscr{I}_m^U(p_k(s)x(s))$ and $\mathscr{I}_m^L(p_k(s)x(s))$ with respect to the product $p_k(s)x(s)$ can be linearly represented by integral terms only with respect to the state $x(s)$. Meanwhile, it is also seen that the polynomial $p_k(s)$ involved in $\mathscr{I}_m^U(p_k(s)x(s))$ or $\mathscr{I}_m^L(p_k(s)x(s))$ just implies multiple integral, and vice versa. In other words, the polynomial and multiple integral can be transformed into each other in this case.

Lemma 3.10 *For an integrable vector function $x(s) : [a, b] \to \mathbb{R}^n$, the following equations hold for any $m \in \mathbb{N}^+$:*

$$\Theta_{0^m} = \mathscr{L}\{\Omega_0, \Omega_{0^2}, \ldots, \Omega_{0^m}\}, \tag{3.26}$$

$$\Omega_{0^m} = \mathscr{L}\{\Theta_0, \Theta_{0^2}, \ldots, \Theta_{0^m}\}. \tag{3.27}$$

Proof First consider the Eq. (3.26). The method of mathematical induction is used. From the definitions, it follows $\Theta_0 = \Omega_0$. With $m = 2$, we have

$$\begin{aligned}\Theta_{0^2} &= \int_a^b \Theta_0(u)du = \int_a^b (\Omega_0 - \Omega_0(u))du\\ &= (b-a)\Omega_0 - \Omega_{0^2}.\end{aligned}$$

Now, assume the following equation

$$\Theta_{0^{m-1}} = c_1\Omega_0 + c_2\Omega_{0^2} + \cdots + c_m\Omega_{0^{m-1}}, \quad m > 3$$

holds, where $c_1, \ldots, c_m$ are constant coefficients. Hence, based on the above assumption we have

$$\begin{aligned}\Theta_{0^m} &= \mathscr{I}_m^L(x(s)) = \mathscr{I}_{m-1}^L(\Theta_0(s))\\ &= c_1\tilde{\Omega}_0 + c_2\tilde{\Omega}_{0^2} + \cdots + c_{m-1}\tilde{\Omega}_{0^{m-1}}\end{aligned}$$

where

$$\tilde{\Omega}_{0^k} = \int_a^b \int_{u_1}^b \cdots \int_{u_{k-1}}^b \Theta_0(u_k)du_k \cdots du_2 du_1, \quad k \in \{1, 2, \ldots, m-1\}.$$

Define $S_k := \mathscr{I}_k^U(1)$. Then, we have $\tilde{\Omega}_{0^k} = S_k\Omega_0 - \Omega_{0^{k+1}}$ since $\Theta_0(u_k) = \Omega_0 - \Omega_0(u_k)$. Therefore, Θ_{0^m} can be linearly represented by $\Omega_0, \Omega_{0^2}, \ldots, \Omega_{0^m}$. In a similar way, Ω_{0^m} can also be linearly represented by $\Theta_0, \Theta_{0^2}, \ldots, \Theta_{0^m}$. This completes the proof. ■

Remark 3.7 From Lemma 3.10, it is found that the integral terms in the upper- and lower-triangular forms can be transformed into each other. In the transformation, the highest degree of integral remains the same. For instance, we have $\Omega_{0,0} = (b-a)\Theta_0 - \Theta_{0,0}$, $\Theta_{0,0,0} = \frac{(b-a)^2}{2}\Omega_0 - (b-a)\Omega_{0,0,} + \Omega_{0,0,0}$. Obviously, $x(s)$ may be any integrable vector function. So, Lemma 3.10 can be stated in a more general version: for any integrable vector function $f(s)$, we have

$$\mathscr{I}_m^L(f) = \mathscr{L}\{\mathscr{I}_1^U(f), \mathscr{I}_2^U(f), \ldots, \mathscr{I}_m^U(f)\},$$
$$\mathscr{I}_m^U(f) = \mathscr{L}\{\mathscr{I}_1^L(f), \mathscr{I}_2^L(f), \ldots, \mathscr{I}_m^L(f)\}.$$

3.2.2.2 Concrete Integral Inequalities

In this subsection, orthogonal polynomials defined in different integral inner spaces are computed. Consequently, various concrete integral inequalities are developed both in the upper- and lower-triangular forms.

Integral Inequalities in the Upper-Triangular Form

(1) $m = 1$

In the single-integral inner space, Legendre polynomials are exactly the orthogonal ones. According to the definition (3.15), the first four Legendre polynomials are listed as follows:

$$\begin{aligned}
L_0(s) &= 1, \\
L_1(s) &= -1 + 2\left(\frac{s-a}{b-a}\right), \\
L_2(s) &= 1 - 6\left(\frac{s-a}{b-a}\right) + 6\left(\frac{s-a}{b-a}\right)^2, \\
L_3(s) &= -1 + 12\left(\frac{s-a}{b-a}\right) - 30\left(\frac{s-a}{b-a}\right)^2 + 20\left(\frac{s-a}{b-a}\right)^3.
\end{aligned}$$

Based on Lemma 3.7, the single-integral inequality

$$\int_a^b x^T(s)Rx(s)ds \geq \sum_{k=0}^{N} \frac{2k+1}{b-a}\Omega_k^T R\Omega_k, \tag{3.28}$$

holds, where $\Omega_k = \int_a^b L_k(s)x(s)ds$. It is seen that the above inequality (3.28) is just the Bessel–Legendre inequality. Through some computations with the four Legendre polynomials, we get the following single-integral inequality from (3.28):

$$\int_a^b x^T(u)Rx(u)du \geq \sum_{i=0}^{4} \frac{2i+1}{b-a}\Omega_i^T R\Omega_i, \tag{3.29}$$

where

$$\begin{aligned}
\Omega_1 &= -\Omega_0 + \frac{2}{b-a}\Omega_{0^2},\\
\Omega_2 &= \Omega_0 - \frac{6}{b-a}\Omega_{0^2} + \frac{12}{(b-a)^2}\Omega_{0^3},\\
\Omega_3 &= -\Omega_0 + \frac{12}{b-a}\Omega_{0^2} - \frac{60}{(b-a)^2}\Omega_{0^3} + \frac{120}{(b-a)^3}\Omega_{0^4},\\
\Omega_4 &= \Omega_0 - \frac{20}{b-a}\Omega_{0^2} + \frac{180}{(b-a)^2}\Omega_{0^3} - \frac{840}{(b-a)^3}\Omega_{0^4} + \frac{1680}{(b-a)^4}\Omega_{0^5}.
\end{aligned}$$

(2) $m = 2$
Based on Lemma 3.7, the double-integral inequality

$$\int_a^b \int_u^b x^T(s)Rx(s)dsdu \geq \sum_{k=0}^{N} \mathscr{I}_2^U(p_k(s)^2)^{-1}\Omega_{0^1,k}^T R\Omega_{0^1,k} \tag{3.30}$$

holds, where $\mathscr{I}_2^U(p_k^2) = \int_a^b \int_u^b p_k(s)^2 dsdu$ and $\Omega_{0^1,k} = \int_a^b \int_u^b p_k(s)x(s)dsdu$. In the upper-triangular double-integral inner space, the orthogonal polynomials are computed as follows with $N = 3$:

$$\begin{aligned}
p_0(s) &= 1,\\
p_1(s) &= -1 + \frac{3}{2}\left(\frac{s-a}{b-a}\right),\\
p_2(s) &= 1 - 4\left(\frac{s-a}{b-a}\right) + \frac{10}{3}\left(\frac{s-a}{b-a}\right)^2,\\
p_3(s) &= -1 + \frac{15}{2}\left(\frac{s-a}{b-a}\right) - 15\left(\frac{s-a}{b-a}\right)^2 + \frac{35}{4}\left(\frac{s-a}{b-a}\right)^3.
\end{aligned}$$

Through some computations, it follows that

$$\begin{aligned}
\dot{p}_0(s) &= 0,\\
\dot{p}_1(s) &= \frac{3}{2(b-a)} = g_0^1 p_0(s),\\
\dot{p}_2(s) &= -\frac{4}{(b-a)} + \frac{20}{3}\frac{(s-a)}{(b-a)^2} = g_0^2 p_0(s) + g_1^2 p_1(s),\\
\dot{p}_3(s) &= \frac{15}{2(b-a)} - 30\frac{(s-a)}{(b-a)^2} + \frac{105}{4}\frac{(s-a)^2}{(b-a)^3} = g_0^3 p_0(s) + g_1^3 p_1(s) + g_2^3 p_2(s),
\end{aligned}$$

where $g_0^1 = \frac{3}{2(b-a)}$, $g_0^2 = \frac{4}{9(b-a)}$, $g_1^2 = \frac{40}{9(b-a)}$, $g_0^3 = \frac{5}{8(b-a)}$, $g_1^3 = \frac{1}{(b-a)}$ and $g_2^3 = \frac{63}{8(b-a)}$.

Additionally, we have $\mathscr{I}_2^U(p_0(s)^2)k = \frac{(b-a)^2}{2}$, $\mathscr{I}_2^U(p_1(s)^2) = \frac{(b-a)^2}{16}$, $\mathscr{I}_2^U(p_2(s)^2) = \frac{(b-a)^2}{54}$ and $\mathscr{I}_2^U(p_3(s)^2) = \frac{(b-a)^2}{128}$.

Therefore, we get the following double-integral inequality from (3.30):

$$\int_a^b \int_u^b x^T(s)Rx(s)dsdu \geq \frac{2}{(b-a)^2}\Omega_{0^2}^T R\Omega_{0^2} + \frac{16}{(b-a)^2}\Omega_{0,1}^T R\Omega_{0,1} + \frac{54}{(b-a)^2}\Omega_{0,2}^T R\Omega_{0,2} + \frac{128}{(b-a)^2}\Omega_{0,3}^T R\Omega_{0,3}, \quad (3.31)$$

where

$$\Omega_{0,1} = -\Omega_{0^2} + \frac{3}{(b-a)}\Omega_{0^3},$$

$$\Omega_{0,2} = \Omega_{0^2} - \frac{8}{(b-a)}\Omega_{0^3} + \frac{20}{(b-a)^2}\Omega_{0^4},$$

$$\Omega_{0,3} = -\Omega_{0^2} + \frac{15}{(b-a)}\Omega_{0^3} - \frac{90}{(b-a)^2}\Omega_{0^4} + \frac{210}{(b-a)^3}\Omega_{0^5}.$$

(3) $m = 3$

Based on Lemma 3.7, the triple-integral inequality

$$\int_a^b \int_u^b \int_r^b x^T(s)Rx(s)dsdrdu \geq \sum_{k=0}^{N} \mathscr{I}_3^U(p_k(s)^2)^{-1}\Omega_{0^2,k}^T R\Omega_{0^2,k} \quad (3.32)$$

holds, where $\Omega_{0^2,k} = \int_a^b \int_u^b \int_r^b p_k(s)x(s)dsdrdu$.

In the upper-triangular triple-integral inner space, the orthogonal polynomials are computed as follows with $N = 2$:

$$p_0(s) = 1,$$

$$p_1(s) = -1 + \frac{4}{3}\left(\frac{s-a}{b-a}\right),$$

$$p_2(s) = 1 - \frac{10}{3}\left(\frac{s-a}{b-a}\right) + \frac{5}{2}\left(\frac{s-a}{b-a}\right)^2.$$

Through some computations with the above three polynomials, we have

$$\dot{p}_0(s) = 0,$$

$$\dot{p}_1(s) = \frac{4}{3(b-a)} = g_0^1 p_0(s),$$

$$\dot{p}_2(s) = -\frac{10}{3(b-a)} + 5\frac{(s-a)}{(b-a)^2} = g_0^2 p_0(s) + g_1^2 p_1(s),$$

where $g_0^1 = \frac{4}{3(b-a)}$, $g_0^2 = \frac{5}{12(b-a)}$ and $g_1^2 = \frac{15}{4(b-a)}$.

Additionally, we have $\mathscr{I}_3^U(p_0(s)^2) = \frac{(b-a)^3}{6}$, $\mathscr{I}_3^U(p_1(s)^2) = \frac{(b-a)^3}{90}$ and $\mathscr{I}_3^U(p_2(s)^2) = \frac{(b-a)^3}{504}$.

Therefore, we get the following triple-integral inequality from (3.32):

$$\begin{aligned}
&\int_a^b \int_{u_1}^b \int_{u_2}^b x^T(u_3)Rx(u_3)du_3du_2du_1 \\
&\geq \frac{6}{(b-a)^3}\Omega_{0^3}^T R\Omega_{0^3} + \frac{90}{(b-a)^3}\Omega_{0^2,1}^T R\Omega_{0^2,1} \\
&+ \frac{504}{(b-a)^3}\Omega_{0^2,2}^T R\Omega_{0^2,2},
\end{aligned} \tag{3.33}$$

where

$$\Omega_{0^2,1} = -\Omega_{0^3} + \frac{4}{(b-a)}\Omega_{0^4},$$

$$\Omega_{0^2,2} = \Omega_{0^3} - \frac{10}{(b-a)}\Omega_{0^4} + \frac{30}{(b-a)^2}\Omega_{0^5}.$$

(4) $m = 4$

Based on Lemma 3.7, the quartic-integral inequality

$$\begin{aligned}
&\int_a^b \int_{u_1}^b \int_{u_2}^b \int_{u_3}^b x^T(u_4)Rx(u_4)du_4du_3du_2du_1 \\
&\geq \sum_{k=0}^{N} \mathscr{I}_4^U(p_k(s)^2)^{-1}\Omega_{0^3,k}^T R\Omega_{0^3,k}
\end{aligned} \tag{3.34}$$

holds. In the upper-triangular quartic-integral inner space, the orthogonal polynomials are computed as follows with $N = 1$:

$$p_0(s) = 1,$$

$$p_1(s) = -1 + \frac{5}{4}\left(\frac{s-a}{b-a}\right).$$

Through some computations, it follows that

$$\dot{p}_0(s) = 0,$$

$$\dot{p}_1(s) = \frac{5}{4(b-a)} = g_0^1 p_0(s),$$

where $g_0^1 = \frac{5}{4(b-a)}$.

Additionally, we have $\mathscr{I}_4^U(p_0(s)^2) = \frac{(b-a)^4}{24}$ and $\mathscr{I}_4^U(p_1(s)^2) = \frac{(b-a)^4}{576}$.

Therefore, we get the following quartic-integral inequality from (3.34):

$$\int_a^b\int_{u_1}^b\int_{u_2}^b\int_{u_3}^b x^T(u_4)Rx(u_4)du_4du_3du_2du_1$$
$$\geq \frac{24}{(b-a)^4}\Omega_{0^4}^T R\Omega_{0^4} + \frac{576}{(b-a)^4}\Omega_{0^3,1}^T R\Omega_{0^3,1}, \tag{3.35}$$

where

$$\Omega_{0^3,1} = -\Omega_{0^4} + \frac{5}{(b-a)}\Omega_{0^5}.$$

Remark 3.8 From the above discussions, it is seen that with a sequence of more orthogonal polynomials considered, a more accurate integral inequality is obtained. There is no doubt that other multiple-integral inequalities can be obtained in a similar way.

With $x(s)$ being replaced by $\dot{x}(s)$ in (3.29), (3.31), (3.33) and (3.35), we have the following corollary.

Corollary 3.5 *For a matrix $R \in \mathbb{S}_+^n$ and a differentiable vector function $x(s)$: $[a, b] \to \mathbb{R}^n$, the following integral inequalities hold:*

$$\int_a^b \dot{x}^T(u_1)R\dot{x}(u_1)du_1$$
$$\geq \frac{1}{b-a}\dot{\Omega}_0^T R\dot{\Omega}_0 + \frac{3}{b-a}\dot{\Omega}_1^T R\dot{\Omega}_1 + \frac{5}{b-a}\dot{\Omega}_2^T R\dot{\Omega}_2 + \frac{7}{b-a}\dot{\Omega}_3^T R\dot{\Omega}_3$$
$$+ \frac{9}{b-a}\dot{\Omega}_4^T R\dot{\Omega}_4,$$
$$\int_a^b\int_{u_1}^b \dot{x}^T(u_2)R\dot{x}(u_2)du_2du_1$$
$$\geq \frac{2}{(b-a)^2}\dot{\Omega}_{0^2}^T R\dot{\Omega}_{0^2} + \frac{16}{(b-a)^2}\dot{\Omega}_{0,1}^T R\dot{\Omega}_{0,1} + \frac{54}{(b-a)^2}\dot{\Omega}_{0,2}^T R\dot{\Omega}_{0,2}$$
$$+ \frac{128}{(b-a)^2}\dot{\Omega}_{0,3}^T R\dot{\Omega}_{0,3},$$
$$\int_a^b\int_{u_1}^b\int_{u_2}^b \dot{x}^T(u_3)R\dot{x}(u_3)du_3du_2du_1$$
$$\geq \frac{6}{(b-a)^3}\dot{\Omega}_{0^3}^T R\dot{\Omega}_{0^3} + \frac{90}{(b-a)^3}\dot{\Omega}_{0^2,1}^T R\dot{\Omega}_{0^2,1} + \frac{504}{(b-a)^3}\dot{\Omega}_{0^2,2}^T R\dot{\Omega}_{0^2,2},$$
$$\int_a^b\int_{u_1}^b\int_{u_2}^b\int_{u_3}^b \dot{x}^T(u_4)R\dot{x}(u_4)du_4du_3du_2du_1$$
$$\geq \frac{24}{(b-a)^4}\dot{\Omega}_{0^4}^T R\dot{\Omega}_{0^4} + \frac{576}{(b-a)^4}\dot{\Omega}_{0^3,1}^T R\dot{\Omega}_{0^3,1},$$

where

$$
\begin{aligned}
\dot{\Omega}_0 &= x(b) - x(a),\\
\dot{\Omega}_{0^2} &= (b-a)x(b) - \Omega_0,\\
\dot{\Omega}_{0^3} &= \frac{(b-a)^2}{2}x(b) - \Omega_{0^2},\\
\dot{\Omega}_{0^4} &= \frac{(b-a)^3}{6}x(b) - \Omega_{0^3},\\
\dot{\Omega}_1 &= x(b) + x(a) - \frac{2}{b-a}\Omega_0,\\
\dot{\Omega}_2 &= x(b) - x(a) + \frac{6}{(b-a)}\Omega_0 - \frac{12}{(b-a)^2}\Omega_{0^2},\\
\dot{\Omega}_3 &= x(b) + x(a) - \frac{12}{(b-a)}\Omega_0 + \frac{60}{(b-a)^2}\Omega_{0^2} - \frac{120}{(b-a)^3}\Omega_{0^3},\\
\dot{\Omega}_4 &= x(b) - x(a) + \frac{20}{(b-a)}\Omega_0 - \frac{180}{(b-a)^2}\Omega_{0^2} + \frac{840}{(b-a)^3}\Omega_{0^3} - \frac{1680}{(b-a)^4}\Omega_{0^4},\\
\dot{\Omega}_{0,1} &= \frac{(b-a)}{2}x(b) + \Omega_0 - \frac{3}{(b-a)}\Omega_{0^2},\\
\dot{\Omega}_{0,2} &= \frac{(b-a)}{3}x(b) - \Omega_0 + \frac{8}{(b-a)}\Omega_{0^2} - \frac{20}{(b-a)^2}\Omega_{0^3},\\
\dot{\Omega}_{0,3} &= \frac{(b-a)}{4}x(b) + \Omega_0 - \frac{15}{(b-a)}\Omega_{0^2} + \frac{90}{(b-a)^2}\Omega_{0^3} - \frac{210}{(b-a)^3}\Omega_{0^4},\\
\dot{\Omega}_{0^2,1} &= \frac{(b-a)^2}{6}x(b) + \Omega_{0^2} - \frac{4}{(b-a)}\Omega_{0^3},\\
\dot{\Omega}_{0^2,2} &= \frac{(b-a)^2}{12}x(b) - \Omega_{0^2} + \frac{10}{(b-a)}\Omega_{0^3} - \frac{30}{(b-a)^2}\Omega_{0^4},\\
\dot{\Omega}_{0^3,1} &= \frac{(b-a)^3}{24}x(b) + \Omega_{0^3} - \frac{5}{(b-a)}\Omega_{0^4}.
\end{aligned}
$$

Integral Inequalities in the Lower-Triangular Form

(1) $m = 2$
The orthogonal polynomials defined in the lower-triangular double-integral inner space are computed as follows with $N = 2$:

$$
\begin{aligned}
p_0(s) &= 1,\\
p_1(s) &= -1 + 3\left(\frac{s-a}{b-a}\right),\\
p_2(s) &= 1 - 8\left(\frac{s-a}{b-a}\right) + 10\left(\frac{s-a}{b-a}\right)^2.
\end{aligned}
$$

Through some computations with the above three polynomials, we get

$$\dot{p}_0(s) = 0, \quad \dot{p}_1(s) = g_0^1 p_0(s), \quad \dot{p}_2(s) = g_0^2 p_0(s) + g_1^2 p_1(s),$$

where $g_0^1 = \frac{3}{(b-a)}$, $g_0^2 = -\frac{4}{3(b-a)}$ and $g_1^2 = \frac{20}{3(b-a)}$.

Additionally, we have $\mathscr{I}_2^L(p_0(s)^2) = \frac{(b-a)^2}{2}$, $\mathscr{I}_2^L(p_1(s)^2) = \frac{(b-a)^2}{4}$ and $\mathscr{I}_2^L(p_2(s)^2) = \frac{(b-a)^2}{6}$.

Therefore, the double-integral inequality is obtained:

$$\int_a^b \int_a^{u_1} x^T(u_2)Rx(u_2)du_2du_1$$
$$\geq \frac{2}{(b-a)^2}\Theta_{0,0}^T R\Theta_{0,0} + \frac{4}{(b-a)^2}\Theta_{0,1}^T R\Theta_{0,1} + \frac{6}{(b-a)^2}\Theta_{0,2}^T R\Theta_{0,2}, \tag{3.36}$$

where

$$\begin{aligned}
\Theta_{0,0} &= (b-a)\Omega_0 - \Omega_{0,0},\\
\Theta_{0,1} &= 2\Theta_{0,0} - \frac{6}{(b-a)}\Theta_{0,0,0}\\
&= -(b-a)\Omega_0 + 4\Omega_{0,0} - \frac{6}{(b-a)}\Omega_{0,0,0},\\
\Theta_{0,2} &= 3\Theta_{0,0} - \frac{24}{(b-a)}\Theta_{0,0,0} + \frac{60}{(b-a)^2}\Theta_{0^4}\\
&= (b-a)\Omega_0 - 9\Omega_{0,0} + \frac{36}{(b-a)}\Omega_{0^3} - \frac{60}{(b-a)^2}\Omega_{0^4}.
\end{aligned}$$

(2) $m = 3$

The orthogonal polynomials defined in the lower-triangular triple-integral inner space are computed as follows with $N = 1$:

$$\begin{aligned}
p_0(s) &= 1,\\
p_1(s) &= -1 + 4\left(\frac{s-a}{b-a}\right).
\end{aligned}$$

Then, based on the above two polynomials, the triple-integral inequality is obtained as follows:

$$\int_a^b \int_a^{u_1} \int_a^{u_2} x^T(u_3)Rx(u_3)du_3du_2du_1$$
$$\geq \frac{6}{(b-a)^3}\Theta_{0^3}^T R\Theta_{0^3} + \frac{10}{(b-a)^3}\Theta_{0^2,1}^T R\Theta_{0^2,1}, \tag{3.37}$$

where

$$\Theta_{0^3} = \frac{(b-a)^2}{2}\Omega_0 - (b-a)\Omega_{0,0} + \Omega_{0,0,0},$$
$$\Theta_{0^2,1} = 3\Theta_{0^3} - \frac{12}{(b-a)}\Theta_{0^4}$$
$$= -\frac{(b-a)^2}{2}\Omega_0 + 3(b-a)\Omega_{0,0} - 9\Omega_{0^3} + \frac{12}{(b-a)}\Omega_{0^4}.$$

(3) $m = 4$
The quartic-integral inequality is obtained from Lemma 3.7:

$$\int_a^b \int_a^{u_1} \int_a^{u_2} \int_a^{u_3} x^T(u_4)Rx(u_4)du_4du_3du_2du_1 \geq \frac{24}{(b-a)^4}\Theta_{0^4}^T R\Theta_{0^4}, \quad (3.38)$$

where

$$\Theta_{0^4} = \frac{(b-a)^3}{6}\Omega_0 - \frac{(b-a)^2}{2}\Omega_{0,0} + (b-a)\Omega_{0,0,0} - \Omega_{0^4}.$$

Now, we have the following corollary with $x(s)$ being replaced by $\dot{x}(s)$ in (3.36), (3.37) and (3.38).

Corollary 3.6 *For a matrix $R \in \mathbb{S}_+^n$ and a differentiate vector function $x(s) : [a, b] \to \mathbb{R}^n$, the following integral inequalities hold:*

$$\int_a^b \int_a^{u_1} \dot{x}^T(u_2)R\dot{x}(u_2)du_2du_1$$
$$\geq \frac{2}{(b-a)^2}\dot{\Theta}_{0,0}^T R\dot{\Theta}_{0,0} + \frac{4}{(b-a)^2}\dot{\Theta}_{0,1}^T R\dot{\Theta}_{0,1} + \frac{6}{(b-a)^2}\dot{\Theta}_{0,2}^T R\dot{\Theta}_{0,2},$$
$$\int_a^b \int_a^{u_1} \int_a^{u_2} \dot{x}^T(u_3)R\dot{x}(u_3)du_3du_2du_1$$
$$\geq \frac{6}{(b-a)^3}\dot{\Theta}_{0^3}^T R\dot{\Theta}_{0^3} + \frac{10}{(b-a)^3}\dot{\Theta}_{0^2,1}^T R\dot{\Theta}_{0^2,1},$$
$$\int_a^b \int_a^{u_1} \int_a^{u_2} \int_a^{u_3} \dot{x}^T(u_4)R\dot{x}(u_4)du_4du_3du_2du_1 \geq \frac{24}{(b-a)^4}\dot{\Theta}_{0^4}^T R\dot{\Theta}_{0^4},$$

where

$$\dot{\Theta}_{0,0} = -(b-a)x(a) + \Omega_0,$$
$$\dot{\Theta}_{0,1} = (b-a)x(a) - 4\Omega_0 + \frac{6}{(b-a)}\Omega_{0,0},$$
$$\dot{\Theta}_{0,2} = -(b-a)x(a) + 9\Omega_0 - \frac{36}{(b-a)}\Omega_{0,0} + \frac{60}{(b-a)^2}\Omega_{0^3},$$

$$\dot{\Theta}_{0^3} = -\frac{(b-a)^2}{2}x(a) + (b-a)\Omega_0 - \Omega_{0,0},$$
$$\dot{\Theta}_{0^2,1} = \frac{(b-a)^2}{2}x(a) - 3(b-a)\Omega_0 + 9\Omega_{0,0} - \frac{12}{(b-a)}\Omega_{0^3},$$
$$\dot{\Theta}_{0^4} = -\frac{(b-a)^3}{6}x(a) + \frac{(b-a)^2}{2}\Omega_0 - (b-a)\Omega_{0,0} + \Omega_{0^3}.$$

3.3 Integral Inequalities with Free Matrices

When applying the L–K functional method to analyze the stability of systems with a time-varying delay, the double-integral functional $\int_{t-h_2}^{t-h_1}\int_u^t \dot{x}(s)^T R\dot{x}(s)dsdu$ is often involved in L–K functionals, where h_1 and h_2 are constant scalars satisfying $0 < h_1 \leq h(t) \leq h_2$. In this case, how to estimate the single-integral term $(h_2 - h_1)\int_{t-h_2}^{t-h_1} \dot{x}(s)^T R\dot{x}(s)ds$ arising in the derivative of L–K functionals becomes an essential issue for obtaining a relaxed stability condition in the form of LMIs. One way is to use an integral inequality without free matrices in combination with the reciprocally convex lemma. For example, applying the Jensen inequality to estimate the single term leads to:

$$\begin{aligned}
&(h_2 - h_1)\int_{t-h_2}^{t-h_1} \dot{x}^T(s)R\dot{x}(s)ds \\
&= (h_2 - h_1)\int_{t-h(t)}^{t-h_1} \dot{x}^T(s)R\dot{x}(s)ds + (h_2 - h_1)\int_{t-h_2}^{t-h(t)} \dot{x}^T(s)R\dot{x}(s)ds \\
&\geq \frac{h_2 - h_1}{h(t) - h_1}(x(t-h_1) - x(t-h(t)))^T R(x(t-h_1) - x(t-h(t))) \\
&\quad + \frac{h_2 - h_1}{h_2 - h(t)}(x(t-h(t)) - x(t-h_2))^T R(x(t-h(t)) - x(t-h_2)) \\
&= \frac{1}{\alpha}\xi(t)^T M_1^T R M_1 \xi(t) + \frac{1}{1-\alpha}\xi(t)^T M_2^T R M_2 \xi(t) \\
&\geq \xi(t)^T \begin{bmatrix} M_1 \\ M_2 \end{bmatrix}^T \begin{bmatrix} R & 0 \\ * & R \end{bmatrix} \begin{bmatrix} M_1 \\ M_2 \end{bmatrix} \xi(t)
\end{aligned} \tag{3.39}$$

where

$$\alpha = \frac{h(t) - h_1}{h_2 - h_1}, \quad M_1 = \begin{bmatrix} I & -I & 0 \end{bmatrix}, \quad M_2 = \begin{bmatrix} 0 & I & -I \end{bmatrix},$$
$$\xi(t) = col\{x(t-h_1), x(t-h(t)), x(t-h_2)\}.$$

Another way is to directly use integral inequalities with free matrices that are also called the free-matrix-based (FMB) integral inequalities. Unlike integral inequalities without free matrices discussed before, the FMB integral inequalities have the term of the integral interval locating in the numerators of the estimated bounds, which makes the reciprocally convex lemma no longer required. In what follows, various FMB inequalities are proposed.

3.3.1 Conventional FMB Inequalities

Lemma 3.11 (The Jensen FMB inequality) *For a matrix $R \in \mathbb{S}_+^n$ and a differentiable vector function $x(s) : [a, b] \to \mathbb{R}^n$, the following integral inequality holds for any matrices $M_1, M_2 \in \mathbb{R}^{n\times n}$:*

$$
\begin{aligned}
-\int_a^b \dot{x}^T(s)R\dot{x}(s)ds \leq{}& \omega^T \begin{bmatrix} M_1^T + M_1 & -M_1^T + M_2 \\ * & -M_2^T - M_2 \end{bmatrix} \omega \\
&+ (b-a)\omega^T \begin{bmatrix} M_1^T \\ M_2^T \end{bmatrix} R^{-1} \begin{bmatrix} M_1 & M_2 \end{bmatrix} \omega
\end{aligned} \tag{3.40}
$$

where $\omega = col\{x(b), x(a)\}$.

Proof Define

$$
M = \begin{bmatrix} M_1 & M_2 \end{bmatrix}, \quad \Phi = \begin{bmatrix} M^T R^{-1} M & M^T \\ M & R \end{bmatrix}.
$$

From the Schur complement, it follows that $\Phi \geq 0$ in view of $R > 0$. Through some computations, we get

$$
\begin{aligned}
\int_a^b \begin{bmatrix} \omega \\ \dot{x}(s) \end{bmatrix}^T \Phi \begin{bmatrix} \omega \\ \dot{x}(s) \end{bmatrix} ds ={}& (b-a)\omega^T M^T R^{-1} M \omega + \omega^T \mathrm{Sym}\left\{M^T \begin{bmatrix} I & -I \end{bmatrix}\right\} \omega \\
&+ \int_a^b \dot{x}^T(s)R\dot{x}(s)ds \geq 0.
\end{aligned}
$$

After a simple arrangement, we get (3.40). This completes the proof. ■

Remark 3.9 Lemma 3.11 is proved by Moon's inequality and the Leibniz–Newton formula in [20]. Here is a relatively simple proof. By setting $M_1 = -\frac{1}{b-a}R$ and $M_2 = \frac{1}{b-a}R$, the inequality (3.40) is reduced to the Jensen inequality (3.2). For this reason, we call it the Jensen FMB inequality. However, unlike the Jensen inequality (3.2), the Jensen FMB inequality (3.40) has the term $(b-a)$ appearing in the numerator of the bound. It is the introduced free matrices M_1 and M_2 that change the elementary construction.

Lemma 3.12 (The Wirtinger-based FMB inequality) *Let $x(s)$ be a differentiable vector function: $[a, b] \to \mathbb{R}^n$. For matrices $R \in \mathbb{S}_+^n$, $Z_{00}, Z_{11} \in \mathbb{S}^{3n}$, $Z_{01} \in \mathbb{R}^{3n\times 3n}$ and $N_0, N_1 \in \mathbb{R}^{3n\times n}$ satisfying*

$$\begin{bmatrix} Z_{00} & Z_{01} & N_0 \\ * & Z_{11} & N_1 \\ * & * & R \end{bmatrix} \geq 0, \tag{3.41}$$

the following inequality holds:

$$\begin{aligned} &-\int_a^b \dot{x}^T(s)R\dot{x}(s)ds \\ &\leq \varpi^T \text{Sym}\{N_0\Pi_0 + N_1\Pi_1\}\varpi + (b-a)\varpi^T \left(Z_{00} + \frac{1}{3}Z_{11}\right)\varpi \end{aligned} \tag{3.42}$$

where

$$\begin{aligned} \varpi &= col\left\{x(b), x(a), \frac{1}{b-a}\int_a^b x(s)ds\right\}, \\ \Pi_0 &= \begin{bmatrix} I & -I & 0 \end{bmatrix}, \quad \Pi_1 = \begin{bmatrix} I & I & -2I \end{bmatrix}. \end{aligned}$$

With the help of the polynomial $p(s) = \frac{b+a-2s}{b-a}$, the Wirtinger-based FMB inequality (3.42) is developed [18]. It is seen that the FMB inequality (3.42) holds under the constraint (3.41) in which several free matrices are introduced. In fact, this constraint can be removed by setting $Z_{ij} = N_i R^{-1} N_j^T, i, j \in \{0, 1\}$. In this case, the constraint (3.41) always holds from the Schur complement. Therefore, we have the following corollary.

Corollary 3.7 (The Wirtinger-based FMB inequality) *Let $x(s)$ be a differentiable vector function: $[a, b] \to \mathbb{R}^n$. For matrices $R \in \mathbb{S}_+^n$ and $N_0, N_1 \in \mathbb{R}^{3n\times n}$, the following inequality holds:*

$$\begin{aligned} -\int_a^b \dot{x}^T(s)R\dot{x}(s)ds &\leq \varpi^T \text{Sym}\{N_0\Pi_0 + N_1\Pi_1\}\varpi \\ &\quad + (b-a)\varpi^T\left(N_0R^{-1}N_0^T + \frac{1}{3}N_1R^{-1}N_1^T\right)\varpi \end{aligned} \tag{3.43}$$

where ϖ, Π_0 and Π_1 are all defined in Lemma 3.12.

Remark 3.10 Some relationships between the two FMB inequalities (3.42) and (3.43) are concluded as follows:

(1) These two FMB inequalities are actually equivalent. In other words, they produce the same tight bounds for the single-integral term $-\int_a^b \dot{x}^T(s)R\dot{x}(s)ds$.

(2) Compared with the inequality (3.42), a smaller number of free matrices are involved in the inequality (3.43) since the free matrices Z_{00}, Z_{01} and Z_{11} are removed.

(3) Unlike the inequality (3.42), the inverse of the matrix R appears in the bound of (3.43), which may result in more rows of LMIs in stability conditions.

Remark 3.11 By setting $N_0 = \frac{1}{b-a}\left[-R \; R \; 0\right]^T$ and $N_1 = \frac{3}{b-a}\left[-R \; -R \; 2R\right]^T$, the FMB inequality (3.43) is just reduced to the Wirtinger-based inequality (3.8). So, we call both of the two inequalities (3.42) and (3.43) the Wirtinger-based FMB inequalities for convenience. It is easily seen that the Wirtinger-based FMB inequality (3.43) includes the Jensen FMB inequality (3.40) as a special case.

Lemma 3.13 (The auxiliary-functional-based FMB inequality) *Let $x(s)$ be a differentiable vector function: $[a, b] \to \mathbb{R}^n$. For matrices $R \in \mathbb{S}_+^n$ and $N_0, N_1, N_2 \in \mathbb{R}^{4n\times n}$, the following integral inequality holds:*

$$-\int_a^b \dot{x}^T(s)R\dot{x}(s)ds \le \varpi^T \Psi \varpi \tag{3.44}$$

where

$$\begin{aligned}\Psi &= (b-a)\left(N_0R^{-1}N_0^T + \frac{1}{3}N_1R^{-1}N_1^T + \frac{1}{5}N_2R^{-1}N_2^T\right)\\ &\quad + \text{Sym}\{N_0\Pi_0 + N_1\Pi_1 + N_2\Pi_2\},\\ \Pi_0 &= \left[I \; -I \; 0 \; 0\right], \quad \Pi_1 = \left[I \; I \; -2I \; 0\right], \quad \Pi_2 = \left[I \; -I \; 6I \; -6I\right],\\ \varpi &= col\left\{x(b), x(a), \frac{1}{b-a}\int_a^b x(s)ds, \frac{2}{(b-a)^2}\int_a^b\int_u^b x(s)dsdu\right\}.\end{aligned}$$

With the help of orthogonal polynomials, the auxiliary-function-based FMB inequality (3.44) is developed by introducing free matrices N_0, N_1 and N_2 [19], which includes the Jensen and Wirtinger-based FMB inequalities.

3.3.2 A Series of Conventional FMB Inequalities

Lemma 3.14 *Let $x(s)$ be a differentiable function: $[a, b] \to \mathbb{R}^n$ and $p_k(s)$, $k \in \mathbb{N}$, be orthogonal polynomials of degree k defined in the integral inner space $\langle p_k, p_l\rangle = \int_a^b p_k(s)p_l(s)ds$. For given integers m, $N \in \mathbb{N}$ satisfying $m \ge N$, matrices $R \in \mathbb{S}_+^n$ and $M_i \in \mathbb{R}^{(m+2)n\times n}$, $i \in \{0, 1, \ldots, N\}$, the following inequality holds:*

$$-\int_a^b \dot{x}^T(s)R\dot{x}(s)ds \le \vartheta_m^T \Phi_N \vartheta_m \tag{3.45}$$

where

$$\begin{aligned}\vartheta_m &= \begin{cases} col\{x(b), x(a), \frac{1}{S_1}\Omega_{0^1}, \ldots, \frac{1}{S_m}\Omega_{0^m}\}, & m \ge 1,\\ col\{x(b), x(a)\}, & m = 0,\end{cases}\\ S_i &= \mathscr{I}_i^U(1), \quad i \in \{1, 2, \ldots, m\},\end{aligned}$$

$$\Phi_N = \sum_{i=0}^{N} \langle p_i, p_i \rangle M_i R^{-1} M_i^T + \sum_{i=0}^{N} \text{Sym}\{M_i \Pi_i\},$$

where Π_i satisfies the following equation

$$\int_a^b p_i(s)\dot{x}(s)ds = \Pi_i \vartheta_m, \quad i \in \{0, 1, \ldots, N\}.$$

Proof Define

$$\begin{aligned} M &= col\{M_0, M_1, \ldots, M_N\}, \\ \varepsilon(s) &= col\{p_0(s)\vartheta_m, p_1(s)\vartheta_m, \ldots, p_N(s)\vartheta_m\}. \end{aligned}$$

From the Schur complement, we have

$$\begin{bmatrix} \varepsilon(s) \\ \dot{x}(s) \end{bmatrix}^T \begin{bmatrix} MR^{-1}M^T & M \\ * & R \end{bmatrix} \begin{bmatrix} \varepsilon(s) \\ \dot{x}(s) \end{bmatrix} \geq 0,$$

which leads to

$$\varepsilon^T(s)MR^{-1}M^T\varepsilon(s) + 2\varepsilon^T(s)M\dot{x}(s) + \dot{x}^T(s)R\dot{x}(s) \geq 0. \tag{3.46}$$

Based on the orthogonal property of $p_k(s)$, we have

$$\begin{aligned} &\int_a^b \varepsilon^T(s)MR^{-1}M^T\varepsilon(s)ds \\ &= \int_a^b \left(\sum_{i=0}^{N} p_i(s)\vartheta_m^T M_i \right) R^{-1} \left(\sum_{i=0}^{N} p_i(s)M_i^T \vartheta_m \right) ds \\ &= \int_a^b \sum_{i=0}^{N} p_i^2(s)\vartheta_m^T M_i R^{-1} M_i^T \vartheta_m ds \\ &= \vartheta_m^T \left\{ \sum_{i=0}^{N} \langle p_i, p_i \rangle M_i R^{-1} M_i^T \right\} \vartheta_m. \end{aligned} \tag{3.47}$$

According to the equation $\int_a^b p_i(s)\dot{x}(s)ds = \Pi_i\vartheta_m$, we have

$$\begin{aligned} \int_a^b 2\varepsilon^T(s)M\dot{x}(s)ds &= 2\sum_{i=0}^{N} \vartheta_m^T M_i \int_a^b p_i(s)\dot{x}(s)ds \\ &= \sum_{i=0}^{N} \vartheta_m^T \text{Sym}\{M_i\Pi_i\}\vartheta_m. \end{aligned} \tag{3.48}$$

Hence, with (3.47) and (3.48), integrating both sides of (3.46) from a to b leads to the inequality (3.45). This completes the proof. ■

Remark 3.12 In Lemma 3.14, the inequality $m \geq N$ is required so that the expression $\int_a^b p_i(s)\dot{x}(s)ds$, $i \in \{0, 1, \ldots, N\}$, can be represented by ϑ_m. On the other hand, if ϑ_m is defined in the lower-triangular form, i.e.,

$$\vartheta_m = \begin{cases} col\{x(b), x(a), \frac{1}{S_1}\Theta_{0^1}, \ldots, \frac{1}{S_m}\Theta_{0^m}\}, & m \geq 1, \\ col\{x(b), x(a)\}, & m = 0, \end{cases}$$

where $S_i = \mathscr{I}_i^L(1)$ and $\Theta_{0^i} = \mathscr{I}_i^L(x(s))$, a series of similar FMB integral inequalities can be obtained.

Remark 3.13 The series of FMB integral inequality (3.45) is general. In fact, with the help of the first one, two and three Legendre polynomials, the Jensen, Wirtinger-based and auxiliary-function-based FMB inequalities are, respectively, recovered from (3.45). The idea involved in Lemma 3.14 can be directly extended to the multiple-integral case.

Lemma 3.15 *Let $x(t)$ be a differentiable function: $[a, b] \to \mathbb{R}^n$ and $p_k(s)$, $k \in \{0, 1, \ldots, N\}$, be orthogonal polynomials defined in the upper-triangular multiple-integral inner space $\langle p_k, p_l\rangle = \mathscr{I}_d^U(p_k(s)p_l(s))$. For given integers $d \in \mathbb{N}^+$, $m, N \in \mathbb{N}$ satisfying $m \geq d - 1 + N$, matrices $R \in \mathbb{S}_+^n$ and $M_i \in \mathbb{R}^{(m+2)n\times n}$, $i \in \{0, 1, \ldots, N\}$, the following inequality holds:*

$$-\mathscr{I}_d^U(\dot{x}^T(s)R\dot{x}(s)) \leq \vartheta_m^T\Phi_N\vartheta_m \tag{3.49}$$

where ϑ_m, Φ_N are defined in Lemma 3.14 with $\langle p_k, p_l\rangle = \mathscr{I}_d^U(p_k(s)p_l(s))$, and Π_i satisfies the following equation:

$$\mathscr{I}_d^U(p_i(s)\dot{x}(s)) = \Pi_i\vartheta_m, \quad i \in \{0, 1, \ldots, N\}.$$

Remark 3.14 When $d = 1$, Lemma 3.15 is reduced to Lemma 3.14. The inequality $m \geq d - 1 + N$ is required so that the expression $\mathscr{I}_d^U(p_i(s)\dot{x}(s))$, $i \in \{0, 1, \ldots, N\}$, can be represented by ϑ_m. Additionally, if ϑ_m is defined in the lower-triangular form, a series of similar multiple-integral inequalities can be developed in a similar way.

3.3.3 Extension of Conventional FMB Inequalities

If the vector ϑ_m is set to different expressions, different kinds of FMB integral inequalities may be developed. Especially, if ϑ_m is chosen as any vector, the following general FMB (GFMB) inequality is developed [23].

Lemma 3.16 *For a matrix $R \in \mathbb{S}_+^n$, any matrices L_0, L_1 with appropriate dimensions, and a differentiable vector function $x(s)$: $[a, b] \to \mathbb{R}^n$, the following inequality holds:*

$$-\int_a^b x^T(s)Rx(s)ds \leq \mathrm{Sym}\{\rho_0^T L_0 \chi_0 + \rho_0^T L_1 \chi_1\} + (b-a)\rho_0^T \left(\frac{3L_0 R^{-1} L_0^T + L_1 R^{-1} L_1^T}{3} \right) \rho_0 \quad (3.50)$$

where ρ_0 is any vector and

$$\chi_0 = \Omega_0,$$
$$\chi_1 = -\Omega_0 + \frac{2}{b-a}\Omega_{0^2}.$$

Remark 3.15 In the Wirtinger-based FMB inequality (3.43), the vector ϑ_m is fixed while in the GFMB inequality (3.50) the vector ρ_0 is a free vector that can be chosen as any vector. Therefore, the GFMB inequality (3.50) provides more freedom to estimate integral terms. Obviously, the GFMB inequality (3.50) encompasses the Wirtinger-based FMB inequality (3.43) as a special case. In what follows, we propose an improved GFMB integral inequality that provides even more freedom than the GFMB inequality (3.50).

Lemma 3.17 (The improved GFMB inequality) *For a matrix $R \in \mathbb{S}_+^n$, any matrices L_0, L_1 with appropriate dimensions and a differentiable vector function $x(s)$: $[a, b] \to \mathbb{R}^n$, the following inequality holds:*

$$-\int_a^b x^T(s)Rx(s)ds \leq \mathrm{Sym}\{\rho_0^T L_0 \chi_0 + \rho_1^T L_1 \chi_1\} + (b-a)\left(\rho_0^T L_0 R^{-1} L_0^T \rho_0 + \frac{1}{3}\rho_1^T L_1 R^{-1} L_1^T \rho_1 \right) \quad (3.51)$$

where ρ_0, ρ_1 are any vectors, and χ_0, χ_1 are defined in Lemma 3.16.

Proof Define

$$p_0(s) = 1, \quad p_1(s) = -1 + 2\frac{s-a}{b-a},$$
$$\Psi = \begin{bmatrix} L_0 R^{-1} L_0^T & L_0 R^{-1} L_1^T & L_0 \\ * & L_1 R^{-1} L_1^T & L_1 \\ * & * & R \end{bmatrix}, \quad \varphi(s) = \begin{bmatrix} \rho_0 p_0(s) \\ \rho_1 p_1(s) \\ x(s) \end{bmatrix}.$$

It follows that $\Psi \geq 0$ from the Schur complement. Through some computations, we have the following equations:

$$\int_a^b p_0(s)p_1(s)ds = 0, \quad \int_a^b p_0(s)^2 ds = b-a,$$
$$\int_a^b p_1(s)^2 ds = \frac{b-a}{3}, \quad \int_a^b p_i(s)x(s)ds = \chi_i, \quad i \in \{0,1\}.$$

Based on the above facts, integrating both sides of $\varphi(s)^T\Psi\varphi(s) \geq 0$ from a to b leads to:

$$0 \leq \text{Sym}\{\rho_0^T L_0\chi_0 + \rho_1^T L_1\chi_1\} + (b-a)\left(\rho_0^T L_0 R^{-1} L_0^T \rho_0 + \frac{1}{3}\rho_1^T L_1 R^{-1} L_1^T \rho_1\right)$$
$$+ \int_a^b x^T(s)Rx(s)ds.$$

After a simple arrangement, we get the improved GFMB inequality (3.51). This completes the proof. ■

Remark 3.16 In the improved GFMB inequality (3.51), the two vectors ρ_0 and ρ_1 are independent from each other. So, they can be chosen freely and independently in case that a stability condition with less conservatism is required or in case that a stability condition with a relatively small number of variables is required. By setting $\rho_1 = \rho_0$, the improved GFMB inequality (3.51) is reduced to the GFMB inequality (3.50). So, compared with (3.50), the improved GFMB inequality (3.51) provides more freedom to estimate integral terms.

With $x(s)$ being replaced by $\dot{x}(s)$, we have the following corollary.

Corollary 3.8 *For a matrix $R \in \mathbb{S}_+^n$, any matrices L_0, L_1 with appropriate dimensions, and a differentiable vector function $x(t)$: $[a,b] \to \mathbb{R}^n$, the following inequality holds:*

$$-\int_a^b \dot{x}^T(s)R\dot{x}(s)ds \leq \text{Sym}\{\rho_0^T L_0\hat{\chi}_0 + \rho_1^T L_1\hat{\chi}_1\}$$
$$+ (b-a)\left(\rho_0^T L_0 R^{-1} L_0^T \rho_0 + \frac{1}{3}\rho_1^T L_1 R^{-1} L_1^T \rho_1\right) \quad (3.52)$$

where ρ_0, ρ_1 are any vectors, and

$$\hat{\chi}_0 = x(b) - x(a),$$
$$\hat{\chi}_1 = x(b) + x(a) - \frac{2}{b-a}\int_a^b x(s)ds.$$

Remark 3.17 By letting $\rho_0 = \hat{\chi}_0$, $\rho_1 = \hat{\chi}_1$, $L_0 = -\frac{R}{b-a}$ and $L_1 = -\frac{3R}{b-a}$, the Wirtinger-based integral inequality (3.8) is recovered from (3.52).

3.4 The Relationship Between Integral Inequalities with and Without Free Matrices

By setting $N_i = -\frac{2i+1}{b-a}\Pi_i^T R$ in Lemma 3.14, we have the following corollary.

Corollary 3.9 *Let $x(t)$ be a differentiable function: $[a, b] \to \mathbb{R}^n$. For a given integer $N \in \mathbb{N}$ and a matrix $R \in \mathbb{S}_+^n$, the following inequality holds:*

$$-\int_a^b \dot{x}^T(s)R\dot{x}(s)ds \leq \vartheta_m^T \Psi_N \vartheta_m \tag{3.53}$$

where

$$\Psi_N = -\sum_{i=0}^{N} \frac{2i+1}{b-a}\Pi_i^T R \Pi_i,$$

and ϑ_m and Π_i are defined in Lemma 3.14.

In what follows, the relationship between the two series of integral inequalities (3.45) and (3.53) is discussed.

Theorem 3.1 *The two series of integral inequalities, respectively, proposed in Lemma 3.14 and Corollary 3.9 are equivalent. In other words, for a given integer N, the two corresponding inequalities with and without free matrices produce the same tight bounds for the integral term $-\int_a^b \dot{x}^T(s)R\dot{x}(s)ds$.*

Proof Firstly, the FMB inequality (3.45) includes the inequality (3.53), which means that (3.45) always produces a tighter bound than (3.53).

Secondly, by considering the two expressions $\Phi_N = \sum_{i=0}^N \langle p_i, p_i \rangle M_i R^{-1} M_i^T + \sum_{i=0}^N \text{Sym}\{M_i \Pi_i\}$ and $\Psi_N = -\sum_{i=0}^N \frac{2i+1}{b-a}\Pi_i^T R \Pi_i$, we have

$$\begin{aligned}
\Phi_N &= \sum_{i=0}^{N} \langle p_i, p_i \rangle M_i R^{-1} M_i^T + \sum_{i=0}^{N} \text{Sym}\{M_i \Pi_i\} \\
&= \sum_{i=0}^{N} \frac{2i+1}{b-a}\left(\Xi_i - \Pi_i^T R \Pi_i\right) \\
&\geq \Psi_N,
\end{aligned}$$

where

$$\Xi_i = \left(\frac{b-a}{2i+1}M_i + \Pi_i^T R\right) R^{-1} \left(\frac{b-a}{2i+1}M_i + \Pi_i^T R\right)^T.$$

This above inequality implies that the integral inequality (3.53) produces a tighter upper bound than (3.45).

Based on the above discussions, it is concluded that for a given integer N, the two corresponding inequalities with free matrices (3.45) and without free matrices (3.53) actually produce the same tight upper bounds. This completes the proof. ■

Remark 3.18 Since the two corresponding integral inequalities with and without free matrices produce the same tight bounds, the conservatism of stability criteria obtained by them should be the same, provided that the same L–K functional is used. However, it is not the case. Owing to different constructions, the integral inequalities without free matrices are usually utilized combined with the reciprocally convex lemma while the integral inequalities with free matrices are used independently. The two different ways to estimate the integral term $\int_{t-h_2}^{t-h_1} \dot{x}(s)^T R\dot{x}(s)ds$ usually lead to different conservative stability conditions.

3.5 Conclusion

This chapter has established a basic framework for integral inequalities. Based on orthogonal polynomials, several series of integral inequalities have been proposed to estimate integral terms which include almost existing integral inequalities. More polynomials considered, tighter bounds produced. According to whether the free matrices are involved or not, integral inequalities are classified into two categories: those without free matrices and those with free matrices. The relationship between them has been briefly discussed.

References

1. Moon YS, Park PG, Kwon WH, Lee YS (2001) Delay-dependent robust stabilization of uncertain state-delayed systems. Int J Control 74:1447–1455
2. Gu K, Kharitonov VL, Chen J (2003) Stability of time-delay systems. Birkhäuser, Boston
3. Ji X, Su H (2015) A note on equivalence between two integral inequalities for time-delay systems. Automatica 53:244–246
4. Kim JH (2011) Note on stability of linear systems with time-varying delay. Automatica 47:2118–2121
5. Gyurkovics E, Takacs T (2019) Comparison of some bounding inequalities applied in stability analysis of time-delay systems. Syst Control Lett 123:40–46
6. Zhang CK, He Y, Jiang L, Wu M, Zeng HB (2016) Stability analysis of systems with time-varying delay via relaxed integral inequalities. Syst Control Lett 92:52–61
7. Chen J, Xu S, Chen W, Zhang B, Ma Q, Zou Y (2016) Two general integral inequalities and their applications to stability analysis for systems with time-varying delay. Int J Robust Nonlinear Control 26:4088–4103
8. Kim JH (2016) Further improvement of Jensen inequality and application to stability of time-delayed systems. Automatica 64:121–125
9. Park PG, Lee WI, Lee SY (2015) Auxiliary function-based integral inequalities for quadratic functions and their applications to time-delay systems. J Frankl Inst 352:1378–1396
10. Park MJ, Kwon OM, Park JH, Lee S, Cha E (2015) Stability of time-delay systems via Wirtinger-based double integral inequality. Automatica 55:204–208

11. Seuret A, Gouaisbaut F (2013) Wirtinger-based integral inequality: application to time-delay systems. Automatica 49:2860–2866
12. Gyurkovics E, Takacs T (2016) Multiple integral inequalities and stability analysis of time delay systems. Syst Control Lett 96:72–80
13. Seuret A, Gouaisbaut F, Fridman E (2015) Stability of discrete-time systems with time-varying delays via a novel summation inequality. Syst Control Lett 81:1–7
14. Seuret A, Gouaisbaut F (2018) Stability of linear systems with time-varying delays using Bessel-Legendre inequalities. IEEE Trans Autom Control 63:225–232
15. Zhang XM, Han QL, Zeng Z (2018) Hierarchical type stability criteria for delayed neural networks via canonical Bessel-Legendre inequalities. IEEE Trans Cybern 48:1660–1671
16. Briat C (2011) Convergence and equivalence results for the Jensen's inequality-application to time-delay and sampled-data systems. IEEE Trans Autom Control 56:1660–1665
17. Gyurkovics E (2015) A note on Wirtinger-type integral inequalities for time-delay systems. Automatica 61:44–46
18. Zeng HB, He Y, Wu M, She J (2015) Free-matrix-based integral inequality for stability analysis of systems with time-varying delay. IEEE Trans Autom Control 60:2768–2772
19. Zeng HB, He Y, Wu M, She J (2015) New results on stability analysis for systems with discrete distributed delay. Automatica 60:189–192
20. Zhang XM, Wu M, She JH, He Y (2005) Delay-dependent stabilization of linear systems with time-varying state and input delays. Automatica 41:1405–1412
21. Chen J, Xu S, Zhang B (2017) Single/multiple integral inequalities with applications to stability analysis of time-delay systems. IEEE Trans Autom Control 62:3488–3493
22. Zhang XM, Lin WJ, Han QL, He Y, Wu M (2018) Global asymptotic stability for delayed neural networks using an integral inequality based on nonorthogonal polynomials. IEEE Trans Neural Netw Learn Syst 29:4487–4493
23. Zhang CK, He Y, Jiang L, Lin WJ, Wu M (2017) Delay-dependent stability analysis of neural networks with time-varying delay: a generalized free-weighting-matrix approach. Appl Math Comput 294:102–120
24. Chen J, Park JH, Xu S (2019) Stability analysis for neural networks with time-varying delay via improved techniques. IEEE Trans Cybern. https://doi.org/10.1109/TCYB.2018.2868136
25. Park PG, Ko JW, Jeong C (2011) Reciprocally convex approach to stability of systems with time-varying delays. Automatica 47:235–238
26. Zhang XM, Han QL, Seuret A, Gouaisbaut F (2017) An improved reciprocally convex inequality and an augmented Lyapunov-Krasovskii functional for stability of linear systems with time-varying delay. Automatica 84:221–226
27. Zhang CK, He Y, Jiang L, Wu M, Wang QG (2017) An extended reciprocally convex matrix inequality for stability analysis of systems with time-varying delay. Automatica 85:481–485

Chapter 4
Summation Inequalities

4.1 Introduction

Summation inequalities can directly estimate summation terms arising in the forward differences of L–K functionals. So just like integral inequalities, summation inequalities have been widely used in the stability analysis of discrete-time systems with time delay [1–14]. In essence, the role of summation inequalities is to transform the summation of the products of vectors into the product of summations of vectors. As a result, tractable stability conditions can be obtained in the form of linear matrix inequalities (LMIs). It is necessarily noted that since the endpoint problem of summation inequalities needs to be carefully addressed, it is a little more difficult to develop summation inequalities than integral ones.

According to whether free matrices are involved or not, summation inequalities can be generally classified into two categories: summation inequalities without free matrices and summation inequalities with free matrices. For the first category, some well-known summation inequalities are reported in the literature such as the Jensen summation inequality [5], the Wirtinger-based summation inequality [7, 10, 12], and the auxiliary-function-based summation inequalities [2, 8]. They can be seen as the discrete counterparts of the corresponding integral inequalities. In this category of summation inequalities, the summation interval appears in the denominator of the bound. So, when using them to estimate summation terms with a time-varying delay, the reciprocally convex lemma is necessarily required. On the other hand, by introducing some free matrices, the second category of summation inequalities are obtained in the literature such as the conventional free-matrix-based (FMB) summation inequality [1] and general FMB (GFMB) summation inequalities [4, 11]. Because the location of the summation interval is changed from the denominator to the numerator by the introduced free matrices, the reciprocally convex lemma is no longer required when applying the second category of summation inequalities. It is seen that the two different categories of summation inequalities may lead to two different ways of estimating summation terms with a time-varying delay.

J. H. Park et al., *Dynamic Systems with Time Delays: Stability and Control*,
https://doi.org/10.1007/978-981-13-9254-2_4

The remainder of this chapter is divided into two parts. In the first part, we mainly study the summation inequalities without free matrices. Some well-known summation inequalities are firstly recalled. Secondly, two series of general summation inequalities without free matrices are, respectively, developed in the upper- and lower-triangular forms with the help of orthogonal polynomials defined in summation inner spaces. Thirdly, various concrete summation inequalities are obtained from the two series of summation inequalities. In the second part, we mainly study the summation inequalities with free matrices. The conventional FMB, GFMB and improved GFMB summation inequalities are successively proposed. Then, based on orthogonal polynomials defined in the single-summation inner product space, a series of GFMB summation inequalities is developed. The above idea can be directly extended to the multiple-summation case.

4.2 Summation Inequalities Without Free Matrices

Before proceeding, the following notations are first defined, $m \in \mathbb{N}$:

$$x_a(i) = x(a+i), \quad y_a(i) = y(a+i), \quad y(i) = x(i+1) - x(i),$$

$$\bar{\Omega}_0 = \sum_{i=0}^{m} x_a(i), \quad \bar{\Omega}_{0^2} = \sum_{i=0}^{m}\sum_{j=i}^{m} x_a(j), \quad \bar{\Omega}_{0^3} = \sum_{i=0}^{m}\sum_{j=i}^{m}\sum_{l=j}^{m} x_a(l).$$

4.2.1 Some Well-Known Summation Inequalities Without Free Matrices

4.2.1.1 The Jensen Summation Inequality

Lemma 4.1 (The Jensen single-summation inequality) *For a matrix $R \in \mathbb{S}_+^n$ and a vector function $\{x(i) \in \mathbb{R}^n | i \in [a, a+m-1]\}$, the following inequality holds:*

$$\sum_{i=0}^{m-1} x_a^T(i) R x_a(i) \geq \frac{1}{m}\left(\sum_{i=0}^{m-1} x_a(i)\right)^T R \left(\sum_{i=0}^{m-1} x_a(i)\right). \tag{4.1}$$

The Jensen single-summation inequality (4.1) is developed by using the Schur complement [5], which can just be seen as the discrete counterpart of the Jensen single-integral inequality. With $x(i)$ being replaced by $y(i)$, we have the following corollary.

Corollary 4.1 (The Jensen single-summation inequality) *For a matrix $R \in \mathbb{S}_+^n$ and a vector function $\{x(i) \in \mathbb{R}^n | i \in [a, a+m]\}$, the following inequality holds:*

$$\sum_{i=0}^{m-1} y_a^T(i) R y_a(i) \geq \frac{1}{m}\Big(x_a(m) - x_a(0)\Big)^T R \Big(x_a(m) - x_a(0)\Big). \tag{4.2}$$

Lemma 4.2 (The Jensen double-summation inequalities) *For a matrix $R \in \mathbb{S}_+^n$ and a vector function $\{x(i) \in \mathbb{R}^n | i \in [a, a+m-1]\}$, the following inequalities hold:*

$$\sum_{i=0}^{m-1}\sum_{j=i}^{m-1} x_a^T(j) R x_a(j) \geq \frac{2}{m(m+1)} \left(\sum_{i=0}^{m-1}\sum_{j=i}^{m-1} x_a(j)\right)^T R \left(\sum_{i=0}^{m-1}\sum_{j=i}^{m-1} x_a(j)\right), \tag{4.3}$$

$$\sum_{i=0}^{m-1}\sum_{j=0}^{i} x_a^T(j) R x_a(j) \geq \frac{2}{m(m+1)} \left(\sum_{i=0}^{m-1}\sum_{j=0}^{i} x_a(j)\right)^T R \left(\sum_{i=0}^{m-1}\sum_{j=0}^{i} x_a(j)\right). \tag{4.4}$$

With $x(i)$ being replaced by $y(i)$, we have the following corollary.

Corollary 4.2 (The Jensen double-summation inequalities) *For a matrix $R \in \mathbb{S}_+^n$ and a vector function $\{x(i) \in \mathbb{R}^n | i \in [a, a+m]\}$, the following inequalities hold:*

$$\sum_{i=0}^{m-1}\sum_{j=i}^{m-1} y_a^T(j) R y_a(j) \geq 2\psi_{u0}^T R \psi_{u0}, \tag{4.5}$$

$$\sum_{i=0}^{m-1}\sum_{j=0}^{i} y_a^T(j) R y_a(j) \geq 2\psi_{l0}^T R \psi_{l0} \tag{4.6}$$

where

$$\psi_{u0} = x_a(m) - \frac{1}{m+1}\bar{\Omega}_0,$$
$$\psi_{l0} = x_a(0) - \frac{1}{m+1}\bar{\Omega}_0.$$

Remark 4.1 According to the summation region, the multiple-summation inequalities are shown in two forms: in the upper-triangular form (e.g., $\sum_{i=0}^{m-1}\sum_{j=i}^{m-1}$ for the double-summation inequality (4.5)) and in the lower-triangular form (e.g., $\sum_{i=0}^{m-1}\sum_{j=0}^{i}$ for the double-summation inequality (4.6)). The two forms of inequalities are similar in the constructions.

Remark 4.2 Compared with integral inequalities, it is a little more difficult to develop summation inequalities since the endpoint problem needs to be carefully addressed. It is worth noting that the summation interval on the left sides of inequalities (4.5) and (4.6) is from a to $a+m-1$ while on the right sides the summation interval becomes from a to $a+m$. Moreover, in order to make the obtained summation inequalities applicable in the stability analysis of time-delay systems, algebraic fractions are often removed from the right sides by using algebraic inequalities. For example, when achieving the two summation inequalities (4.5) and (4.6), the algebraic inequality, $\frac{m+1}{m} \geq 1$, is considered so that the algebraic fraction $\frac{m+1}{m}$ is removed.

4.2.1.2 The Wirtinger-Based Summation Inequality

Lemma 4.3 (The Wirtinger-based summation inequality) *For a matrix $R \in \mathbb{S}_+^n$ and a vector function $\{x(i) \in \mathbb{R}^n | i \in [a, a+m]\}$, the following inequality holds:*

$$\sum_{i=0}^{m-1} y_a^T(i) R_a y(i) \geq \frac{1}{m} \vartheta_0^T R \vartheta_0 + \frac{3}{m} \vartheta_1^T R \vartheta_1 \tag{4.7}$$

where

$$\vartheta_0 = x_a(m) - x_a(0),$$
$$\vartheta_1 = x_a(m) + x_a(0) - \frac{2}{m+1} \bar{\Omega}_0.$$

Based on orthogonal polynomials [7, 10] or Abel lemma [12], the Wirtinger-based summation inequality (4.7) is developed, which can just be seen as the discrete counterpart of the Wirtinger-based integral inequality. Obviously, the Wirtinger-based summation inequality (4.7) includes the Jensen summation inequality (4.2) as a special case.

Remark 4.3 If the vector function $x(i)$ is a constant vector, the summation term $\sum_{i=0}^{m-1} y_a(i)^T R y_a(i)$ is equal to zero. In this case, every summand on the right side of the inequality (4.7) must be equal to zero so that it holds. It is found that both ϑ_0 and ϑ_1 are zero vectors, which verifies the validity of the inequality (4.7). In fact, every summation inequality presented in this chapter has this property.

4.2.1.3 The Auxiliary-Function-Based Summation Inequality

Lemma 4.4 (The auxiliary-function-based single-summation inequality) *For a matrix $R \in \mathbb{S}_+^n$ and a vector function $\{x(i) \in \mathbb{R}^n | i \in [a, a+m]\}$, the following inequality holds:*

$$\sum_{i=0}^{m-1} y_a^T(i) R y_a(i) \geq \frac{1}{m} \vartheta_0^T R \vartheta_0 + \frac{3}{m} \vartheta_1^T R \vartheta_1 + \frac{5}{m} \vartheta_2^T R \vartheta_2 \tag{4.8}$$

where ϑ_0, ϑ_1 are defined in Lemma 4.3, and

$$\vartheta_2 = x_a(m) - x_a(0) + \frac{6}{m+1}\bar{\Omega}_0 - \frac{12}{(m+1)(m+2)}\bar{\Omega}_{0^2}.$$

With the help of two auxiliary functions, the auxiliary-function-based single-summation inequality (4.8) is obtained [8]. In a similar way, the following double-summation inequalities are obtained, respectively, in the upper- and lower-triangular forms.

Lemma 4.5 *For a matrix $R \in \mathbb{S}_+^n$ and a vector function $\{x(i)|i \in [a, a+m]\}$, the following double-summation inequalities hold:*

$$\sum_{i=0}^{m-1}\sum_{j=i}^{m-1} y_a^T(j)Ry_a(j) \geq 2\psi_{u0}^T R\psi_{u0} + 4\psi_{u1}^T R\psi_{u1}, \tag{4.9}$$

$$\sum_{i=0}^{m-1}\sum_{j=0}^{i} y_a^T(j)Ry_a(j) \geq 2\psi_{l0}^T R\psi_{l0} + 4\psi_{l1}^T R\psi_{l1} \tag{4.10}$$

where ψ_{u0}, ψ_{l0} are defined in Lemma 4.2, and

$$\psi_{u1} = x_a(m) + \frac{2}{m+1}\bar{\Omega}_0 - \frac{6}{(m+1)(m+2)}\bar{\Omega}_{0^2},$$
$$\psi_{l1} = x_a(0) - \frac{4}{m+1}\bar{\Omega}_0 + \frac{6}{(m+1)(m+2)}\bar{\Omega}_{0^2}.$$

Obviously, the auxiliary-function-based double-summation inequalities (4.9) and (4.10) include the Jensen double-summation inequalities (4.5) and (4.6), respectively.

4.2.2 Two Series of General Summation Inequalities

Let $p_k(i)$ be a polynomial function of degree k, $i \in \mathbb{Z}$, $k \in \mathbb{N}$, and define $p_{ka}(i) = p_k(i+a)$. Further, the following two notations are defined for any function $f(i)$:

$$\mathscr{S}_d^U(f(i)) = \sum_{i_1=0}^{m-1}\sum_{i_2=i_1}^{m-1}\cdots\sum_{i_d=i_{d-1}}^{m-1} f_a(i_d),$$

$$\mathscr{S}_d^L(f(i)) = \sum_{i_1=0}^{m-1}\sum_{i_2=0}^{i_1}\cdots\sum_{i_d=0}^{i_{d-1}} f_a(i_d).$$

Lemma 4.6 *For given integers $N \in \mathbb{N}$, $d \in \mathbb{N}^+$, a matrix $R \in \mathbb{S}_+^n$ and a vector function $\{x(i)|i \in [a, a+m-1]\}$, the following summation inequalities hold for any scalar polynomial $p_k(i)$:*

$$\mathscr{S}_d^U(x_a^T(i)Rx_a(i)) \geq \sum_{k=0}^{N} \mathscr{S}_d^U(p_k(i)^2)^{-1}\Big(\mathscr{S}_d^U(p_{ka}(i)x_a(i))\Big)^T R\Big(\mathscr{S}_d^U(p_{ka}(i)x_a(i))\Big), \tag{4.11}$$

$$\mathscr{S}_d^L(x_a^T(i)Rx_a(i)) \geq \sum_{k=0}^{N} \mathscr{S}_d^L(p_k(i)^2)^{-1}\Big(\mathscr{S}_d^L(p_{ka}(i)x_a(i))\Big)^T R\Big(\mathscr{S}_d^L(p_{ka}(i)x_a(i))\Big) \tag{4.12}$$

satisfying

$$\mathscr{S}_d^U(p_k(i)p_l(i))\begin{cases} = 0, & k \neq l, \\ \neq 0, & k = l, \end{cases} \tag{4.13}$$

$$\mathscr{S}_d^L(p_k(i)p_l(i))\begin{cases} = 0, & k \neq l, \\ \neq 0, & k = l, \end{cases} \tag{4.14}$$

for (4.11) and (4.12), respectively.

Proof Suppose polynomials $p_k(i)$, $k \in \{0, 1, \ldots, N\}$, satisfy (4.13). Define

$$W(i) = \begin{bmatrix} x_a(i) & p_{0a}(i)I & p_{1a}(i)I & \cdots & p_{Na}(i)I \end{bmatrix},$$
$$\Psi(i) = W(i)^T RW(i).$$

By considering the orthogonal property of $p_{ka}(i)$, we get

$$\mathscr{S}_d^U(\Psi(i)) = \begin{bmatrix} \mathscr{S}_d^U(x_a^T(i)Rx_a(i)) & \mathscr{S}_d^U(p_{0a}(i)x_a^T(i))R & \mathscr{S}_d^U(p_{1a}(i)x_a^T(i))R & \cdots & \mathscr{S}_d^U(p_{Na}(i)x_a^T(i))R \\ * & \mathscr{S}_d^U(p_{0a}^2(i))R & 0 & \cdots & 0 \\ * & * & \mathscr{S}_d^U(p_{1a}^2(i))R & \cdots & 0 \\ * & * & * & \ddots & \vdots \\ * & * & * & * & \mathscr{S}_d^U(p_{Na}^2(i))R \end{bmatrix}.$$

From the fact $\Psi(i) \geq 0$, we get $\mathscr{S}_d^U(\Psi(i)) \geq 0$. Then, applying the Schur complement to the inequality $\mathscr{S}_d^U(\Psi(i)) \geq 0$ directly leads to (4.11). Additionally, the inequality (4.12) can be obtained in a similar way. This completes the proof. ■

Remark 4.4 In Lemma 4.6, the sequence of the polynomials $p_k(i)$, $k \in \mathbb{N}$, forms an orthogonal basis with respect to the summation inner product $\langle p_k, p_l\rangle = \mathscr{S}_d^U(p_k(i)p_l(i))$ or $\langle p_k, p_l\rangle = \mathscr{S}_d^L(p_k(i)p_l(i))$. Thanks to these orthogonal polynomials, the two series of general summation inequalities (4.11) and (4.12) are developed, respectively, in the upper- and lower-triangular forms. The well-known summation inequalities presented previously can be recovered from them. Moreover, for a given d, the larger N, the tighter bound.

Remark 4.5 For given d and N, the orthogonal polynomials can be obtained by the Schmidt orthogonal method with a basis which can be initially chosen as $\{1, (i-a), \ldots, (i-a)^N\}$ in any summation inner product space. However, provided that the orthogonal polynomials are involved in them, it is still difficult to directly apply the two series of inequalities (4.11) and (4.12) to the stability analysis of discrete-time systems with a time-varying delay. So, how to get concrete summation inequalities by eliminating the involved polynomials becomes an essential and urgent issue.

4.2.3 Concrete Summation Inequalities

4.2.3.1 Properties of Two Summation Terms

The two summation terms $\mathscr{S}_d^U(p_{ka}(i)x_a(i))$ and $\mathscr{S}_d^L(p_{ka}(i)x_a(i))$ are involved in the two series of general summation inequalities (4.11) and (4.12). Their properties play an important role in eliminating the orthogonal polynomials, which are consequently studied in this subsection. Before proceeding, the following notations are defined for convenience, $d \in \mathbb{N}^+$, $k \in \mathbb{N}$:

$$\Omega_k(j) := \sum_{l=j}^{m-1} p_{ka}(l)x_a(l), \quad \Omega_{k,0^d}(j) := \sum_{l=j}^{m-1} p_{ka}(l)\Omega_{0^d}(l),$$

$$\Omega_{0^d,k} := \sum_{i_1=0}^{m-1}\sum_{i_2=i_1}^{m-1}\cdots\sum_{i_d=i_{d-1}}^{m-1} \Omega_k(i_d),$$

$$\Theta_k(j) := \sum_{l=0}^{j} p_{ka}(l)x_a(l), \quad \Theta_{k,0^d}(j) := \sum_{l=0}^{j} p_{ka}(l)\Theta_{0^d}(l),$$

$$\Theta_{0^d,k} := \sum_{i_1=0}^{m-1}\sum_{i_2=0}^{i_1}\cdots\sum_{i_d=0}^{i_{d-1}} \Theta_k(i_d),$$

where 0^d is the abbreviation of $\underbrace{0,\ldots,0}_{d}$. Particularly, we define $\Omega_k := \Omega_k(0)$, $\Omega_{k,0^d} := \Omega_{k,0^d}(0)$, $\Theta_k := \Theta_k(m-1)$ and $\Theta_{k,0^d} := \Theta_{k,0^d}(m-1)$.

We assume $p_0(i) = 1$ in any inner product space without loss of generality. According to the above definitions, we enumerate some expressions for clarity:

$$\Omega_0 = \sum_{i=0}^{m-1} x_a(i), \quad \Omega_{0^3} = \sum_{i=0}^{m-1}\sum_{j=i}^{m-1}\sum_{l=j}^{m-1} x_a(l),$$

$$\Omega_{0^3}(i) = \sum_{i_1=i}^{m-1}\sum_{i_2=i_1}^{m-1}\sum_{i_3=i_2}^{m-1} x_a(i_3), \quad \Omega_{2,0^3}(j) = \sum_{l=j}^{m-1} p_{2a}(l)\Omega_{0^3}(l),$$

$$\Theta_{0^3} = \sum_{i=0}^{m-1}\sum_{j=0}^{i}\sum_{l=0}^{j} x_a(l), \quad \Theta_0(j) = \sum_{l=0}^{j} x_a(l), \quad \Theta_{2,0^3}(j) = \sum_{l=0}^{j} p_{2a}(l)\Theta_{0^3}(l),$$

$$\Omega_{0,k} = \sum_{i_1=0}^{m-1}\sum_{i_2=i_1}^{m-1} p_{ka}(i_2)x_a(i_2), \quad \Theta_{0,k} = \sum_{i_1=0}^{m-1}\sum_{i_2=0}^{i_1} p_{ka}(i_2)x_a(i_2).$$

Lemma 4.7 *For an integer $m \in \mathbb{N}^+$ and a vector function $\{x(i)|i \in [a, a+m-1]\}$, the following equations hold:*

$$\sum_{i=0}^{m-1} i x_a(i) = \Omega_{0,0} - \Omega_0, \tag{4.15}$$

$$\sum_{i=0}^{m-1} i^2 x_a(i) = 2\Omega_{0,0,0} - 3\Omega_{0,0} + \Omega_0, \tag{4.16}$$

$$\sum_{i=0}^{m-1} i^3 x_a(i) = 6\Omega_{0^4} - 12\Omega_{0,0,0} + 7\Omega_{0,0} - \Omega_0, \tag{4.17}$$

$$\sum_{i=0}^{m-1}\sum_{j=i}^{m-1} j x_a(j) = 2\Omega_{0,0,0} - 2\Omega_{0,0}, \tag{4.18}$$

$$\sum_{i=0}^{m-1}\sum_{j=i}^{m-1} j^2 x_a(j) = 6\Omega_{0^4} - 10\Omega_{0,0,0} + 4\Omega_{0,0}, \tag{4.19}$$

$$\sum_{i=0}^{m-1}\sum_{j=i}^{m-1}\sum_{l=j}^{m-1} l x_a(l) = 3\Omega_{0^4} - 3\Omega_{0,0,0}. \tag{4.20}$$

Proof Here, the proofs for 6 equalities above are given one by one.
Proof of (4.15). By some computations, we have

$$\begin{aligned}
\sum_{i=0}^{m-1} i x_a(i) &= \sum_{i=0}^{m-1}\sum_{j=0}^{i} x_a(i) - \sum_{i=0}^{m-1} x_a(i) \\
&= \sum_{j=0}^{m-1}\sum_{i=j}^{m-1} x_a(i) - \sum_{i=0}^{m-1} x_a(i) \\
&= \Omega_{0,0} - \Omega_0.
\end{aligned}$$

Proof of (4.16). By applying (4.15), we have

$$\begin{aligned}\sum_{i=0}^{m-1} i^2 x_a(i) &= \sum_{i=0}^{m-1}\sum_{j=i}^{m-1} j x_a(j) - \sum_{i=0}^{m-1} i x_a(i)\\ &= \sum_{i=0}^{m-1}\left(\sum_{j=0}^{m-1} j x_a(j) - \sum_{j=0}^{i-1} j x_a(j)\right) - \sum_{i=0}^{m-1} i x_a(i)\\ &= (m-1)(\Omega_{0,0} - \Omega_0) - \sum_{i=0}^{m-1}\sum_{j=0}^{i-1} j x_a(j). \qquad (4.21)\end{aligned}$$

Since

$$\begin{aligned}\sum_{j=0}^{i-1} j x_a(j) &= \sum_{j=0}^{i-1}\sum_{l=j}^{i-1} x_a(l) - \sum_{j=0}^{i-1} x_a(j)\\ &= \left(\sum_{j=0}^{m-1} - \sum_{j=i}^{m-1}\right)\left(\sum_{l=j}^{m-1} - \sum_{l=i}^{m-1}\right) x_a(l) - \left(\sum_{j=0}^{m-1} - \sum_{j=i}^{m-1}\right) x_a(j)\\ &= \Big(\Omega_{0,0} - m\Omega_0(i) - \Omega_{0,0}(i) + (m-i)\Omega_0(i)\Big) - \Omega_0 + \Omega_0(i)\\ &= \Omega_{0,0} - \Omega_0 - (i-1)\Omega_0(i) - \Omega_{0,0}(i),\end{aligned}$$

we have

$$\begin{aligned}\sum_{i=0}^{m-1} i^2 x_a(i) &= (m-1)(\Omega_{0,0} - \Omega_0) - \sum_{i=0}^{m-1}\Big(\Omega_{0,0} - \Omega_0 - (i-1)\Omega_0(i) - \Omega_{0,0}(i)\Big)\\ &= -(\Omega_{0,0} - \Omega_0) + \sum_{i=0}^{m-1} i\Omega_0(i) - \sum_{i=0}^{m-1}\Omega_0(i) + \sum_{i=0}^{m-1}\Omega_{0,0}(i)\\ &= 2\Omega_{0,0,0} - 3\Omega_{0,0} + \Omega_0.\end{aligned}$$

Note that the following equations have been considered:

$$\sum_{i=0}^{m-1}\Omega_0(i) = \Omega_{0,0}, \quad \sum_{i=0}^{m-1}\Omega_{0,0}(i) = \Omega_{0,0,0}, \quad \sum_{i=0}^{m-1} i\Omega_0(i) = \Omega_{0,0,0} - \Omega_{0,0}.$$

Proof of (4.17). Similarly to (4.21), we have

$$\sum_{i=0}^{m-1} i^3 x_a(i) = (m-1)\sum_{j=0}^{m-1} j^2 x_a(j) - \sum_{i=0}^{m-1}\sum_{j=0}^{i-1} j^2 x_a(j). \qquad (4.22)$$

By applying (4.16), we have

$$\sum_{j=0}^{i-1} j^2 x_a(j) = 2\Omega_{0,0,0}^i - 3\Omega_{0,0}^i + \Omega_0^i, \tag{4.23}$$

where

$$\Omega_0^i = \sum_{i_1=0}^{i-1} x_a(i_1), \quad \Omega_{0,0}^i = \sum_{i_1=0}^{i-1}\sum_{i_2=i_1}^{i-1} x_a(i_2), \quad \Omega_{0,0,0}^i = \sum_{i_1=0}^{i-1}\sum_{i_2=i_1}^{i-1}\sum_{i_3=i_2}^{i-1} x_a(i_3).$$

By some computations, we have

$$\Omega_0^i = \left(\sum_{i_1=0}^{m-1} - \sum_{i_1=i}^{m-1}\right) x_a(i_1) = \Omega_0 - \Omega_0(i), \tag{4.24}$$

$$\begin{aligned}
\Omega_{0,0}^i &= \sum_{i_1=0}^{i-1}\left(\sum_{i_2=i_1}^{m-1} - \sum_{i_2=i}^{m-1}\right) x_a(i_2) \\
&= \left(\sum_{i_1=0}^{m-1} - \sum_{i_1=i}^{m-1}\right)\Big(\Omega_0(i_1) - \Omega_0(i)\Big) \\
&= \Omega_{0,0} - \Omega_{0,0}(i) - i\Omega_0(i),
\end{aligned} \tag{4.25}$$

$$\begin{aligned}
\Omega_{0,0,0}^i &= \sum_{i_1=0}^{i-1}\sum_{i_2=i_1}^{i-1}\left(\sum_{i_3=i_2}^{m-1} - \sum_{i_3=i}^{m-1}\right) x_a(i_3) \\
&= \sum_{i_1=0}^{i-1}\sum_{i_2=i_1}^{i-1}\sum_{i_3=i_2}^{m-1} x_a(i_3) - \frac{i(i+1)}{2}\Omega_0(i) \\
&= \left(\sum_{i_1=0}^{m-1} - \sum_{i_1=i}^{m-1}\right)\left(\sum_{i_2=i_1}^{m-1} - \sum_{i_2=i}^{m-1}\right)\Omega_0(i_2) - \frac{i(i+1)}{2}\Omega_0(i) \\
&= \Omega_{0,0,0} - \Omega_{0,0,0}(i) - i\Omega_{0,0}(i) - \frac{i(i+1)}{2}\Omega_0(i).
\end{aligned} \tag{4.26}$$

Then, substituting (4.24), (4.25) and (4.26) into (4.23) yields

$$\begin{aligned}
\sum_{j=0}^{i-1} j^2 x_a(j) = {} & 2\Omega_{0,0,0} - 3\Omega_{0,0} + \Omega_0 - 2\Omega_{0,0,0}(i) \\
& - (2i-3)\Omega_{0,0}(i) - (i^2 - 2i + 1)\Omega_0(i).
\end{aligned} \tag{4.27}$$

Substituting (4.27) into (4.22) yields

$$\begin{aligned}
\sum_{i=0}^{m-1} i^3 x_a(i) &= (m-1)\Big(2\Omega_{0,0,0} - 3\Omega_{0,0} + \Omega_0\Big) - \sum_{i=0}^{m-1}\Big(2\Omega_{0,0,0} - 3\Omega_{0,0} + \Omega_0 \\
&\quad - 2\Omega_{0,0,0}(i) - (2i-3)\Omega_{0,0}(i) - (i^2 - 2i + 1)\Omega_0(i)\Big) \\
&= \sum_{i=0}^{m-1}\Big(2\Omega_{0,0,0}(i) + (2i-3)\Omega_{0,0}(i) + (i^2 - 2i + 1)\Omega_0(i)\Big) \\
&\quad - \Big(2\Omega_{0,0,0} - 3\Omega_{0,0} + \Omega_0\Big).
\end{aligned} \tag{4.28}$$

As a result, we get (4.17) from (4.28) by considering the following equations:

$$\begin{aligned}
&\sum_{i=0}^{m-1} i\Omega_0(i) = \Omega_{0,0,0} - \Omega_{0,0}, \quad \sum_{i=0}^{m-1} i\Omega_{0,0}(i) = \Omega_{0^4} - \Omega_{0,0,0}, \\
&\sum_{i=0}^{m-1} i^2\Omega_0(i) = 2\Omega_{0^4} - 3\Omega_{0,0,0} + \Omega_{0,0}.
\end{aligned}$$

Proof of (4.18). From (4.21), we have

$$\begin{aligned}
\sum_{i=0}^{m-1}\sum_{j=i}^{m-1} j x_a(j) &= \sum_{i=0}^{m-1} i^2 x_a(i) + \sum_{i=0}^{m-1} i x_a(i) \\
&= 2\Omega_{0,0,0} - 2\Omega_{0,0}.
\end{aligned}$$

Proof of (4.19). From (4.15), we have

$$\sum_{i=0}^{m-1} i^3 x_a(i) = \sum_{i=0}^{m-1}\sum_{j=i}^{m-1} j^2 x_a(j) - \sum_{i=0}^{m-1} i^2 x_a(i).$$

So with (4.16) and (4.17), we get

$$\begin{aligned}
\sum_{i=0}^{m-1}\sum_{j=i}^{m-1} j^2 x_a(j) &= \sum_{i=0}^{m-1} i^3 x_a(i) + \sum_{i=0}^{m-1} i^2 x_a(i) \\
&= 6\Omega_{0^4} - 10\Omega_{0,0,0} + 4\Omega_{0,0}.
\end{aligned}$$

Proof of (4.20). From (4.18), we have

$$\sum_{i=0}^{m-1}\sum_{j=i}^{m-1} j^2 x_a(j) = 2\sum_{i=0}^{m-1}\sum_{j=i}^{m-1}\sum_{l=j}^{m-1} l x_a(l) - 2\sum_{i=0}^{m-1}\sum_{j=i}^{m-1} j x_a(j).$$

So with (4.18) and (4.19), we get

$$\begin{aligned}\sum_{i=0}^{m-1}\sum_{j=i}^{m-1}\sum_{l=j}^{m-1} l x_a(l) &= \frac{1}{2}\sum_{i=0}^{m-1}\sum_{j=i}^{m-1} j^2 x_a(j) + \sum_{i=0}^{m-1}\sum_{j=i}^{m-1} j x_a(j)\\ &= 3\Omega_{0^4} - 5\Omega_{0,0,0} + 2\Omega_{0,0} + 2\Omega_{0,0,0} - 2\Omega_{0,0}\\ &= 3\Omega_{0^4} - 3\Omega_{0,0,0}.\end{aligned}$$

This completes the proof. ■

Remark 4.6 From the proof of Lemma 4.7, we conclude that the single-summation term $\sum_{i=0}^{m-1} i^k x_a(i)$ for all $k \in \mathbb{N}^+$ can be represented as a linear combination of the sequence of summation terms $\Omega_{0^{k+1}}$, Ω_{0^k}, ..., and Ω_0. Note that there is no monomial involved in the summation term Ω_{0^k}. Since $p_{ka}(i)$ is a polynomial function of degree k, the summation term Ω_k can also be linearly represented by the sequence of summation terms $\Omega_{0^{k+1}}$, Ω_{0^k}, ..., and Ω_0. In fact, the preceding conclusion can be further extended to the multiple-summation terms such as $\Omega_{0^d,k}$, $\Omega_{k,0^d}$, $\Theta_{0^d,k}$ and $\Theta_{k,0^d}$, that is, any of the summation terms $\Omega_{0^d,k}$, $\Omega_{k,0^d}$, $\Theta_{0^d,k}$ and $\Theta_{k,0^d}$ can be linearly represented by the sequence of summation terms such as Ω_0, Ω_{0^2} and so on. Additionally, it is worth pointing out that the two groups of summation terms $\{\Omega_{0^{k+1}}, \Omega_{0^k}, \ldots, \Omega_0\}$ and $\{\Theta_{0^{k+1}}, \Theta_{0^k}, \ldots, \Theta_0\}$ can be linearly represented by each other.

4.2.3.2 Summation Inequalities in the Upper-Triangular Form

In this subsection, various concrete single- and multiple-summation inequalities are developed in the upper-triangular forms.

(1) $m = 1$
In the single-summation inner product space with the inner product defined as $\langle p_k, p_l\rangle := \sum_{i=a}^{a+m-1} p_k(i)p_l(i)$, the orthogonal polynomials are computed as follows with $N = 3$:

$$\begin{aligned}p_0(i) &= 1,\\ p_1(i) &= (i-a) - \frac{(m-1)}{2},\\ p_2(i) &= (i-a)^2 - (m-1)(i-a) + \frac{(m-1)(m-2)}{6},\end{aligned}$$

$$p_3(i) = (i-a)^3 - \frac{3(m-1)}{2}(i-a)^2 \frac{(6m^2-15m+11)}{10}(i-a) \\ - \frac{(m-1)(m-2)(m-3)}{20}.$$

Through some computations with the above orthogonal polynomials, we have

$$\begin{aligned}
\langle p_0, p_0 \rangle &= m, \\
\langle p_1, p_1 \rangle &= \frac{m(m-1)(m+1)}{12}, \\
\langle p_2, p_2 \rangle &= \frac{m(m-1)(m+1)(m-2)(m+2)}{180}, \\
\langle p_3, p_3 \rangle &= \frac{m(m-1)(m+1)(m-2)(m+2)(m-3)(m+3)}{2800}.
\end{aligned}$$

From Lemma 4.7, the following equations hold:

$$\begin{aligned}
\Omega_1 &= \sum_{i=0}^{m-1} p_{1a}(i)x_a(i) = \Omega_{0,0} - \Omega_0 - \frac{m-1}{2}\Omega_0 \\
&= \Omega_{0,0} - \frac{(m+1)}{2}\Omega_0, \\
\Omega_2 &= \sum_{i=0}^{m-1} p_{2a}(i)x_a(i) \\
&= 2\Omega_{03} - 3\Omega_{02} + \Omega_0 - (m-1)(\Omega_{02} - \Omega_0) + \frac{(m-1)(m-2)}{6}\Omega_0 \\
&= 2\Omega_{03} - (m+2)\Omega_{02} + \frac{(m+1)(m+2)}{6}\Omega_0, \\
\Omega_3 &= \sum_{i=0}^{m-1} p_{3a}(i)x_a(i) \\
&= 6\Omega_{04} - 3(m+3)\Omega_{03} + \frac{3(m+2)(m+3)}{5}\Omega_{0,0} - \frac{(m+1)(m+2)(m+3)}{20}\Omega_0.
\end{aligned}$$

As a result, the following single-summation inequality holds from Lemma 4.6:

$$\sum_{i=0}^{m-1} x_a^T(i)Rx_a(i) \geq \sum_{k=0}^{3} \frac{1}{\langle p_k, p_k \rangle}\Omega_k^T R\Omega_k. \tag{4.29}$$

(2) $m = 2$

In the double-summation inner product space with the inner product defined as $\langle p_k, p_l \rangle := \sum_{i=a}^{a+m-1}\sum_{j=i}^{a+m-1} p_k(j)p_l(j)$, the orthogonal polynomials are computed as follows with $N = 2$:

$$\begin{aligned}
p_0(i) &= 1, \\
p_1(i) &= (i-a) - \frac{2(m-1)}{3}, \\
p_2(i) &= (i-a)^2 - \frac{(6m-7)}{5}(i-a) + \frac{3(m-1)(m-2)}{10}.
\end{aligned}$$

Through some computations with the above orthogonal polynomials, we have

$$\begin{aligned}
\langle p_0, p_0 \rangle &= \frac{m(m+1)}{2}, \\
\langle p_1, p_1 \rangle &= \frac{m(m-1)(m+1)(m+2)}{36}, \\
\langle p_2, p_2 \rangle &= \frac{m(m-1)(m+1)(m-2)(m+2)(m+3)}{600}.
\end{aligned}$$

The following equations hold from Lemma 4.7:

$$\begin{aligned}
\Omega_{0,1} &= \sum_{i=0}^{m-1}\sum_{j=i}^{m-1} p_{1a}(j)x_a(j) = 2\Omega_{03} - 2\Omega_{02} - \frac{2(m-1)}{3}\Omega_{02} \\
&= 2\Omega_{03} - \frac{2(m+2)}{3}\Omega_{0,0}, \\
\Omega_{0,2} &= \sum_{i=0}^{m-1}\sum_{j=i}^{m-1} p_{2a}(j)x_a(j) \\
&= 6\Omega_{04} - 10\Omega_{03} + 4\Omega_{0,0} - \frac{(6m-7)}{5}(2\Omega_{03} - 2\Omega_{02}) + \frac{3(m-1)(m-2)}{10}\Omega_{02} \\
&= 6\Omega_{04} - \frac{12(m+3)}{5}\Omega_{03} + \frac{3(m+2)(m+3)}{10}\Omega_{0,0}.
\end{aligned}$$

As a result, the following double-summation inequality holds from Lemma 4.6:

$$\sum_{i=0}^{m-1}\sum_{j=i}^{m-1} x_a^T(j)Rx_a(j) \geq \sum_{i=0}^{2} \frac{1}{\langle p_i, p_i \rangle}\Omega_{0,i}^T R\Omega_{0,i}. \tag{4.30}$$

(3) $m = 3$
In the triple-summation inner product space with the inner product defined as $\langle p_k, p_l \rangle := \mathscr{S}_3^U(p_k(i)p_l(i))$, the orthogonal polynomials are computed as follows with $N = 1$:

$$\begin{aligned}
p_0(i) &= 1, \\
p_1(i) &= (i-a) - \frac{3(m-1)}{4}.
\end{aligned}$$

Based on the above polynomials, the following triple-summation inequality is developed from Lemmas 4.6 and 4.7:

$$\sum_{i_1=0}^{m-1}\sum_{i_2=i_1}^{m-1}\sum_{i_3=i_2}^{m-1} x_a^T(i_3)Rx_a(i_3) \geq \frac{1}{\langle p_0, p_0\rangle}\Omega_{0^3}^T R\Omega_{0^3} + \frac{1}{\langle p_1, p_1\rangle}\Omega_{0,0,1}^T R\Omega_{0,0,1}, \tag{4.31}$$

where

$$\begin{aligned}
\langle p_0, p_0\rangle &= \frac{m(m+1)(m+2)}{6}, \\
\langle p_1, p_1\rangle &= \frac{m(m-1)(m+1)(m+2)(m+3)}{160}, \\
\Omega_{0,0,1} &= \sum_{i_1=0}^{m-1}\sum_{i_2=i_1}^{m-1}\sum_{i_3=i_2}^{m-1}\left(i - \frac{3(m-1)}{4}\right)x_a(i_3) \\
&= 3\Omega_{0^4} - 3\Omega_{0^3} - \frac{3(m-1)}{4}\Omega_{0^3} \\
&= 3\Omega_{0^4} - \frac{3(m+3)}{4}\Omega_{0^3}.
\end{aligned}$$

(4) $m = 4$ (in the upper-triangular form)
In the quartic-summation inner product space with the inner product defined as $\langle p_k, p_l\rangle := \mathscr{S}_4^U(p_k(i)p_l(i))$, the following quartic-summation inequality is developed:

$$\sum_{i_1=0}^{m-1}\sum_{i_2=i_1}^{m-1}\sum_{i_3=i_2}^{m-1}\sum_{i_4=i_3}^{m-1} x_a^T(i_4)Rx_a(i_4) \geq \frac{1}{\langle p_0, p_0\rangle}\Omega_{0^4}^T R\Omega_{0^4}, \tag{4.32}$$

where

$$\langle p_0, p_0\rangle = \frac{m(m+1)(m+2)(m+3)}{24}.$$

In the stability analysis for time-delay systems, the commonly encountered problem is to estimate summation terms such as $\mathscr{S}_d^U(y^T(i)Ry(i))$. So, with $x(i)$ being replaced by $y(i)$ in (4.29), (4.30), (4.31) and (4.32) we get the following corollary.

Corollary 4.3 *For a matrix $R \in \mathbb{S}_+^n$ and a vector function $\{x(i)|i \in [a, a+m]\}$, the following summation inequalities hold:*

$$\sum_{i=0}^{m-1} y_a^T(i)Ry_a(i) \geq \sum_{i=0}^{3}\frac{2i+1}{m}\dot{\Omega}_i^T R\dot{\Omega}_i, \tag{4.33}$$

$$\sum_{i=0}^{m-1}\sum_{j=i}^{m-1} y_a^T(j)Ry_a(j) \geq 2\dot{\Omega}_{0^2}^T R\dot{\Omega}_{0^2} + 4\dot{\Omega}_{0,1}^T R\dot{\Omega}_{0,1} + 6\dot{\Omega}_{0,2}^T R\dot{\Omega}_{0,2}, \tag{4.34}$$

$$\sum_{i=0}^{m-1}\sum_{j=i}^{m-1}\sum_{l=j}^{m-1} y_a^T(l)Ry_a(l) \geq \frac{3(m+2)}{2}\dot{\Omega}_{0^3}^T R\dot{\Omega}_{0^3} + \frac{5(m+3)}{2}\dot{\Omega}_{0^2,1}^T R\dot{\Omega}_{0^2,1}, \tag{4.35}$$

$$\sum_{i_1=0}^{m-1}\sum_{i_2=i_1}^{m-1}\sum_{i_3=i_2}^{m-1}\sum_{i_4=i_3}^{m-1} y_a^T(i_4)Ry_a(i_4) \geq \frac{2(m+2)(m+3)}{3}\dot{\Omega}_{0^4}^T R\dot{\Omega}_{0^4} \tag{4.36}$$

where

$$\begin{aligned}
\dot{\Omega}_0 &= x_a(m) - x_a(0),\\
\dot{\Omega}_1 &= x_a(m) + x_a(0) - \frac{2}{m+1}\bar{\Omega}_0,\\
\dot{\Omega}_2 &= x_a(m) - x_a(0) + \frac{6}{(m+1)}\bar{\Omega}_0 - \frac{12}{(m+1)(m+2)}\bar{\Omega}_{0^2},\\
\dot{\Omega}_3 &= x_a(m) + x_a(0) - \frac{12}{(m+1)}\bar{\Omega}_0 + \frac{60}{(m+1)(m+2)}\bar{\Omega}_{0^2}\\
&\quad - \frac{120}{(m+1)(m+2)(m+3)}\bar{\Omega}_{0^3},\\
\dot{\Omega}_{0^2} &= x_a(m) - \frac{1}{(m+1)}\bar{\Omega}_0,\\
\dot{\Omega}_{0,1} &= x_a(m) + \frac{2}{(m+1)}\bar{\Omega}_0 - \frac{6}{(m+1)(m+2)}\bar{\Omega}_{0^2},\\
\dot{\Omega}_{0,2} &= x_a(m) - \frac{3}{(m+1)}\bar{\Omega}_0 + \frac{24}{(m+1)(m+2)}\bar{\Omega}_{0^2} - \frac{60}{(m+1)(m+2)(m+3)}\bar{\Omega}_{0^3},\\
\dot{\Omega}_{0^3} &= x_a(m) - \frac{2}{(m+1)(m+2)}\bar{\Omega}_{0^2},\\
\dot{\Omega}_{0^2,1} &= x_a(m) + \frac{6}{(m+1)(m+2)}\bar{\Omega}_{0^2} - \frac{24}{(m+1)(m+2)(m+3)}\bar{\Omega}_{0^3},\\
\dot{\Omega}_{0^4} &= x_a(m) - \frac{6}{(m+1)(m+2)(m+3)}\bar{\Omega}_{0^3}.
\end{aligned}$$

Remark 4.7 When developing Corollary 4.3, some algebraic fractions have been removed for convenience of applications of these summation inequalities in the stability analysis of time-delay systems by considering inequalities such as $\frac{m+1}{m-1} > 1$, $\frac{(m+1)(m+2)}{(m-1)(m-2)} > 1$ and $\frac{(m+1)(m+2)(m+3)}{(m-1)(m-2)(m-3)} > 1$. It is worth pointing out that the summation intervals in the bounds of all summation inequalities proposed in Corollary 4.3 are from a to $a+m$, not from a to $a+m-1$. Meanwhile, it is observed that any summation term in the bounds is a zero vector such as $\dot{\Omega}_{0^3}$ and $\dot{\Omega}_{0^2,1}$ if the vector function $x(i)$ is assumed to be constant. This interesting phenomenon verifies the correctness of the proposed summation inequalities.

Remark 4.8 It is seen from Corollary 4.3 that the single-summation inequality (4.33) includes the Jensen (4.2), the Wirtinger-based (4.7) and the auxiliary-function-based summation inequalities (4.8). It is also found that the double-summation inequality (4.34) includes the Jensen (4.5) and auxiliary-function-based double-summation inequalities (4.9). More accurate single- and multiple-summation inequalities can be obtained in a similar way if more orthogonal polynomials are considered.

4.2.3.3 Summation Inequalities in the Lower-Triangular Form

(1) $m = 2$

In the double-summation inner product space with the inner product defined as $\langle p_k, p_l\rangle = \sum_{i=0}^{m-1}\sum_{j=0}^{i} p_k(j)p_l(j)$, the orthogonal polynomials are computed with $N = 2$:

$$
\begin{aligned}
p_0(i) &= 1, \\
p_1(i) &= (i-a) - \frac{(m-1)}{3}, \\
p_2(i) &= (i-a)^2 - \frac{(4m-3)}{5}(i-a) + \frac{(m-1)(m-2)}{10}.
\end{aligned}
$$

Through some computations with the above polynomials, we have

$$
\begin{aligned}
\langle p_0, p_0\rangle &= \frac{m(m+1)}{2}, \\
\langle p_1, p_1\rangle &= \frac{m(m-1)(m+1)(m+2)}{36}, \\
\langle p_2, p_2\rangle &= \frac{m(m-1)(m+1)(m-2)(m+2)(m+3)}{600}.
\end{aligned}
$$

Additionally, the following equations hold from Lemma 4.7:

$$
\begin{aligned}
\Theta_{0,0} =& -\Omega_{0,0} + (m+1)\Omega_0, \\
\Theta_{0,1} =& -2\Omega_{0,0,0} + \frac{4(m+2)}{3}\Omega_{0,0} - \frac{(m+1)(m+2)}{3}\Omega_0, \\
\Theta_{0,2} =& -6\Omega_{0^4} + \frac{18(m+3)}{5}\Omega_{0,0,0} - \frac{9(m+2)(m+3)}{10}\Omega_{0,0} \\
& + \frac{(m+1)(m+2)(m+3)}{10}\Omega_0.
\end{aligned}
$$

Hence, the following double-summation inequality holds from Lemma 4.6:

$$\sum_{i=0}^{m-1}\sum_{j=0}^{i} x_a^T(j)Rx_a(j) \geq \sum_{k=0}^{2} \frac{1}{\langle p_k, p_k\rangle}\Theta_{0,k}^T R\Theta_{0,k}. \tag{4.37}$$

(2) $m = 3$
In the triple-summation inner product space with the inner product defined as $\langle p_k, p_l\rangle := \mathscr{S}_3^L(p_k(i)p_l(i))$, the orthogonal polynomials are computed as follows with $N = 1$:

$$p_0(i) = 1,$$
$$p_1(i) = (i-a) - \frac{(m-1)}{4}.$$

Based on the above orthogonal polynomials, the following triple-summation inequality

$$\sum_{i_1=0}^{n-1}\sum_{i_2=0}^{i_1-1}\sum_{i_3=0}^{i_2-1} x_a^T(i_3)Rx_a(i_3) \geq \frac{1}{\langle p_0, p_0\rangle}\Theta_{0^3}^T R\Theta_{0^3} + \frac{1}{\langle p_1, p_1\rangle}\Theta_{0,0,1}^T R\Theta_{0,0,1} \tag{4.38}$$

holds from Lemmas 4.6 and 4.7, where

$$\begin{aligned}
\langle p_0, p_0\rangle &= \frac{m(m+1)(m+2)}{6},\\
\langle p_1, p_1\rangle &= \frac{m(m-1)(m+1)(m+2)(m+3)}{160},\\
\Theta_{0,0,0} &= \frac{(m+1)(m+2)}{2}\Omega_0 - (m+2)\Omega_{0,0} + \Omega_{0,0,0},\\
\Theta_{0,0,1} &= -\frac{(m+1)(m+2)(m+3)}{8}\Omega_0 + \frac{3(m+2)(m+3)}{4}\Omega_{0,0}\\
&\quad - \frac{9(m+3)}{4}\Omega_{0,0,0} + 3\Omega_{0^4}.
\end{aligned}$$

(3) $m = 4$ (in the lower-triangular form)
In the quartic-summation inner product space with the inner product defined as $\langle p_k, p_l\rangle := \mathscr{S}_4^L(p_k(i)p_l(i))$, the following quartic-summation inequality

$$\sum_{i_1=0}^{n-1}\sum_{i_2=0}^{i_1-1}\sum_{i_3=0}^{i_2-1}\sum_{i_4=0}^{i_3-1} x_a^T(i_4)Rx_a(i_4) \geq \frac{1}{\langle p_0, p_0\rangle}\Theta_{0^4}^T R\Theta_{0^4} \tag{4.39}$$

holds, where

$$\langle p_0, p_0\rangle = \frac{m(m+1)(m+2)(m+3)}{24}.$$

With $x_a(i)$ being replaced by $y_a(i)$ in (4.37), (4.38) and (4.39), we get the following lemma.

Lemma 4.8 *For a matrix $R \in \mathbb{S}_+^n$ and a vector function $\{x(i)|i \in [a, a+m]\}$, the following summation inequalities hold:*

$$\sum_{i_1=0}^{m-1}\sum_{i_2=0}^{i_1} y_a^T(i_2)Ry_a(i_2) \geq 2\dot{\Theta}_{0^2}^T R\dot{\Theta}_{0^2} + 4\dot{\Theta}_{0,1}^T R\dot{\Theta}_{0,1} + 6\dot{\Theta}_{0,2}^T R\dot{\Theta}_{0,2}, \tag{4.40}$$

$$\sum_{i_1=0}^{m-1}\sum_{i_2=0}^{i_1}\sum_{i_3=0}^{i_2} y_a^T(i_3)Ry_a(i_3) \geq \frac{3(m+2)}{2}\dot{\Theta}_{0^3}^T R\dot{\Theta}_{0^3} + \frac{5(m+3)}{2}\dot{\Theta}_{0^2,1}^T R\dot{\Theta}_{0^2,1}, \tag{4.41}$$

$$\sum_{i_1=0}^{m-1}\sum_{i_2=0}^{i_1}\sum_{i_3=0}^{i_2}\sum_{i_4=0}^{i_3} y_a^T(i_4)Ry_a(i_4) \geq \frac{2(m+2)(m+3)}{3}\dot{\Theta}_{0^4}^T R\dot{\Theta}_{0^4} \tag{4.42}$$

where

$$\begin{aligned}
\dot{\Theta}_{0,0} &= -x_a(0) + \frac{1}{(m+1)}\bar{\Omega}_0, \\
\dot{\Theta}_{0,1} &= x_a(0) - \frac{4}{(m+1)}\bar{\Omega}_0 + \frac{6}{(m+1)(m+2)}\bar{\Omega}_{0,0}, \\
\dot{\Theta}_{0,2} &= -x_a(0) + \frac{9}{(m+1)}\bar{\Omega}_0 - \frac{36}{(m+1)(m+2)}\bar{\Omega}_{0,0} + \frac{60}{(m+1)(m+2)(m+3)}\bar{\Omega}_{0^3}, \\
\dot{\Theta}_{0^3} &= -x_a(0) + \frac{2}{(m+1)}\bar{\Omega}_0 - \frac{2}{(m+1)(m+2)}\bar{\Omega}_{0,0}, \\
\dot{\Theta}_{0^2,1} &= x_a(0) - \frac{6}{(m+1)}\bar{\Omega}_0 + \frac{18}{(m+1)(m+2)}\bar{\Omega}_{0,0} - \frac{24}{(m+1)(m+2)(m+3)}\bar{\Omega}_{0^3}, \\
\dot{\Theta}_{0^4} &= -x_a(0) + \frac{3}{(m+1)}\bar{\Omega}_0 - \frac{6}{(m+1)(m+2)}\bar{\Omega}_{0,0}.
\end{aligned}$$

4.3 Summation Inequalities with Free Matrices

In this section, various summation inequalities with free matrices are proposed, including conventional FMB, GFMB and improved GFMB summation inequalities. Especially, based on orthogonal polynomials defined in the single-summation inner product space, a series of GFMB summation inequalities is developed.

4.3.1 Conventional FMB Inequalities

Lemma 4.9 (The conventional FMB summation inequality) *For matrices* $R \in \mathbb{S}_{+}^{n}$, $Z_{00}, Z_{11} \in \mathbb{S}_{+}^{3n}$, $Z_{01} \in \mathbb{R}^{3n\times 3n}$, $N_0, N_1 \in \mathbb{R}^{3n\times n}$ *and a vector function* $\{x(i)|i \in [a, a+m]\}$ *such that*

$$\Phi = \begin{bmatrix} Z_{00} & Z_{01} & N_0 \\ * & Z_{11} & N_1 \\ * & * & R \end{bmatrix} \geq 0, \tag{4.43}$$

the following inequality holds:

$$-\sum_{i=a}^{a+m-1} y^T(i)Ry(i) \leq \vartheta^T \Psi \vartheta \tag{4.44}$$

where

$$\begin{aligned}
\Psi &= m\left(Z_{00} + \frac{1}{3}Z_{11}\right) + \text{Sym}\{N_0 e_{12} + N_1 e_{123}\}, \\
\vartheta &= col\left\{x(a+m), x(a), \frac{1}{m+1}\bar{\Omega}_0\right\}, \\
e_{12} &= \begin{bmatrix} I & -I & 0 \end{bmatrix}, \quad e_{123} = \begin{bmatrix} I & I & -2I \end{bmatrix}.
\end{aligned}$$

Proof Define

$$f(i) = \frac{i - a - \frac{m-1}{2}}{\frac{m+1}{2}}, \quad \varepsilon(i) = col\{\vartheta, f(i)\vartheta, y(i)\}.$$

By using Lemma 4.7, we get

$$\sum_{i=a}^{a+m-1} f(i) = 0, \tag{4.45}$$

$$\begin{aligned}
\sum_{i=a}^{a+m-1} f^2(i) &= \frac{\sum_{i=a}^{a+m-1}\left(i - a - \frac{m-1}{2}\right)^2}{\frac{(m+1)^2}{4}} \\
&= \frac{4}{(m+1)^2} \times \left\{\frac{m(m+1)^2}{4} + \frac{m(m+1)(2m+1)}{6} - \frac{m(m+1)^2}{2}\right\} \\
&= \frac{4}{(m+1)^2} \times \frac{1}{12}m(m+1)(m-1) \\
&= \frac{m(m-1)}{3(m+1)},
\end{aligned} \tag{4.46}$$

$$\begin{aligned}
\sum_{i=a}^{a+m-1} f(i)y(i) &= \frac{2}{m+1}\sum_{i=a}^{a+m-1}\sum_{j=i}^{a+m-1} y(j) - \sum_{i=a}^{a+m-1} y(i) \\
&= \frac{2}{m+1}\sum_{i=a}^{a+m-1}(x(a+m)-x(i)) - (x(a+m)-x(a)) \\
&= x(a+m)+x(a) - \frac{2}{m+1}\sum_{i=a}^{a+m} x(i) \\
&= e_{123}\vartheta. \qquad (4.47)
\end{aligned}$$

Summating $\varepsilon^T(i)\Phi\varepsilon(i)$ from a to $a+m-1$ with (4.45)–(4.47) leads to

$$\begin{aligned}
\sum_{i=a}^{a+m-1}\varepsilon^T(i)\Phi\varepsilon(i) = m\vartheta^T Z_{00}\vartheta + \frac{m(m-1)}{3(m+1)}\vartheta^T Z_{11}\vartheta + 2\vartheta^T N_0 e_{12}\vartheta \\
+2\vartheta^T N_1 e_{123}\vartheta + \sum_{i=a}^{a+m-1} y^T(i)Ry(i).
\end{aligned}$$

Considering the fact that $\varepsilon^T(i)\Phi\varepsilon(i) \geq 0$, we get

$$\begin{aligned}
&-\sum_{i=a}^{a+m-1} y^T(i)Ry(i) \\
&\leq m\vartheta^T Z_{00}\vartheta + \frac{m(m-1)}{3(m+1)}\vartheta^T Z_{11}\vartheta + 2\vartheta^T N_0 e_{12}\vartheta + 2\vartheta^T N_1 e_{123}\vartheta \\
&\leq \vartheta^T \Psi\vartheta. \qquad (4.48)
\end{aligned}$$

Note that the faction $\frac{m-1}{m+1}$ is removed from the FMB summation inequality since this term may bring some difficulty in applying the summation inequality to stability analysis for time-delay systems. This completes the proof. ■

Remark 4.9 By introducing free matrices such as Z_{00}, Z_{11}, Z_{01}, N_0, N_1, the conventional FMB summation inequality (4.44) is developed in Lemma 4.9, which includes the Wirtinger-based summation inequality (4.7). Actually, by setting $N_0 = \frac{1}{m}[-R,\ R,\ 0]^T$, $N_1 = \frac{3}{m}[R,\ R,\ -2R]^T$, $Z_{00} = N_0R^{-1}N_0^T$, $Z_{01} = N_0R^{-1}N_1^T$ and $Z_{11} = N_1R^{-1}N_1^T$, the Wirtinger-based summation inequality (4.7) is recovered. Especially, when the FMB inequality (4.44) is applied to the stability analysis of time-delay systems, the reciprocally convex lemma is no longer required.

It is necessarily noted that the free matrix Z_{01} does not appear in the inequality (4.44). It is only used to satisfy the inequality constraint (4.43). If we set $Z_{00} = N_0R^{-1}N_0^T$, $Z_{01} = N_0R^{-1}N_1^T$ and $Z_{11} = N_1R^{-1}N_1^T$, the constraint (4.43) always holds. In this case, the three free matrices Z_{00}, Z_{01} and Z_{11} can be eliminated from the conventional FMB inequality (4.44).

Corollary 4.4 (The conventional FMB summation inequality) *For matrices $R \in \mathbb{S}_+^n$, $N_0, N_1 \in \mathbb{R}^{3n\times n}$ and a vector function $\{x(i)|i \in [a, a+m]\}$, the following inequality holds:*

$$-\sum_{i=a}^{a+m-1} y^T(i)Ry(i) \leq \vartheta^T \Psi \vartheta \tag{4.49}$$

where ϑ, e_{12} and e_{123} are defined in Lemma 4.9, and

$$\Psi = m\left(N_0R^{-1}N_0^T + \frac{1}{3}N_1R^{-1}N_1^T\right) + \mathrm{Sym}\{N_0e_{12} + N_1e_{123}\}.$$

Remark 4.10 It is worth pointing out that the chosen polynomial $f(i)$ is orthogonal with the polynomial 1 with respect to the single-summation inner product $\langle p_k, p_l\rangle = \sum_{i=a}^{a+m-1} p_k(i)p_l(i)$. Naturally, if a sequence of such orthogonal polynomials is considered, one series of conventional FMB can be obtained in a similar way.

4.3.2 GFMB Inequalities

If the augmented vector ϑ is set to be an arbitrary vector in the conventional FMB inequality (4.44), the GFMB inequality is obtained.

Lemma 4.10 (The GFMB summation inequality) *For a matrix $R \in \mathbb{S}_+^n$, any matrices M_0, M_1 with appropriate dimensions and a vector function $\{x(i)|i \in [a, a+m]\}$, the following inequality holds:*

$$\begin{aligned}-\sum_{i=0}^{m-1} x_a^T(i)Rx_a(i) \leq \eta^T\left(mM_0R^{-1}M_0^T + \frac{m}{3}M_1R^{-1}M_1^T\right)\eta \\ + \eta^T\mathrm{Sym}\{M_0\chi_0 + M_1\chi_1\}\end{aligned} \tag{4.50}$$

where η is an arbitrary vector, and

$$\chi_0 = \sum_{i=0}^{m-1} x_a(i),$$

$$\chi_1 = -\chi_0 + \frac{2}{m+1}\sum_{i=0}^{m-1}\sum_{j=i}^{m-1} x_a(j).$$

Proof For an arbitrary vector η, we define

$$
\begin{aligned}
&f_0(i) = 1, \quad f_1(i) = \frac{2i-(m-1)}{m+1},\\
&\zeta(i) = col\{f_0(i)\eta, f_1(i)\eta, x_a(i)\},\\
&M = col\{M_0, M_1\}, \quad \Phi = \begin{bmatrix} MR^{-1}M^T & M \\ * & R \end{bmatrix}.
\end{aligned}
$$

Through some computations, it follows that

$$
\begin{aligned}
&\sum_{i=0}^{m-1} f_0(i)f_1(i) = 0,\\
&\sum_{i=0}^{m-1} f_0(i)^2 = m, \quad \sum_{i=0}^{m-1} f_1(i)^2 = \frac{m(m-1)}{3(m+1)},\\
&\sum_{i=0}^{m-1} f_0(i)x_a(i) = \chi_0, \quad \sum_{i=0}^{m-1} f_1(i)x_a(i) = \chi_1.
\end{aligned}
$$

Then, based on the above results, summating $\zeta(i)^T\Phi\zeta(i)$ from 0 to $m-1$ leads to

$$
\begin{aligned}
\sum_{i=0}^{m-1}\zeta^T(i)\Phi\zeta(i) = {}& m\eta^T M_0R^{-1}M_0^T\eta + \frac{m(m-1)}{3(m+1)}\eta^T M_1R^{-1}M_1^T\eta\\
&+\eta^T\sum_{p=0}^{1}\text{Sym}\left\{M_p\chi_p\right\} + \sum_{i=0}^{m-1}x_a^T(i)Rx_a(i).
\end{aligned}
\tag{4.51}
$$

It follows $\sum_{i=0}^{m-1}\zeta(i)^T\Phi\zeta(i) \geq 0$ from the Schur complement. Then the GFMB inequality (4.50) is directly obtained by considering the fact $\frac{m(m-1)}{3(m+1)} \leq \frac{m}{3}$. This completes the proof. ■

With $x_a(i)$ being replaced by $y_a(i)$, the following corollary is obtained.

Corollary 4.5 (The GFMB summation inequality) *For a matrix $R \in \mathbb{S}_+^n$, any matrices M_0, M_1 with appropriate dimensions and a vector function $\{x(i)|i \in [a, a+m]\}$, the following inequality holds:*

$$
\begin{aligned}
-\sum_{i=0}^{m-1}y_a^T(i)Ry_a(i) \leq {}& \eta^T\left(mM_0R^{-1}M_0^T + \frac{m}{3}M_1R^{-1}M_1^T\right)\eta\\
&+\eta^T\text{Sym}\{M_0\hat{\chi}_0 + M_1\hat{\chi}_1\}
\end{aligned}
\tag{4.52}
$$

where η is an arbitrary vector, and

$$\hat{\chi}_0 = x_a(m) - x_a(0),$$
$$\hat{\chi}_1 = x_a(m) + x_a(0) - \frac{2}{m+1}\bar{\Omega}_0.$$

Remark 4.11 Since η is an arbitrary vector, the GFMB inequality (4.52) definitely provides more freedom to estimate summation terms than the conventional FMB inequality (4.44). Particularly, when η is fixed by setting $\eta = col\{x_a(m), x_a(0), \frac{1}{m+1}\bar{\Omega}_0\}$, the conventional FMB inequality (4.44) is recovered.

It is noted that in the GFMB inequality (4.52), the vector η in the two terms $\eta^T N_0 R^{-1} N_0^T \eta$ and $\eta^T N_1 R^{-1} N_1^T \eta$ is identical. If they are chosen differently from each other, more freedom should be got.

Lemma 4.11 *For a matrix $R \in \mathbb{S}^n_+$, any matrices M_0, M_1 with appropriate dimensions, and a vector function $\{x_a(i)|i \in [0, m]\}$, the following inequalities hold:*

$$-\sum_{i=0}^{m-1} x_a^T(i) R x_a(i) \leq \text{Sym}\{\eta_0^T M_0 \chi_0 + \eta_1^T M_1 \chi_1\}$$
$$+ m\eta_0^T M_0 R^{-1} M_0^T \eta_0 + \frac{m}{3}\eta_1^T M_1 R^{-1} M_1^T \eta_1, \qquad (4.53)$$

$$-\sum_{i=0}^{m-1} y_a^T(i) R y_a(i) \leq \text{Sym}\{\eta_0^T M_0 \hat{\chi}_0 + \eta_1^T M_1 \hat{\chi}_1\}$$
$$+ m\eta_0^T M_0 R^{-1} M_0^T \eta_0 + \frac{m}{3}\eta_1^T M_1 R^{-1} M_1^T \eta_1 \qquad (4.54)$$

where η_0, η_1 are arbitrary vectors, χ_0, χ_1 are defined in Lemma 4.10 and $\hat{\chi}_0$, $\hat{\chi}_1$ are defined in Corollary 4.5.

Remark 4.12 In the GFMB summation inequalities (4.50) and (4.52), the two vectors η_0 and η_1 are set to the same, which definitely restricts their flexibility. However, in the improved GFMB summation inequalities (4.53) and (4.54), the expressions of η_0 and η_1 can be chosen freely and independently when applying them. So, compared with the previous summation inequalities, the improved GFMB summation inequalities (4.53) and (4.54) provide more freedom to estimate summation terms.

4.3.3 A Series of GFMB Summation Inequalities

In the previous subsection, we have discussed the properties of summation terms such as $\mathscr{S}_d^U(p_{ka}(i)x_a(i))$ and $\mathscr{S}_d^L(p_{ka}(i)x_a(i))$. Here, we further study the properties of the term $\mathscr{S}_d^U(p_{ka}(i)x_a(i))$. Before proceeding, the following notations are defined:

$$\bar{S}_d := \sum_{i_1=0}^{m}\sum_{i_2=i_1}^{m}\cdots\sum_{i_d=i_{d-1}}^{m} 1, \quad \hat{\Omega}_k =: \sum_{i=0}^{m-1} p_{ka}(i)y_a(i)$$

$$\hat{\Omega}_{0^d} := \sum_{i_1=0}^{m-1}\sum_{i_2=i_1}^{m-1}\cdots\sum_{i_d=i_{d-1}}^{m-1} y_a(i_d), \quad \bar{\Omega}_{0^d} := \sum_{i_1=0}^{m}\sum_{i_2=i_1}^{m}\cdots\sum_{i_d=i_{d-1}}^{m} x_a(i_d).$$

Lemma 4.12 *For any vector function $\{x_a(i) \in \mathbb{R}^n | i \in [0, m]\}$, the following equations hold:*

$$\mathscr{S}_d^U(ix_a(i)) = d(\Omega_{0^{d+1}} - \Omega_{0^d}), \quad d \geq 1, \tag{4.55}$$

$$\mathscr{S}_1^U(i^l x_a(i)) = \mathscr{L}\{\Omega_{0^1}, \Omega_{0^2}, \ldots, \Omega_{0^{l+1}}\}, \quad l \geq 0, \tag{4.56}$$

$$\hat{\Omega}_{0^{d+1}} = \begin{cases} x_a(m) - x_a(0), & d = 0, \\ \bar{S}_d x_a(m) - \bar{\Omega}_{0^d}, & d > 0, \end{cases} \tag{4.57}$$

$$\mathscr{S}_1^U(i^l y_a(i)) = \begin{cases} \mathscr{L}\{x_a(m), x_a(0)\}, & l = 0, \\ \mathscr{L}\{x_a(m), \bar{\Omega}_{0^1}, \ldots, \bar{\Omega}_{0^l}\}, & l > 0. \end{cases} \tag{4.58}$$

Proof The four equalities above are proved in turn below.
Proof of (4.55). We use mathematics induction. When $d = 1$, the Eq. (4.55) holds from Lemma 4.7. Let us assume the following equation holds:

$$\mathscr{S}_{d-1}^U(ix_a(i)) = (d-1)(\Omega_{0^d} - \Omega_{0^{d-1}}), \quad d \geq 3. \tag{4.59}$$

Through some computations, we have

$$\begin{aligned} \mathscr{S}_d^U(ix_a(i)) &= \sum_{i_1=0}^{m-1}\cdots\sum_{i_{d-1}=i_{d-2}}^{m-1}\left(\sum_{i_d=0}^{m-1} - \sum_{i_d=0}^{i_{d-1}-1}\right) i_d x_a(i_d) \\ &= \sum_{i_1=0}^{m-1}\cdots\sum_{i_{d-1}=i_{d-2}}^{m-1}\sum_{i_d=0}^{m-1} i_d x_a(i_d) \\ &\quad - \sum_{i_1=0}^{m-1}\cdots\sum_{i_{d-1}=i_{d-2}}^{m-1}\sum_{i_d=0}^{i_{d-1}-1} i_d x_a(i_d). \end{aligned} \tag{4.60}$$

It follows from Lemma 4.7 that

$$\begin{aligned} \sum_{j=0}^{i-1} jx_a(j) &= \sum_{j=0}^{i-1}\sum_{l=j}^{i-1} x_a(l) - \sum_{j=0}^{i-1} x_a(j) \\ &= \Omega_{0^2} - \Omega_{0^1} - \Omega_{0^2}(i) + \Omega_{0^1}(i) - i\Omega_{0^1}(i). \end{aligned} \tag{4.61}$$

Then with the assumption (4.59), substituting (4.61) into (4.60) leads to

$$\begin{aligned}\mathscr{S}_d^U(ix_a(i)) &= \sum_{i_1=0}^{m-1}\cdots\sum_{i_{d-1}=i_{d-2}}^{m-1}\Big(i_{d-1}\Omega_{0^1}(i_{d-1})+\Omega_{0^2}(i_{d-1})-\Omega_{0^1}(i_{d-1})\Big)\\ &= \mathscr{S}_{d-1}^U(i\Omega_{0^1}(i))+\mathscr{S}_{d-1}^U(\Omega_{0^2}(i))-\mathscr{S}_{d-1}^U(\Omega_{0^1}(i))\\ &= (d-1)(\Omega_{0^{d+1}}-\Omega_{0^d})+\Omega_{0^{d+1}}-\Omega_{0^d}\\ &= d(\Omega_{0^{d+1}}-\Omega_{0^d}).\end{aligned}$$

Proof of (4.56). We still use mathematics induction. When $l = 0, 1$, the Eq. (4.56) obviously holds. Let us assume the following equation holds:

$$\mathscr{S}_1^U(i^{l-1}x_a(i)) = \mathscr{L}\{\Omega_{0^1}, \Omega_{0^2}, \ldots, \Omega_{0^l}\}, \quad l \geq 3. \tag{4.62}$$

With the assumption (4.62), it follows that

$$\mathscr{S}_1^U(i^l x_a(i)) = \sum_{i=0}^{m-1} i^{l-1} i x_a(i) = \mathscr{L}\{\mathscr{S}_1^U(ix_a(i)), \mathscr{S}_2^U(ix_a(i)), \ldots, \mathscr{S}_l^U(ix_a(i))\}.$$

According to the Eq. (4.55), we get $\mathscr{S}_d^U(ix_a(i)) = \mathscr{L}\{\Omega_{0^d}, \Omega_{0^{d+1}}\}$ for all $d \in \mathbb{N}^+$. Therefore, the Eq. (4.56) holds for all $l \in \mathbb{N}$.
Proof of (4.57). Obviously, $\hat{\Omega}_{0^1} = x_a(m) - x_a(0)$. Then with $d \in \mathbb{N}^+$, we have

$$\begin{aligned}\hat{\Omega}_{0^{d+1}} &= \mathscr{S}_{d+1}^U(y_a(i)) = \mathscr{S}_d^U(x_a(m) - x_a(i))\\ &= \sum_{i_1=0}^{m-1}\cdots\sum_{i_d=i_{d-1}}^{m-1}\big(x_a(m)-x_a(i_d)\big)\\ &= \sum_{i_1=0}^{m}\cdots\sum_{i_d=i_{d-1}}^{m}\big(x_a(m)-x_a(i_d)\big)\\ &= \bar{S}_d x_a(m) - \bar{\Omega}_{0^d}.\end{aligned}$$

Proof of (4.58). When $l = 0$, the equation $\mathscr{S}_1^U(y_a(i)) = \mathscr{L}\{x_a(m), x_a(0)\}$ obviously holds. When $l > 0$, it follows from (4.56) and (4.57) that

$$\begin{aligned}\mathscr{S}_1^U(i^l y_a(i)) &= \mathscr{L}\{\hat{\Omega}_{0^1}, \hat{\Omega}_{0^2}, \ldots, \hat{\Omega}_{0^{l+1}}\}\\ &= \mathscr{L}\{x_a(m), \bar{\Omega}_{0^1}, \ldots, \bar{\Omega}_{0^l}\}.\end{aligned}$$

This completes the proof. ■

Now let us consider the summation term $\hat{\Omega}_k = \sum_{i=0}^{m-1} p_{ka}(i)y_a(i)$. Since $p_{ka}(i)$ is a polynomial function of degree k, the following corollary is easily obtained from Lemma 4.12.

Corollary 4.6 *For any polynomial function $p_{ka}(i)$ of degree k and any vector function $\{x_a(i) \in \mathbb{R}^n | i \in [0, m]\}$, there exists a constant vector $\Pi_k \in \mathbb{R}^{n\times(k+2)n}$ such that the following equation holds:*

$$\hat{\Omega}_k = \Pi_k \vartheta_k, \quad k \in \mathbb{N} \tag{4.63}$$

where

$$\vartheta_k := \begin{cases} col\{x_a(m), x_a(0)\}, & k = 0, \\ col\{x_a(m), x_a(0), \frac{1}{\bar{S}_1}\bar{\Omega}_{0^1}, \ldots, \frac{1}{\bar{S}_k}\bar{\Omega}_{0^k}\}, & k \geq 1. \end{cases} \tag{4.64}$$

According to Lemma 4.3, the first three expressions of Π_k are listed as follows:

$$\Pi_0 = \begin{bmatrix} I & -I \end{bmatrix}, \quad \Pi_1 = \begin{bmatrix} I & I & -2I \end{bmatrix}, \quad \Pi_2 = \begin{bmatrix} I & -I & 6I & -6I \end{bmatrix}.$$

Suppose polynomial functions $p_{ka}(i)$, $k \in \mathbb{N}$, form an orthogonal basis in the single-summation inner product space with the inner product defined as $\langle p_k, p_l \rangle := \sum_{i=a}^{a+m-1} p_k(i) p_l(i)$. Now based on the sequence of the orthogonal polynomials satisfying the following property:

$$\langle p_k, p_l \rangle \begin{cases} = 0, & k \neq l, \\ \neq 0, & k = l, \end{cases} \tag{4.65}$$

we develop a series of GFMB summation inequalities by introducing some free matrices.

Lemma 4.13 *For a given integer $N \in \mathbb{N}$, orthogonal polynomials $p_{ka}(i)$, $k \in \mathbb{N}$, satisfying (4.65), a matrix $R \in \mathbb{S}_+^n$, any matrices M_k, $k \in \{0, 1, \ldots, N\}$, with appropriate dimensions, and any vector function $\{x_a(i) \in \mathbb{R}^n | i \in [0, m]\}$, the following summation inequality holds:*

$$-\sum_{i=0}^{m-1} y_a^T(i) R y_a(i) \leq \sum_{k=0}^{N} \langle p_k, p_k \rangle \eta_k^T M_k R^{-1} M_k^T \eta_k + \text{Sym}\left\{\sum_{k=0}^{N} \eta_k^T M_k \Pi_k \vartheta_k\right\} \tag{4.66}$$

where η_k, $k \in \{0, 1, \ldots, N\}$, are arbitrary vectors, ϑ_k is defined in (4.64), and Π_k satisfies the Eq. (4.63).

Proof For orthogonal polynomials $p_{ka}(i)$ and arbitrary vectors η_k, we define

$$\zeta(i) = col\{p_{0a}(i)\eta_0, p_{1a}(i)\eta_1, \ldots, p_{Na}(i)\eta_N, y_a(i)\},$$
$$M = col\{M_0, M_1, \ldots, M_N\}, \quad \Phi = \begin{bmatrix} MR^{-1}M^T & M \\ * & R \end{bmatrix}.$$

With Corollary 4.6, summating $\zeta^T(i)\Phi\zeta(i)$ from 0 to $m-1$ leads to

$$\sum_{i=0}^{m-1} \zeta^T(i)\Phi\zeta(i) = \sum_{k=0}^{N} \langle p_k, p_k\rangle \eta_k^T M_k R^{-1} M_k^T \eta_k + \sum_{k=0}^{N} \mathrm{Sym}\left\{\eta_k^T M_k \Pi_k \vartheta_k\right\} + \sum_{i=0}^{m-1} y_a^T(i) R y_a(i). \tag{4.67}$$

From the Schur complement, it is known that $\sum_{i=0}^{m-1} \zeta^T(i)\Phi\zeta(i) \geq 0$. Therefore, by a simple arrangement, we get (4.66). This completes the proof. ■

Remark 4.13 The series of GFMB inequality (4.66) is exactly the extension of the improved GFMB inequality (4.52). If the terms η_k and ϑ_k are fixed by setting $\eta_k = \vartheta_k = \vartheta_N$, $k \in \{0, 1, \ldots, N\}$, another series of FMB summation inequalities is developed in the following corollary.

Corollary 4.7 *For a given integer $N \in \mathbb{N}$, orthogonal polynomials $p_{ka}(i)$, $k \in \mathbb{N}$, satisfying (4.65), a matrix $R \in \mathbb{S}_+^n$, any matrices $M_k \in \mathbb{R}^{(N+2)n\times n}$, $k \in \{0, 1, \ldots, N\}$, and any vector function $\{x_a(i) \in \mathbb{R}^n | i \in [0, m]\}$, the following summation inequality holds:*

$$-\sum_{i=0}^{m-1} y_a^T(i) R y_a(i) \leq \vartheta_N^T \Phi_N \vartheta_N \tag{4.68}$$

where ϑ_N is defined in (4.64),

$$\Phi_N = \sum_{l=0}^{N} \langle p_l, p_l\rangle M_l R^{-1} M_l^T + \sum_{l=0}^{N} \mathrm{Sym}\{M_l \tilde{\Pi}_l\},$$

and Π_l satisfies the Eq. (4.63).

Remark 4.14 The series of FMB summation inequality (4.68) is exactly an extension of the conventional FMB summation inequality (4.44). The larger N, the tighter bound. By setting $N = 2$ in Corollary 4.7, a special FMB summation inequality is obtained in the following corollary, which includes the FMB summation (4.44) as a special case.

Corollary 4.8 *For matrices $R \in \mathbb{S}_+^n$, $M_0, M_1, M_2 \in \mathbb{R}^{4n\times n}$, and any vector function $\{x_a(i) \in \mathbb{R}^n | i \in [0, m]\}$, the following inequality holds:*

$$-\sum_{i=0}^{m-1} y_a^T(i) R y_a(i) \leq \vartheta_2^T \Phi_2 \vartheta_2 \tag{4.69}$$

where

$$\Phi_2 = \sum_{l=0}^{2} \frac{m}{2l+1} M_l R^{-1} M_l^T + \sum_{l=0}^{2} \text{Sym}\{M_l \tilde{\Pi}_l\},$$

$$\vartheta_2 = col \left\{ x_a(m), x_a(0), \frac{1}{\bar{S}_1} \bar{\Omega}_{0^1}, \frac{1}{\bar{S}_2} \bar{\Omega}_{0^2} \right\},$$

$$\bar{S}_1 = m+1, \quad \bar{S}_2 = \frac{(m+1)(m+2)}{2},$$

$$\tilde{\Pi}_0 = \begin{bmatrix} I & -I & 0 & 0 \end{bmatrix}, \quad \tilde{\Pi}_1 = \begin{bmatrix} I & I & -2I & 0 \end{bmatrix}, \quad \tilde{\Pi}_2 = \begin{bmatrix} I & -I & 6I & -6I \end{bmatrix}.$$

By setting $M_l = -\langle p_l, p_l \rangle^{-1} \tilde{\Pi}_l^T R$ in (4.68), another series of summation inequalities is obtained in which no free matrix is involved.

Corollary 4.9 *For a given integer $N \in \mathbb{N}$ and any vector function $\{x_a(i) \in \mathbb{R}^n | i \in [0, m]\}$, the following summation inequality holds:*

$$-\sum_{i=0}^{m-1} y_a^T(i) R y_a(i) \leq \vartheta_N^T \Psi_N \vartheta_N \tag{4.70}$$

where ϑ_N is defined in (4.64),

$$\Psi_N = -\sum_{l=0}^{N} \langle p_l, p_l \rangle^{-1} \tilde{\Pi}_l^T R^{-1} \tilde{\Pi}_l,$$

and $\tilde{\Pi}_l$ satisfies $\hat{\Omega}_l = \tilde{\Pi}_l \vartheta_N$, $l \in \{0, 1, \ldots, N\}$.

Remark 4.15 The series of summation inequality (4.70) is exactly an extension of the Jensen and Wirtinger-based summation inequalities. Further, it can also be seen as the discrete counterpart of the Bessel–Legendre inequality. The above idea can be directly extended to the multiple-summation case. The series of multiple-summation FMB inequalities can be obtained in a similar way.

Remark 4.16 It can be proved that the two series of summation inequalities (4.68) and (4.70) are equivalent. In other words, with a given integral N they produce the same tight bounds.

4.4 Conclusions

This chapter has intensively investigated the topic of summation inequalities. On one hand, the summation inequalities without free matrices have been studied. With the help of orthogonal polynomials defined in summation inner product spaces, two series of general summation inequalities have been developed. From them, various concrete summation inequalities have been obtained, including the well-known Jensen and Wirtinger-based ones. On the other hand, the summation inequalities with free matrices have been studied. Several free-matrix-based (FMB) summation inequalities have been proposed, including the conventional and general FMB ones. The relationship between two series of summation inequalities with and without free matrices has been briefly discussed.

References

1. Chen J, Lu J, Xu S (2016) Summation inequality and its application to stability analysis for time-delay systems. IET Control Theory Appl 10:391–395
2. Chen J, Xu S, Ma Q, Li Y, Chu Y, Zhang Z (2017) Two novel general summation inequalities to discrete-time systems with time-varying delay. J Frankl Inst 354:5537–5558
3. Chen J, Xu S, Jia X, Zhang B (2017) Novel summation inequalities and their applications to stability analysis for systems with time-varying delay. IEEE Trans Autom Control 62:2470–2475
4. Chen J, Park JH, Xu S (2019) Stability analysis of discrete-time neural networks with an interval-like time-varying delay. Neurocomputing 329:248–254
5. Gu K, Kharitonov VL, Chen J (2003) Stability of time-delay systems. Springer Science & Business Media, Berlin
6. Gyurkovics E, Kiss K, Nagy I, Takacs T (2017) Multiple summation inequalities and their application to stability analysis of discrete-time delay systems. J Frankl Inst 354:123–144
7. Nam PT, Pathirana PN, Trinh H (2015) Discrete Wirtinger-based inequality and its application. J Frankl Inst 352:1893–1905
8. Nam PT, Trinh H, Pathirana PN (2015) Discrete inequalities based on multiple auxiliary functions and their applications to stability analysis of time-delay systems. J Frankl Inst 352:5810–5831
9. Lee WI, Park PG, Lee SY, Newcomb RW (2015) Auxiliary function-based summation inequalities for quadratic functions and their application to discrete-time delay systems. In: Proceedings of 12th IFAC workshop on time delay systems, pp 203–208
10. Seuret A, Gouaisbaut F, Fridman E (2015) Stability of discrete-time systems with time-varying delays via a novel summation inequality. IEEE Trans Autom Control 60:2740–2745
11. Xiao S, Xu L, Zeng HB, Teo KL (2018) Improved stability criteria for discrete-time delay systems via novel summation inequalities. Int J Control Autom Syst 16:1592–1602
12. Zhang XM, Han QL (2015) Abel lemma-based finite-sum inequality and its application to stability analysis for linear discrete time-delay systems. Automatica 57:199–202
13. Zhang CK, He Y, Jiang YL, Wu M (2016) An improved summation inequality to discrete-time systems with time-varying delay. Automatica 74:10–15
14. Zhang CK, He Y, Jiang L, Wu M, Zeng HB (2017) Summation inequalities to bounded real lemmas of discrete-time systems with time-varying delay. IEEE Trans Autom Control 62:2582–2588

Chapter 5
Stability Analysis for Linear Systems with Time-Varying Delay

5.1 Introduction

It is well known that for any practical control system, stability is the primary condition for its normal operation. On the other hand, time delay widely exists in the real world which is usually a source of undesirable dynamics like performance degradation and even instability [1]. So, the stability research of time-delay systems, especially linear continuous- and discrete-time systems with a time-varying delay, has been attracting considerable attention [1–3]. Many relevant research results have been reported in the literature in recent years, e.g., the complete L–K functional [1], the descriptor model approach [4], the delay-partitioning method [5, 6], the free-weighting-matrix method [7, 8], the bounding inequality method [9–23]. Due to its simplicity, the bounding inequality method has been widely utilized in the stability analysis of time-delay systems.

When studying the stability problem of linear systems with a time-varying delay, the main goal is to develop a more relaxed stability criterion in the form of linear matrix inequalities (LMIs) so that the considered system is ensured to be stable with the time delay varying in an interval as large as possible. To develop a less conservative condition, constructing a proper L–K functional is very fundamental and essential. Generally speaking, more information of the system involved in the L–K functional, less conservatism achieved by the stability criterion. So, in order to reduce the conservatism continually, various double-, triple- and even quartic-integral/summation terms multiple-integral/summation terms have been included in the L–K functionals [12, 24]. Besides adding multiple-integral/summation terms, some novel L–K functionals have also been proposed in the literature [20, 23, 25–28]. For example, along with the idea of the free-matrix-based method, a novel L–K functional is proposed in [26] that does not require the quadratic matrix to be positive definite. By studying the role of L–K functionals in reducing the conservatism of stability conditions, the delay-product L–K functional is proposed in [20].

Another effective way to reduce the conservatism is to develop tighter bounding inequalities, including either an integral/summation inequality or a matrix inequality

J. H. Park et al., *Dynamic Systems with Time Delays: Stability and Control*,
https://doi.org/10.1007/978-981-13-9254-2_5

or both. It is believed that a more accurate inequality ought to be helpful in reducing the conservatism. However, this is not the case. If the L–K functional is not appropriately constructed, a more accurate integral/summation inequality may not lead to a less conservative condition [20, 23]. Therefore, how to construct a proper L–K functional to take full advantage of the interest of a more accurate integral/summation inequality becomes an important topic. On the other hand, when bounding single-integral/summation terms with a time-varying delay, the matrix inequality, i.e., the reciprocally convex lemma (RCL), is often utilized. The problem is firstly studied in [29], in which the conventional RCL is proposed. Recently, improved RCLs are successively proposed in [18, 23] by introducing some free matrices. Since the information of the time-varying delay is considered, the improved RCLs produce tighter bounds than the conventional one.

This chapter focuses on the stability analysis of linear systems with a time-varying delay. The remainder is divided into two parts.

The first part is concerned with the stability problem of the continuous-time systems. Based on integral inequalities proposed in Chap. 3, two types of stability conditions are, respectively, developed. One type is stability conditions obtained via the combination of integral inequalities without free matrices and RCL. The other type is stability conditions obtained via integral inequalities with free matrices. One conclusion can be made that if the L–K functional is appropriately constructed, a tighter bounding inequality, no matter an integral inequality or a matrix inequality, leads to a more relaxed stability condition.

The second part is concerned with the stability problem of discrete-time systems with a time-varying delay. Based on several typical ways of estimating the single-summation term proposed in Chap. 4, some stability conditions are obtained and some meaningful conclusions are consequently made.

5.2 Stability Analysis of Continuous-Time Systems with a Time-Varying Delay

Consider the following linear system with a time-varying delay:

$$\begin{cases} \dot{x}(t) = Ax(t) + A_d x(t - h(t)), & t \geq 0, \\ x(t) = \phi(t), & t \in [-h_M, 0], \end{cases} \tag{5.1}$$

where $x(t) \in \mathbb{R}^n$ is the state vector; A and A_d are constant system matrices; $\phi(t)$ is the initial condition; $h(t)$ is the time-varying delay satisfying the following constraints:

$$0 \leq h(t) \leq h_M, \quad d_m \leq \dot{h}(t) \leq d_M, \quad \forall t \geq 0, \tag{5.2}$$

where h_M, d_m and d_M are constant scalars.

When we obtain time-dependent stability conditions for system (5.1), the double-integral term $h_M \int_{t-h_M}^{t} \int_{\theta}^{t} \dot{x}^T(s)R\dot{x}(s)dsd\theta$ is commonly contained in the L–K functional candidate. The term is very helpful in reducing the conservatism of the obtained stability condition. In this case, how to estimate the single-integral term

$$\psi(t) = h_M \int_{t-h_M}^{t} \dot{x}^T(s)R\dot{x}(s)ds \tag{5.3}$$

arising in the L–K functional becomes very important since different ways may lead to different types of stability conditions that are of different conservatism.

In the early research, the model transformations and free-weighting-matrix methods are often used [2, 4, 7, 8]. Nowadays the bounding inequality method is widely employed because of its effectiveness and simplicity.

In the following, we further study the stability problem for system (5.1) via the L–K functional method. The well-known inequalities such as the Jensen, Wirtinger-based, auxiliary-function-based and Bessel–Legendre (B–L) ones are, respectively, employed to develop stability conditions and in the meanwhile, corresponding L–K functionals are deliberately constructed.

Before proceeding, the following notations are defined:

$$\begin{aligned}
&h_t = h(t), \quad h_{Mt} = h_M - h_t, \quad \alpha = \frac{h_t}{h_M},\\
&\zeta(t) = col\{x(t), x(t-h_t), x(t-h_M)\},\\
&u_0(t) = \int_{t-h_t}^{t} x(s)ds, \quad v_0(t) = \int_{t-h_M}^{t-h_t} x(s)ds,\\
&u_1(t) = \frac{1}{h_t}\int_{t-h_t}^{t}\int_{u}^{t} x(s)dsdu, \quad v_1(t) = \frac{1}{h_{Mt}}\int_{t-h_M}^{t-h_t}\int_{u}^{t-h_t} x(s)dsdu,\\
&V_1(t) = \int_{t-h_t}^{t} x^T(s)Q_1x(s)ds + \int_{t-h_M}^{t} x^T(s)Q_2x(s)ds\\
&\qquad + h_M \int_{-h_M}^{0}\int_{t+\theta}^{t} \dot{x}^T(s)R\dot{x}(s)dsd\theta.
\end{aligned}$$

5.2.1 *Via Integral Inequalities Without Free Matrices*

5.2.1.1 Via the Jensen Integral Inequality

The L–K functional candidate is constructed as follows:

$$V_J(t) = x^T(t)Px(t) + V_1(t). \tag{5.4}$$

Based on the simple L–K functional candidate (5.4), we obtain the following stability condition via the Jensen integral inequality.

Theorem 5.1 *For given h_M, d_m and d_M satisfying (5.2), matrices $P, Q_1, Q_2, R \in \mathbb{S}_+^n$ and $X \in \mathbb{R}^{n\times n}$, system (5.1) is asymptotically stable if the following LMIs hold for $\dot{h}_t \in \{d_m, d_M\}$:*

$$\begin{bmatrix} R & X \\ * & R \end{bmatrix} \geq 0, \tag{5.5}$$

$$\Sigma_1(\dot{h}_t) - \Sigma_2 < 0 \tag{5.6}$$

where

$$\begin{aligned} \Sigma_1(\dot{h}_t) =& \text{Sym}\{e_1^T P e_s\} + e_1^T Q_1 e_1 - (1-\dot{h}_t) e_2^T Q_1 e_2 \\ &+ e_1^T Q_2 e_1 - e_3^T Q_2 e_3 + h_M^2 e_s^T R e_s, \end{aligned} \tag{5.7}$$

$$\Sigma_2 = \begin{bmatrix} E_{12} \\ E_{23} \end{bmatrix}^T \begin{bmatrix} R & X \\ * & R \end{bmatrix} \begin{bmatrix} E_{12} \\ E_{23} \end{bmatrix}, \tag{5.8}$$

with

$$\begin{aligned} E_{12} &= e_1 - e_2, \quad E_{23} = e_2 - e_3, \\ e_i &= \begin{bmatrix} 0_{n\times(i-1)n} & I_{n\times n} & 0_{n\times(3-i)n} \end{bmatrix}, \quad i \in \{1, 2, 3\}, \\ e_s &= Ae_1 + A_d e_2. \end{aligned}$$

Proof Owing to the fact $P, Q_1, Q_2, R > 0$, there exists a sufficient small scalar $\varepsilon_1 > 0$ such that $V_J(t) \geq \varepsilon_1 \|x(t)\|^2 > 0$ for any $x(t) \neq 0$.

Now, differentiating $V_J(t)$ along the trajectory of (5.1) yields:

$$\begin{aligned} \dot{V}_J(t) =& 2x^T(t)P\dot{x}(t) + x^T(t)Q_1x(t) - (1-\dot{h}_t)x^T(t-h_t)Q_1x(t-h_t) \\ &+ x^T(t)Q_2x(t) - x^T(t-h_M)Q_2x(t-h_M) \\ &+ h_M^2\dot{x}^T(t)R\dot{x}(t) - h_M\int_{t-h_M}^t \dot{x}^T(s)R\dot{x}(s)ds \\ =& \zeta^T(t)\Sigma_1(\dot{h}_t)\zeta(t) - h_M\int_{t-h_M}^t \dot{x}^T(s)R\dot{x}(s)ds \end{aligned} \tag{5.9}$$

where $\Sigma_1(\dot{h}_t)$ is defined in (5.7).

With (5.5), applying the Jensen inequality and RCL to the integral term in (5.9) leads to:

$$h_M\int_{t-h_M}^t \dot{x}^T(s)R\dot{x}(s)ds$$

$$
\begin{aligned}
&= h_M \int_{t-h_t}^{t} \dot{x}^T(s)R\dot{x}(s)ds + h_M \int_{t-h_M}^{t-h_t} \dot{x}^T(s)R\dot{x}(s)ds \\
&\geq \frac{h_M}{h_t}\zeta^T(t)E_{12}^T R E_{12}\zeta(t) + \frac{h_M}{h_{Mt}}\zeta^T(t)E_{23}^T R E_{23}\zeta(t) \\
&= \zeta^T(t)\begin{bmatrix} E_{12} \\ E_{23} \end{bmatrix}^T \begin{bmatrix} \frac{1}{\alpha}R & 0 \\ * & \frac{1}{1-\alpha}R \end{bmatrix} \begin{bmatrix} E_{12} \\ E_{23} \end{bmatrix}\zeta(t) \\
&\geq \zeta^T(t)\Sigma_2\zeta(t),
\end{aligned}
$$

where Σ_2 is defined in (5.8). According to the above discussions, we have

$$
\dot{V}_J(t) \leq \zeta^T(t)(\Sigma_1(\dot{h}_t) - \Sigma_2)\zeta(t).
$$

Hence, from (5.6), there exists a sufficient small scalar $\varepsilon_2 > 0$ such that $\dot{V}_J(t) \leq -\varepsilon_2\|x(t)\|^2 < 0$ for any $x(t) \neq 0$. In summation, it is concluded that system (5.1) subject to (5.2) is asymptotically stable. This completes the proof. ■

Remark 5.1 The L–K functional $V_J(t)$ defined in (5.4) is composed of the quadratic term, the single- and double-integral terms. This is the basic structural form commonly used in the literature. Since the Jensen integral inequality is used, the simple quadratic term $x^T(t)Px(t)$ is constructed, in which no integral vector with respect to the state is involved. When estimating the integral term $h_M \int_{t-h_M}^{t} \dot{x}^T(s)R\dot{x}(s)ds$, the combination of the Jensen inequality and conventional RCL is employed. If the improved RCL is used, a more relaxed condition is naturally expected. In this case, the variable α is inevitably involved in the obtained stability condition. Additionally, if the variation of the time-varying delay is unknown, the single-integral term $\int_{t-h_t}^{t} x^T(s)Q_1x(s)ds$ should be removed from the L–K functional candidate.

5.2.1.2 Via the Wirtinger-Based Integral Inequality

The L–K functional candidate is constructed as follows:

$$
V_W(t) = \chi_0^T(t)P\chi_0(t) + V_1(t), \tag{5.10}
$$

where

$$
\chi_0(t) = col\{x(t), u_0(t), v_0(t)\}.
$$

Based on the augmented L–K functional candidate (5.10), we obtain the following stability condition via the Wirtinger-based integral inequality.

Theorem 5.2 *For given h_M, d_m and d_M satisfying (5.2), matrices $P \in \mathbb{S}_+^{3n}$, $Q_1, Q_2, R \in \mathbb{S}_+^n$ and $X \in \mathbb{R}^{2n\times 2n}$, system (5.1) is asymptotically stable if the following LMIs hold for any $h_t \times \dot{h}_t \in \{0, h_M\} \times \{d_m, d_M\}$:*

$$\begin{bmatrix} \tilde{R} & X \\ * & \tilde{R} \end{bmatrix} \geq 0, \tag{5.11}$$

$$\Upsilon_1(h_t, \dot{h}_t) - \Upsilon_2 < 0 \tag{5.12}$$

where

$$\begin{aligned} \Upsilon_1(h_t, \dot{h}_t) &= \text{Sym}\{e_{\chi_0}(h_t)^T P e_{\dot{\chi}_0}(\dot{h}_t)\} + e_1^T Q_1 e_1 - (1 - \dot{h}_t) e_2^T Q_1 e_2 \\ &\quad + e_1^T Q_2 e_1 - e_3^T Q_2 e_3 + h_M^2 e_s^T R e_s, \end{aligned} \tag{5.13}$$

$$\Upsilon_2 = \begin{bmatrix} E_{124} \\ E_{235} \end{bmatrix}^T \begin{bmatrix} \tilde{R} & X \\ * & \tilde{R} \end{bmatrix} \begin{bmatrix} E_{124} \\ E_{235} \end{bmatrix}, \tag{5.14}$$

with

$$\begin{aligned} &E_{124} = \begin{bmatrix} e_1 - e_2 \\ e_1 + e_2 - 2e_4 \end{bmatrix}, \quad E_{235} = \begin{bmatrix} e_2 - e_3 \\ e_2 + e_3 - 2e_5 \end{bmatrix}, \\ &e_i = \begin{bmatrix} 0_{n\times(i-1)n} & I_{n\times n} & 0_{n\times(5-i)n} \end{bmatrix}, \quad i \in \{1, \ldots, 5\}, \\ &e_{\chi_0}(h_t) = col\{e_1, h_t e_4, h_{Mt} e_5\}, \\ &e_{\dot{\chi}_0}(\dot{h}_t) = col\{e_s, e_1 - (1 - \dot{h}_t) e_2, (1 - \dot{h}_t) e_2 - e_3\}, \\ &e_s = A e_1 + A_d e_2, \quad \tilde{R} = \text{diag}\{R, 3R\}. \end{aligned}$$

Proof Differentiating $V_W(t)$ along the trajectory of (5.1) yields:

$$\dot{V}_W(t) = \xi_0^T(t) \Upsilon_1(h_t, \dot{h}_t) \xi_0(t) - h_M \int_{t-h_M}^{t} \dot{x}^T(s) R \dot{x}(s) ds, \tag{5.15}$$

where $\Upsilon_1(h_t, \dot{h}_t)$ is defined in (5.13), and

$$\xi_0(t) =: col\left\{\zeta(t), \frac{1}{h_t} u_0(t), \frac{1}{h_{Mt}} v_0(t)\right\}.$$

Note that

$$\chi_0(t) = e_{\chi_0}(h_t)\xi_0(t), \quad \dot{\chi}_0(t) = e_{\dot{\chi}_0}(\dot{h}_t)\xi_0(t).$$

With (5.11), applying the Wirtinger-based inequality and RCL to the integral term in (5.15) leads to:

$$\begin{aligned} &h_M \int_{t-h_M}^{t} \dot{x}^T(s) R \dot{x}(s) ds \\ &= h_M \int_{t-h_t}^{t} \dot{x}^T(s) R \dot{x}(s) ds + h_M \int_{t-h_M}^{t-h_t} \dot{x}^T(s) R \dot{x}(s) ds \end{aligned}$$

$$
\begin{aligned}
&\geq \frac{h_M}{h_t}\xi_0^T(t)E_{124}^T\tilde{R}E_{124}\xi_0(t) + \frac{h_M}{h_{Mt}}\xi_0^T(t)E_{235}^T\tilde{R}E_{235}\xi_0(t) \\
&= \xi_0^T(t)\begin{bmatrix} E_{124} \\ E_{235} \end{bmatrix}^T \begin{bmatrix} \frac{1}{\alpha}\tilde{R} & 0 \\ * & \frac{1}{1-\alpha}\tilde{R} \end{bmatrix} \begin{bmatrix} E_{124} \\ E_{235} \end{bmatrix} \xi_0(t) \\
&\geq \xi_0^T(t)\Upsilon_2\xi_0(t),
\end{aligned}
$$

where Υ_2 is defined in (5.14).

According to the above discussions, we have

$$
\dot{V}_W(t) \leq \xi_0{}^T(t)(\Upsilon_1(h_t, \dot{h}_t) - \Upsilon_2)\xi_0(t).
$$

It is noted that $\Upsilon_1(h_t, \dot{h}_t)$ is affine with both h_t and $\dot{h}_t$. So from (5.12), there exists a sufficient small scalar $\varepsilon > 0$ such that $\dot{V}_W(t) \leq -\varepsilon\|x(t)\|^2 < 0$ for any $x(t) \neq 0$. In summation, it is concluded that system (5.1) subject to (5.2) is asymptotically stable. This completes the proof. ■

Remark 5.2 In order to coordinate with the use of the Wirtinger-based inequality, the L–K functional $V_W(t)$ is constructed [15], in which the vector $\chi_0(t)$ is augmented by adding two integral terms $\int_{t-h_t}^{t} x(s)ds$ and $\int_{t-h_M}^{t-h_t} x(s)ds$. Otherwise, the stability condition obtained by the Wirtinger-based integral inequality is equivalent to that obtained by the Jensen integral inequality. That is, they are of the same conservatism. When differentiating the augmented quadratic term $\chi_0^T(t)P\chi_0(t)$, the cross-product terms with respect to the vectors $x(t)$, $u_0(t)$ and $v_0(t)$ arise that are helpful in reducing the conservatism.

5.2.1.3 Via the Auxiliary-Function-Based Integral Inequality

The L–K functional candidate is constructed as follows:

$$
V_{AF}(t) = \chi_1^T(t)P\chi_1(t) + V_1(t), \tag{5.16}
$$

where

$$
\chi_1(t) = col\left\{x(t), u_0(t), v_0(t), u_1(t), v_1(t)\right\}.
$$

Based on the augmented L–K functional candidate (5.16), we obtain the following stability condition via the auxiliary-function-based integral inequality.

Theorem 5.3 *For given h_M, d_m and d_M satisfying (5.2), matrices $P \in \mathbb{S}_+^{5n}$, Q_1, Q_2, $R \in \mathbb{S}_+^n$ and $X \in \mathbb{R}^{3n\times 3n}$, system (5.1) is asymptotically stable if the following LMIs hold for $h_t \times \dot{h}_t \in \{0, h_M\} \times \{d_m, d_M\}$:*

$$\begin{bmatrix} \tilde{R} & X \\ * & \tilde{R} \end{bmatrix} \geq 0, \tag{5.17}$$

$$\Xi_1(h_t, \dot{h}_t) - \Xi_2 < 0 \tag{5.18}$$

where

$$\begin{aligned} \Xi_1(h_t, \dot{h}_t) &= \mathrm{Sym}\{e_{\chi_1}(h_t)^T P e_{\dot{\chi}_1}(\dot{h}_t)\} + e_1^T Q_1 e_1 - (1 - \dot{h}_t) e_2^T Q_1 e_2 \\ &\quad + e_1^T Q_2 e_1 - e_3^T Q_2 e_3 + h_M^2 e_s^T R e_s, \end{aligned} \tag{5.19}$$

$$\Xi_2 = \begin{bmatrix} E_{1246} \\ E_{2357} \end{bmatrix}^T \begin{bmatrix} \tilde{R} & X \\ * & \tilde{R} \end{bmatrix} \begin{bmatrix} E_{1246} \\ E_{2357} \end{bmatrix}, \tag{5.20}$$

with

$$E_{1246} = \begin{bmatrix} e_1 - e_2 \\ e_1 + e_2 - 2e_4 \\ e_1 - e_2 + 6e_4 - 12e_6 \end{bmatrix}, \quad E_{2357} = \begin{bmatrix} e_2 - e_3 \\ e_2 + e_3 - 2e_5 \\ e_2 - e_3 + 6e_5 - 12e_7 \end{bmatrix},$$

$$\begin{aligned} e_i &= \begin{bmatrix} 0_{n\times(i-1)n} & I_{n\times n} & 0_{n\times(7-i)n} \end{bmatrix}, \quad i \in \{1, \ldots, 7\}, \\ e_{\chi_1} &= col\{e_1, h_t e_4, h_{Mt} e_5, h_t e_6, h_{Mt} e_7\}, \\ e_{\dot{u}_1} &= e_1 - (1 - \dot{h}_t) e_4 - \dot{h}_t e_6, \quad e_{\dot{v}_1} = (1 - \dot{h}_t) e_2 - e_5 + \dot{h}_t e_7, \\ e_{\dot{\chi}_1} &= col\{e_s, e_1 - (1 - \dot{h}_t) e_2, (1 - \dot{h}_t) e_2 - e_3, e_{\dot{u}_1}, e_{\dot{v}_1}\}, \\ e_s &= A e_1 + A_d e_2, \quad \tilde{R} = \mathrm{diag}\{R, 3R, 5R\}. \end{aligned}$$

Proof Differentiating $V_{AF}(t)$ along the trajectory of (5.1) yields:

$$\dot{V}_{AF}(t) = \xi_1^T(t) \Xi_1(h_t, \dot{h}_t) \xi_1(t) - h_M \int_{t-h_M}^{t} \dot{x}^T(s) R \dot{x}(s) ds, \tag{5.21}$$

where $\Xi_1(h_t, \dot{h}_t)$ is defined in (5.19), and

$$\xi_1(t) := col \left\{ \zeta(t), \frac{1}{h_t} u_0(t), \frac{1}{h_{Mt}} v_0(t), \frac{1}{h_t} u_1(t), \frac{1}{h_{Mt}} v_1(t) \right\}.$$

With (5.17), applying the auxiliary-function-based inequality and RCL to the integral term in (5.21) leads to:

$$\begin{aligned} & h_M \int_{t-h_M}^{t} \dot{x}^T(s) R \dot{x}(s) ds \\ &= h_M \int_{t-h_t}^{t} \dot{x}^T(s) R \dot{x}(s) ds + h_M \int_{t-h_M}^{t-h_t} \dot{x}^T(s) R \dot{x}(s) ds \\ &\geq \frac{h_M}{h_t} \xi_1^T(t) E_{1246}^T \tilde{R} E_{1246} \xi_1(t) + \frac{h_M}{h_{Mt}} \xi_1^T(t) E_{2357}^T \tilde{R} E_{2357} \xi_1(t) \end{aligned}$$

$$= \xi_1^T(t) \begin{bmatrix} E_{1246} \\ E_{2357} \end{bmatrix}^T \begin{bmatrix} \frac{1}{\alpha}\tilde{R} & 0 \\ * & \frac{1}{1-\alpha}\tilde{R} \end{bmatrix} \begin{bmatrix} E_{1246} \\ E_{2357} \end{bmatrix} \xi_1(t)$$
$$\geq \xi_1^T(t) \Xi_2 \xi_1(t),$$

where Ξ_2 is defined in (5.20).

According to the above discussions, we have

$$\dot{V}_{AF}(t) \leq \xi_1(t)^T (\Xi_1(h_t, \dot{h}_t) - \Xi_2)\xi_1(t).$$

The remainder is similar to that in the proof of Theorem 5.2 and therefore, omitted for brevity. This completes the proof. ∎

Remark 5.3 In order to coordinate with the auxiliary-function-based integral inequality, the augmented L–K functional $V_{AF}(t)$ is constructed, in which the vector $\chi_1(t)$ is further augmented by adding two double-integral terms $u_1(t)$ and $v_1(t)$. Otherwise, the stability condition obtained by the auxiliary-function-based inequality is equivalent to that obtained by the Wirtinger-based one.

Remark 5.4 It can be proved in theory that Theorem 5.3 is more relaxed than Theorem 5.2 and that Theorem 5.2 is more relaxed than Theorem 5.1. So, it can be concluded that a more accurate integral inequality leads to a more relaxed stability condition if the corresponding L–K functional candidate is suitably constructed. Furthermore, the B–L inequality can lead to a hierarchy of stability conditions for system (5.1), which include Theorems 5.1, 5.2 and 5.3 as special cases.

5.2.1.4 Via the B–L Inequality

Before proceeding, the following notations are defined first:

$$g_{(a,b)}^{[i]} := \int_a^b \left(\frac{s-a}{b-a}\right)^i x(s)ds,$$
$$u_i(t) := g_{(t-h_t,t)}^{[i]}, \quad v_i(t) := g_{(t-h_M,t-h_t)}^{[i]}. \tag{5.22}$$

Lemma 5.1 *For scalar functions $a(t)$ and $b(t)$ satisfying $a < b$ and any vector function $x(t) : [a, b] \to \mathbb{R}^n$, the equations hold for any $i \in \mathbb{N}$:*

$$\Omega_{0^{i+1}} = \frac{(b-a)^i}{i!} g_{(a,b)}^{[i]},$$
$$\frac{d}{dt} g_{(a,b)}^{[i]} = \begin{cases} \dot{b}x(b) - \dot{a}x(a), & i = 0, \\ \dot{b}x(b) - \frac{i\dot{a}}{b-a} g_{(a,b)}^{[i-1]} - \frac{i(\dot{b}-\dot{a})}{b-a} g_{(a,b)}^{[i]}, & i \geq 1. \end{cases}$$

From the above lemma, we get

$$u_0(t) = \int_{t-h_t}^{t} x(s)ds, \quad v_0(t) = \int_{t-h_M}^{t-h_t} x(s)ds,$$
$$u_1(t) = \int_{t-h_t}^{t} \left(\frac{s-t+h_t}{h_t}\right) x(s)ds = \frac{1}{h_t}\int_{t-h_t}^{t}\int_{u}^{t} x(s)dsdu,$$
$$v_1(t) = \int_{t-h_M}^{t-h_t} \left(\frac{s-t+h_M}{h_{Mt}}\right) x(s)ds = \frac{1}{h_{Mt}}\int_{t-h_M}^{t-h_t}\int_{u}^{t-h_t} x(s)dsdu.$$

It is seen that the definition of (5.22) is consistent with the previous definitions of $u_i(t)$ and $v_i(t)$, $i = 0, 1$.

Now, the L–K functional candidate is constructed as follows:

$$V_{BL}(t) = \chi_N^T(t) P \chi_N(t) + V_1(t), \tag{5.23}$$

where

$$\chi_N(t) = col\left\{x(t), u_0(t), v_0(t), u_1(t), v_1(t), \ldots, u_N(t), v_N(t)\right\}. \tag{5.24}$$

Based on the augmented L–K functional candidate (5.23), we obtain a hierarchy of stability conditions via the B–L integral inequality.

Theorem 5.4 *For given h_M, d_m and d_M satisfying (5.2), matrices $P \in \mathbb{S}_+^{(2N+3)n}$, $Q_1, Q_2, R \in \mathbb{S}_+^n$ and $X \in \mathbb{R}^{(N+2)n\times(N+2)n}$, system (5.1) is asymptotically stable if the following LMIs hold for any $h_t \times \dot{h}_t \in \{0, h_M\} \times \{d_m, d_M\}$:*

$$\begin{bmatrix} \tilde{R}_{N+1} & X \\ * & \tilde{R}_{N+1} \end{bmatrix} \geq 0, \tag{5.25}$$
$$\Gamma_1(h_t, \dot{h}_t) - \Gamma_2 < 0 \tag{5.26}$$

where

$$\Gamma_1(h_t, \dot{h}_t) = \mathrm{Sym}\{e_{\chi_N}(h_t)^T P e_{\dot{\chi}_N}(\dot{h}_t)\} + e_1^T Q_1 e_1 - (1-\dot{h}_t) e_2^T Q_1 e_2$$
$$+ e_1^T Q_2 e_1 - e_3^T Q_2 e_3 + h_M^2 e_s^T R e_s, \tag{5.27}$$
$$\Gamma_2 = \begin{bmatrix} \bar{\Pi}_{N+1} T_{N1} \\ \bar{\Pi}_{N+1} T_{N2} \end{bmatrix}^T \begin{bmatrix} \tilde{R}_{N+1} & X \\ * & \tilde{R}_{N+1} \end{bmatrix} \begin{bmatrix} \bar{\Pi}_{N+1} T_{N1} \\ \bar{\Pi}_{N+1} T_{N2} \end{bmatrix}, \tag{5.28}$$

with

$$T_{N1} = col\{e_1, e_2, e_4, 2e_6, 3e_8, \ldots, (N+1)e_{2N+4}\},$$
$$T_{N2} = col\{e_2, e_3, e_5, 2e_7, 3e_9, \ldots, (N+1)e_{2N+5}\},$$
$$e_i = \left[\, 0_{n\times(i-1)n} \; I_{n\times n} \; 0_{n\times(2N+5-i)n} \,\right], \quad i \in \{1, \ldots, 2N+5\},$$

$$\begin{aligned}
&e_{u_i} = h_t e_{2i+4}, \quad e_{v_i} = h_{Mt} e_{2i+5},\\
&e_{\chi_N}(h_t) = col\{e_1, e_{u_0}, e_{v_0}, \ldots, e_{u_N}, e_{v_N}\},\\
&e_{\dot{u}_i} = \begin{cases} e_1 - (1-\dot{h}_t)e_2, & i = 0,\\ e_1 - i(1-\dot{h}_t)e_{2i+2} - i\dot{h}_t e_{2i+4}, & i \geq 1,\end{cases}\\
&e_{\dot{v}_i} = \begin{cases} (1-\dot{h}_t)e_2 - e_3, & i = 0,\\ (1-\dot{h}_t)e_2 - ie_{2i+3} + i\dot{h}_t e_{2i+5}, & i \geq 1,\end{cases}\\
&e_{\dot{\chi}_N}(\dot{h}_t) = col\{e_s, e_{\dot{u}_0}, e_{\dot{v}_0}, \ldots, e_{\dot{u}_N}, e_{\dot{v}_N}\},\\
&e_s = Ae_1 + A_d e_2, \quad \tilde{R}_{N+1} = \text{diag}\{R, 3R, \ldots, (2N+3)R\},\\
&\bar{\Pi}_{N+1} = col\{\Pi_0^{N+1}, \Pi_1^{N+1}, \ldots, \Pi_{N+1}^{N+1}\},
\end{aligned}$$

in which Π_i^{N+1} satisfies the following equation

$$\int_a^b L_i(s)\dot{x}(s)ds = \Pi_i^{N+1}\vartheta_{N+1}, \quad i \in \{0, 1, \ldots, N+1\}.$$

Proof Define

$$\xi_N(t) := col\left\{\zeta(t), \frac{u_0(t)}{h_t}, \frac{v_0(t)}{h_{Mt}}, \frac{u_1(t)}{h_t}, \frac{v_1(t)}{h_{Mt}}, \ldots, \frac{u_N(t)}{h_t}, \frac{v_N(t)}{h_{Mt}}\right\}.$$

Differentiating $V_{BL}(t)$ defined in (5.23) along the trajectory of system (5.1) yields:

$$\dot{V}(t) = \xi_N^T(t)\Gamma_1(h_t, \dot{h}_t)\xi_N(t) - h_M\int_{t-h_M}^t \dot{x}^T(s)R\dot{x}(s)ds, \tag{5.29}$$

where $\Gamma_1(h_t, \dot{h}_t)$ is defined in (5.27).

Obviously, we have the fact $\chi_N(t) = e_{\chi_N}(h_t)\xi_N(t)$.

Meanwhile, from Lemma 5.1 the equation holds:

$$\dot{\chi}_N(t) = e_{\dot{\chi}_N}(\dot{h}_t)\xi_N(t).$$

Now, applying the B–L integral inequality to $h_t\int_{t-h_t}^t \dot{x}^T(s)R\dot{x}(s)ds$ leads to

$$\begin{aligned}
h_t\int_{t-h_t}^t \dot{x}^T(s)R\dot{x}(s)ds &\geq \vartheta_{N+1}^T\bar{\Pi}_{N+1}^T\tilde{R}_{N+1}\bar{\Pi}_{N+1}\vartheta_{N+1}\\
&= \xi_N(t)^T T_{N1}^T\bar{\Pi}_{N+1}^T\tilde{R}_{N+1}\bar{\Pi}_{N+1}T_{N1}\xi_N(t),
\end{aligned}$$

where

$$\vartheta_{N+1} = col\left\{x(b), x(a), \frac{1}{S_1}\Omega_0, \frac{1}{S_2}\Omega_{0^2}, \ldots, \frac{1}{S_{N+1}}\Omega_{0^{N+1}}\right\}$$

$$
\begin{aligned}
&= col\left\{x(b), x(a), \frac{1}{b-a}\Omega_0, \frac{2!}{(b-a)^2}\Omega_{0^2}, \ldots, \frac{(N+1)!}{(b-a)^{N+1}}\Omega_{0^{N+1}}\right\} \\
&= col\left\{x(t), x(t-h_t), \frac{u_0(t)}{h_t}, \frac{2u_1(t)}{h_t}, \ldots, \frac{(N+1)u_N(t)}{h_t}\right\} \\
&= T_{N1}\xi_N(t).
\end{aligned}
$$

Similarly, applying the B–L integral inequality to $h_{Mt}\int_{t-h_M}^{t-h_t}\dot{x}^T(s)R\dot{x}(s)ds$ leads to

$$
\begin{aligned}
h_{Mt}\int_{t-h_M}^{t-h_t}\dot{x}^T(s)R\dot{x}(s)ds &\geq \vartheta_{N+1}^T\bar{\Pi}_{N+1}^T\tilde{R}_{N+1}\bar{\Pi}_{N+1}\vartheta_{N+1} \\
&= \xi_N^T(t)T_{N2}^T\bar{\Pi}_{N+1}^T\tilde{R}_{N+1}\bar{\Pi}_{N+1}T_{N2}\xi_N(t)
\end{aligned}
$$

where

$$
\begin{aligned}
\vartheta_{N+1} &= col\left\{x(b), x(a), \frac{1}{b-a}\Omega_0, \frac{2!}{(b-a)^2}\Omega_{0^2}, \ldots, \frac{(N+1)!}{(b-a)^{N+1}}\Omega_{0^{N+1}}\right\} \\
&= col\left\{x(t-h_t), x(t-h_M), \frac{v_0(t)}{h_{Mt}}, \frac{2v_1(t)}{h_{Mt}}, \ldots, \frac{(N+1)v_N(t)}{h_{Mt}}\right\} \\
&= T_{N2}\xi_N(t).
\end{aligned}
$$

Then, by applying the RCL with (5.25) we have

$$
\begin{aligned}
& h_M\int_{t-h_M}^{t}\dot{x}^T(s)R\dot{x}(s)ds \\
&= h_M\int_{t-h_t}^{t}\dot{x}^T(s)R\dot{x}(s)ds + h_M\int_{t-h_M}^{t-h_t}\dot{x}^T(s)R\dot{x}(s)ds \\
&\geq \frac{h_M}{h_t}\xi_N^T(t)T_{N1}^T\bar{\Pi}_{N+1}^T\tilde{R}_{N+1}\bar{\Pi}_{N+1}T_{N1}\xi_N(t) \\
&\quad + \frac{h_M}{h_{Mt}}\xi_N^T(t)T_{N2}^T\bar{\Pi}_{N+1}^T\tilde{R}_{N+1}\bar{\Pi}_{N+1}T_{N2}\xi_N(t) \\
&= \xi_N^T(t)\begin{bmatrix}\bar{\Pi}_{N+1}T_{N1} \\ \bar{\Pi}_{N+1}T_{N2}\end{bmatrix}^T\begin{bmatrix}\frac{1}{\alpha}\tilde{R}_{N+1} & 0 \\ * & \frac{1}{1-\alpha}\tilde{R}_{N+1}\end{bmatrix}\begin{bmatrix}\bar{\Pi}_{N+1}T_{N1} \\ \bar{\Pi}_{N+1}T_{N2}\end{bmatrix}\xi_N(t) \\
&\geq \xi_N^T(t)\Gamma_2\xi_N(t)
\end{aligned}
$$

where Γ_2 is defined in (5.28).

According to the above discussions, we have

$$
\dot{V}_{BL}(t) \leq \xi_N^T(t)(\Gamma_1(h_t, \dot{h}_t) - \Gamma_2)\xi_N(t).
$$

The remainder is similar to that in the proof of Theorem 5.2 and therefore, omitted for brevity. This completes the proof. ■

Remark 5.5 Based on the augmented L–K functional (5.23) with a parameter N, a hierarchy of stability conditions is developed for system (5.1) in Theorem 5.4. It can be proved that the stability condition with a larger N embraces that with a smaller N. By setting $N = 0, 1$, Theorem 5.4 is, respectively, reduced to Theorems 5.2 and 5.3. And by setting $\xi_N(t) = \zeta(t)$, Theorem 5.4 is reduced to Theorem 5.1.

Remark 5.6 When $N = 3$, we can obtain the following results from Chap. 3:

$$\begin{aligned}
&\Pi_0^4 = \begin{bmatrix} I & -I & 0 & 0 & 0 & 0 \end{bmatrix}, \quad \Pi_1^4 = \begin{bmatrix} I & I & -2I & 0 & 0 & 0 \end{bmatrix},\\
&\Pi_2^4 = \begin{bmatrix} I & -I & 6I & -6I & 0 & 0 \end{bmatrix}, \quad \Pi_3^4 = \begin{bmatrix} I & I & -12I & 30I & -20I & 0 \end{bmatrix},\\
&\Pi_4^4 = \begin{bmatrix} I & -I & 20I & -90I & 140I & -70I \end{bmatrix}.
\end{aligned}$$

It can be seen that there is an interesting phenomenon, that is, the sum of each element of Π_i^4 is zero matrix. Obviously, when $N = 1$, we have the following results:

$$\Pi_0^1 = \begin{bmatrix} I & -I & 0 \end{bmatrix}, \quad \Pi_1^1 = \begin{bmatrix} I & I & -2I \end{bmatrix}.$$

5.2.1.5 Numerical Examples

In this subsection, two commonly-used numerical examples are given to illustrate the effectiveness of the proposed approach.

The maximum allowable upper bounds (MAUBs) and the numbers of decision variables (NVs) are, respectively, computed to compare the conservatism and computation burden of different stability conditions.

Example 1 Consider system (5.1) with

$$A = \begin{bmatrix} -2 & 0 \\ 0 & -0.9 \end{bmatrix}, \quad A_d = \begin{bmatrix} -1 & 0 \\ -1 & -1 \end{bmatrix}.$$

The MAUBs and NVs are, respectively, computed by Theorems 5.1, 5.2, 5.3 and 5.4 with $N = 2, 3$. It is seen that the MAUBs obtained by Theorem 5.2 are, respectively, larger than those obtained by Theorem 5.1 and that the MAUBs obtained by Theorem 5.3 are, respectively, larger than those obtained by Theorem 5.2, and so on. Therefore, we can conclude that a more accurate integral inequality can lead to a more relaxed stability condition provided that the corresponding L–K functional is suitably constructed. Of course, the reduction of conservatism is achieved at the price of the computation burden, which can be clearly seen from Table 5.1.

On the other hand, as the value of N increases, the improvement of MAUBs becomes very limited. From the balance between the conservatism and computation burden, Theorem 4 (N = 2) may be a good choice.

Table 5.1 The MAUBs h_M for $d_M = -d_m$ in Example 1

d_M	0.1	0.5	0.8	NVs
Theorem 5.1	3.658	2.337	1.934	$3n^2 + 2n$
Theorem 5.2	4.703	2.420	2.137	$10n^2 + 3n$
Theorem 5.3	4.713	2.570	2.281	$23n^2 + 4n$
Theorem 5.4 (N = 2)	4.797	2.711	2.327	$42n^2 + 5n$
Theorem 5.4 (N = 3)	4.828	2.725	2.329	$67n^2 + 6n$

Table 5.2 The MAUBs h_M for $d_M = -d_m$ in Example 2

d_M	0.1	0.2	0.5	0.8
Theorem 5.1	5.551	3.392	1.268	1.067
Theorem 5.2	6.590	3.672	1.411	1.275
Theorem 5.3	6.604	3.715	1.573	1.342
Theorem 5.4 (N = 2)	6.685	3.813	1.607	1.354
Theorem 5.4 (N = 3)	6.721	3.818	1.608	1.354

Example 2 Consider system (5.1) with

$$A = \begin{bmatrix} 0 & 1 \\ -1 & -2 \end{bmatrix}, \quad A_d = \begin{bmatrix} 0 & 0 \\ -1 & 1 \end{bmatrix}.$$

The comparative results among Theorems 5.1, 5.2, 5.3 and 5.4 with $N = 2, 3$ are listed in Table 5.2. It is still observed that the MAUBs obtained by Theorem 5.2 are, respectively, larger than those obtained by Theorem 5.1 and that the MAUBs obtained by Theorem 5.3 are, respectively, larger than those obtained by Theorem 5.2, and so on. Similar conclusions can also be made.

5.2.2 Via Integral Inequalities with Free Matrices

In this subsection, we develop stability conditions for system (5.1) by employing the Jensen, Wirtinger-based, auxiliary-function-based and B–L FMB integral inequalities, respectively. In this case, the RCL is not required.

5.2.2.1 Via the Jensen FMB Inequality

Theorem 5.5 *For given h_M, d_m and d_M satisfying (5.2), matrices $P, Q_1, Q_2, R \in \mathbb{S}_+^n$ and $N_0, M_0 \in \mathbb{R}^{2n\times n}$, system (5.1) is asymptotically stable if the following LMIs hold for $\dot{h}_t \in \{d_m, d_M\}$:*

$$\begin{bmatrix} \Sigma_1(\dot{h}_t) + h_M \tilde{\Sigma}_2 & h_M T_2^T M_0 \\ * & -R \end{bmatrix} < 0, \tag{5.30}$$

$$\begin{bmatrix} \Sigma_1(\dot{h}_t) + h_M \tilde{\Sigma}_2 & h_M T_1^T N_0 \\ * & -R \end{bmatrix} < 0, \tag{5.31}$$

where $\Sigma_1(\dot{h}_t)$, e_s and e_i, $i \in \{1, 2, 3\}$, are all defined in Theorem 5.1, and

$$\begin{aligned} &\tilde{\Sigma}_2 = T_1^T \mathrm{Sym}\{N_0 \Pi_0\} T_1 + T_2^T \mathrm{Sym}\{M_0 \Pi_0\} T_2, \\ &T_1 = col\{e_1, e_2\}, \quad T_2 = col\{e_2, e_3\}, \quad \Pi_0 = \begin{bmatrix} I & -I \end{bmatrix}. \end{aligned} \tag{5.32}$$

Proof Let us consider the L–K function (5.4) again. Its time derivative is computed as follows from the proof of Theorem 5.1:

$$\dot{V}_J(t) = \zeta^T(t) \Sigma_1(\dot{h}_t) \zeta(t) - h_M \int_{t-h_M}^{t} \dot{x}^T(s) R \dot{x}(s) ds. \tag{5.33}$$

Applying the Jensen FMB integral inequality to the above integral term leads to

$$\begin{aligned} &- \int_{t-h_M}^{t} \dot{x}^T(s) R \dot{x}(s) ds \\ &= - \int_{t-h_t}^{t} \dot{x}^T(s) R \dot{x}(s) ds - \int_{t-h_M}^{t-h_t} \dot{x}^T(s) R \dot{x}(s) ds \\ &\leq \zeta^T(t) T_1^T \left(\mathrm{Sym}\{N_0 \Pi_0\} + h_t N_0 R^{-1} N_0^T \right) T_1 \zeta(t) \\ &\quad + \zeta^T(t) T_2^T \left(\mathrm{Sym}\{M_0 \Pi_0\} + h_{Mt} M_0 R^{-1} M_0^T \right) T_2 \zeta(t) \\ &= \zeta^T(t) (\tilde{\Sigma}_2 + \Sigma_3(h_t)) \zeta(t), \end{aligned}$$

where $\tilde{\Sigma}_2$ is defined in (5.32), and

$$\Sigma_3(h_t) = h_t T_1^T N_0 R^{-1} N_0^T T_1 + h_{Mt} T_2^T M_0 R^{-1} M_0^T T_2.$$

As a result, we have

$$\dot{V}_J(t) \leq \zeta^T(t) \Sigma(h_t) \zeta(t),$$

where $\Sigma(h_t) = \Sigma_1(\dot{h}_t) + h_M \tilde{\Sigma}_2 + h_M \Sigma_3(h_t)$. Since $\Sigma(h_t)$ is affine with the delay h_t, the inequality $\Sigma(h_t) < 0$ for any $h_t \in [0, h_M]$ is ensured by the two inequalities $\Sigma(0) < 0$ and $\Sigma(h_M) < 0$.

On the other hand, from the Schur complement the two inequalities (5.30) and (5.31) are, respectively, equivalent to $\Sigma(0) < 0$ and $\Sigma(h_M) < 0$. Therefore, there exists a sufficiently small scalar $\varepsilon > 0$ such that $\dot{V}_J(t) \leq -\varepsilon \|x(t)\| < 0$ for any

$x(t) \neq 0$, which guarantees the asymptotic stability of system (5.1). This completes the proof. ■

5.2.2.2 Via the Wirtinger-Based FMB Inequality

Theorem 5.6 *For given h_M, d_m and d_M satisfying (5.2), matrices $P \in \mathbb{S}_+^{3n}$, Q_1, Q_2, $R \in \mathbb{S}_+^n$ and $N_0, N_1, M_0, M_1 \in \mathbb{R}^{3n \times n}$, system (5.1) is asymptotically stable if the following LMIs hold for $\dot{h}_t \in \{d_m, d_M\}$:*

$$\begin{bmatrix} \Upsilon_1(0, \dot{h}_t) + h_M \tilde{\Upsilon}_2 & h_M T_2^T \bar{M} \\ * & -\tilde{R} \end{bmatrix} < 0, \tag{5.34}$$

$$\begin{bmatrix} \Upsilon_1(h_M, \dot{h}_t) + h_M \tilde{\Upsilon}_2 & h_M T_1^T \bar{N} \\ * & -\tilde{R} \end{bmatrix} < 0, \tag{5.35}$$

where

$$\begin{aligned} &\tilde{\Upsilon}_2 = T_1^T \mathrm{Sym}\{\bar{N}\bar{\Pi}\} T_1 + T_2^T \mathrm{Sym}\{\bar{M}\bar{\Pi}\} T_2, \\ &\Pi_0 = \begin{bmatrix} I & -I & 0 \end{bmatrix}, \quad \Pi_1 = \begin{bmatrix} I & I & -2I \end{bmatrix}, \quad \bar{\Pi} = col\{\Pi_0, \Pi_1\}, \\ &T_1 = col\{e_1, e_2, e_4\}, \quad T_2 = col\{e_2, e_3, e_5\}, \\ &\bar{N} = \begin{bmatrix} N_0 & N_1 \end{bmatrix}, \quad \bar{M} = \begin{bmatrix} M_0 & M_1 \end{bmatrix}, \end{aligned} \tag{5.36}$$

and other notations are all defined in Theorem 5.2.

Proof Let us consider the L–K functional (5.10). Its time derivative is computed from the proof of Theorem 5.2:

$$\dot{V}_W(t) = \xi_0^T(t)\Upsilon_1(h_t, \dot{h}_t)\xi_0(t) - h_M \int_{t-h_M}^{t} \dot{x}^T(s) R \dot{x}(s) ds.$$

Applying the Wirtinger-based FMB integral inequality to the above integral term leads to

$$\begin{aligned} &- \int_{t-h_M}^{t} \dot{x}^T(s) R \dot{x}(s) ds \\ &= - \int_{t-h_t}^{t} \dot{x}^T(s) R \dot{x}(s) ds - \int_{t-h_M}^{t-h_t} \dot{x}^T(s) R \dot{x}(s) ds \\ &\leq \xi_0^T(t) T_1^T \left(\mathrm{Sym}\{\bar{N}\bar{\Pi}\} + h_t \bar{N} \tilde{R}^{-1} \bar{N}^T \right) T_1 \xi_0(t) \\ &\quad + \xi_0^T(t) T_2^T \left(\mathrm{Sym}\{\bar{M}\bar{\Pi}\} + h_{Mt} \bar{M} \tilde{R}^{-1} \bar{M}^T \right) T_2 \xi_0(t) \\ &= \xi_0^T(t)(\tilde{\Upsilon}_2 + \Upsilon_3(h_t))\xi_0(t), \end{aligned}$$

where $\tilde{\Upsilon}_2$ is defined in (5.36), and

$$\Upsilon_3(h_t) = h_t T_1^T \bar{N} \tilde{R}^{-1} \bar{N}^T T_1 + h_{Mt} T_2^T \bar{M} \tilde{R}^{-1} \bar{M}^T T_2.$$

As a result, we have

$$\dot{V}_W(t) \leq \xi_0^T(t) \Upsilon(h_t, \dot{h}_t) \xi_0(t),$$

where $\Upsilon(h_t, \dot{h}_t) = \Upsilon_1(h_t, \dot{h}_t) + h_M \tilde{\Upsilon}_2 + h_M \Upsilon_3(h_t)$. It is noted that $\Upsilon(h_t, \dot{h}_t)$ is affine with both h_t and $\dot{h}_t$. The rest is similar to that in the proof of Theorem 5.5 and omitted for brevity. This completes the proof. ■

5.2.2.3 Via the Auxiliary-Function-Based FMB Inequality

Theorem 5.7 *For given h_M, d_m and d_M satisfying (5.2), matrices $P \in \mathbb{S}_+^{5n}$, Q_1, Q_2, $R \in \mathbb{S}_+^n$ and $N_i, M_i \in \mathbb{R}^{4n \times n}$, $i \in \{0, 1, 2\}$, system (5.1) is asymptotically stable if the following LMIs hold for $\dot{h}_t \in \{d_m, d_M\}$:*

$$\begin{bmatrix} \Xi_1(0, \dot{h}_t) + h_M \tilde{\Xi}_2 & h_M T_2^T \bar{M} \\ * & -\tilde{R} \end{bmatrix} < 0, \tag{5.37}$$

$$\begin{bmatrix} \Xi_1(h_M, \dot{h}_t) + h_M \tilde{\Xi}_2 & h_M T_1^T \bar{N} \\ * & -\tilde{R} \end{bmatrix} < 0, \tag{5.38}$$

where $\Xi_1(h_t, \dot{h}_t)$ is defined in Theorem 5.3, and

$$\begin{aligned}
&\tilde{\Xi}_2 = T_1^T \mathrm{Sym}\{\bar{N}\bar{\Pi}\} T_1 + T_2^T \mathrm{Sym}\{\bar{M}\bar{\Pi}\} T_2, \\
&\Pi_0 = \begin{bmatrix} I & -I & 0 & 0 \end{bmatrix}, \quad \Pi_1 = \begin{bmatrix} I & I & -2I & 0 \end{bmatrix}, \quad \Pi_2 = \begin{bmatrix} I & -I & 6I & -6I \end{bmatrix}, \\
&T_1 = col\{e_1, e_2, e_4, 2e_6\}, \quad T_2 = col\{e_2, e_3, e_5, 2e_7\}, \quad \bar{\Pi} = col\{\Pi_0, \Pi_1, \Pi_2\}, \\
&\bar{N} = \begin{bmatrix} N_0 & N_1 & N_2 \end{bmatrix}, \quad \bar{M} = \begin{bmatrix} M_0 & M_1 & M_2 \end{bmatrix}, \quad \tilde{R} = \mathrm{diag}\{R, 3R, 5R\},
\end{aligned}$$

and other notations are all defined in Theorem 5.3.

5.2.2.4 Via the B–L FMB Inequality

Theorem 5.8 *For given h_M, d_m and d_M satisfying (5.2), matrices $P \in \mathbb{S}_+^{(2N+3)n}$, $Q_1, Q_2, R \in \mathbb{S}_+^n$ and $N_i, M_i \in \mathbb{R}^{(N+3)n \times n}$, $i \in \{0, 1, \ldots, N+1\}$, system (5.1) is asymptotically stable if the following LMIs hold for $\dot{h}_t \in \{d_m, d_M\}$:*

$$\begin{bmatrix} \Gamma_1(0, \dot{h}_t) + h_M \tilde{\Gamma}_2 & h_M T_{N2}^T \bar{M}_{N+1} \\ * & -\tilde{R}_{N+1} \end{bmatrix} < 0, \tag{5.39}$$

$$\begin{bmatrix} \Gamma_1(h_M, \dot{h}_t) + h_M \tilde{\Gamma}_2 & h_M T_{N1}^T \bar{N}_{N+1} \\ * & -\tilde{R}_{N+1} \end{bmatrix} < 0 \tag{5.40}$$

where

$$\tilde{\Gamma}_2 = T_{N1}^T \text{Sym}\{\bar{N}_{N+1}\bar{\Pi}_{N+1}\}T_{N1} + T_{N2}^T \text{Sym}\{\bar{M}_{N+1}\bar{\Pi}_{N+1}\}T_{N2}, \tag{5.41}$$
$$\bar{N}_{N+1} = \begin{bmatrix} N_0 & N_1 & \cdots & N_{N+1} \end{bmatrix}, \quad \bar{M}_{N+1} = \begin{bmatrix} M_0 & M_1 & \cdots & M_{N+1} \end{bmatrix},$$

and other notations are all defined in Theorem 5.4.

Proof From the proof of Theorem 5.4, the time derivative of $V_{BL}(t)$ defined in (5.23) is computed as follows:

$$\dot{V}_{BL}(t) = \xi_N^T(t)\Gamma_1(h_t, \dot{h}_t)\xi_N(t) - h_M \int_{t-h_M}^{t} \dot{x}^T(s)R\dot{x}(s)ds. \tag{5.42}$$

Applying the B–L FMB integral inequality to $-\int_{t-h_t}^{t} \dot{x}(s)^T R\dot{x}(s)ds$ leads to

$$\begin{aligned}
&-\int_{t-h_t}^{t} \dot{x}^T(s)R\dot{x}(s)ds \\
&\leq \vartheta_{N+1}^T\Big(h_t \bar{N}_{N+1}\tilde{R}_{N+1}^{-1}\bar{N}_{N+1}^T + \text{Sym}\{\bar{N}_{N+1}\bar{\Pi}_{N+1}\}\Big)\vartheta_{N+1} \\
&= \xi_N^T(t)T_{N1}^T\Big(h_t \bar{N}_{N+1}\tilde{R}_{N+1}^{-1}\bar{N}_{N+1}^T + \text{Sym}\{\bar{N}_{N+1}\bar{\Pi}_{N+1}\}\Big)T_{N1}\xi_N(t).
\end{aligned}$$

Similarly, applying the B–L FMB integral inequality to $-\int_{t-h_M}^{t-h_t} \dot{x}(s)^T R\dot{x}(s)ds$ leads to

$$\begin{aligned}
&-\int_{t-h_M}^{t-h_t} \dot{x}^T(s)R\dot{x}(s)ds \\
&\leq \vartheta_{N+1}^T\Big(h_{Mt} \bar{M}_{N+1}\tilde{R}_{N+1}^{-1}\bar{M}_{N+1}^T + \text{Sym}\{\bar{M}_{N+1}\bar{\Pi}_{N+1}\}\Big)\vartheta_{N+1} \\
&= \xi_N^T(t)T_{N2}^T\Big(h_{Mt} \bar{M}_{N+1}\tilde{R}_{N+1}^{-1}\bar{M}_{N+1}^T + \text{Sym}\{\bar{M}_{N+1}\bar{\Pi}_{N+1}\}\Big)T_{N2}\xi_N(t).
\end{aligned}$$

According to the above discussions, we have

$$\dot{V}_{BL}(t) \leq \xi_N^T(t)\Big(\Gamma_1(h_t, \dot{h}_t) + h_M\tilde{\Gamma}_2 + h_M\Gamma_3(h_t)\Big)\xi_N(t),$$

where $\Gamma_1(h_t, \dot{h}_t)$ is defined in Theorem 5.4, $\tilde{\Gamma}_2$ is defined in (5.41), and

$$\Gamma_3(h_t) = h_t T_{N1}^T \bar{N}_{N+1}\tilde{R}_{N+1}^{-1}\bar{N}_{N+1}^T T_{N1} + h_{Mt} T_{N2}^T \bar{M}_{N+1}\tilde{R}_{N+1}^{-1}\bar{M}_{N+1}^T T_{N2}.$$

Table 5.3 The MAUBs h_M for $d_M = -d_m$ in Example 3

d_M	0.1	0.5	0.8	NVs
Theorem 5.5	3.623	2.259	1.892	$6n^2 + 2n$
Theorem 5.6	4.710	2.459	2.212	$18n^2 + 3n$
Theorem 5.7	4.714	2.655	2.485	$38n^2 + 4n$
Theorem 5.8 (N = 2)	4.786	2.880	2.592	$66n^2 + 5n$
Theorem 5.8 (N = 3)	4.802	2.914	2.605	$102n^2 + 6n$

The rest is similar to that in the proof of Theorem 5.5 and therefore, omitted for brevity. This completes the proof. ■

Remark 5.7 Based on the Jensen, Wirtinger-based and auxiliary-function-based FMB inequalities, three stability conditions are, respectively, obtained in Theorems 5.5, 5.6 and 5.7. In the process, the RCL is not needed. It can be proved theoretically that Theorem 5.7 includes Theorem 5.6 that further includes Theorem 5.5. Based on the B–L FMB inequality, a hierarch of general stability conditions is obtained in Theorem 5.8 with a parameter N. By setting $N = 0$ and $N = 1$, Theorem 5.8 is, respectively, reduced to Theorems 5.6 and 5.7. And by setting $\xi_N(t) = \zeta(t)$, Theorem 5.8 is reduced to Theorem 5.5.

5.2.3 *Continuous-Time Numerical Examples*

In order to illustrate the effectiveness of the proposed approach, the following two commonly-used numerical examples are presented, in which the maximum allowable upper bounds (MAUBs) are carefully compared, including the numbers of decision variables (NVs).

Example 3 Consider system (5.1) with

$$A = \begin{bmatrix} -2 & 0 \\ 0 & -0.9 \end{bmatrix}, \quad A_d = \begin{bmatrix} -1 & 0 \\ -1 & -1 \end{bmatrix}.$$

The MAUBs and NVs are, respectively, computed by Theorems 5.5, 5.6, 5.7 and 5.8 with $N = 2, 3$. It is seen that the MAUBs obtained by Theorem 5.6 are, respectively, larger than those obtained by Theorem 5.5 and that the MAUBs obtained by Theorem 5.7 are, respectively, larger than those obtained by Theorem 5.6, and so on. Therefore, we can conclude that a more accurate integral inequality with free matrices can lead to a more relaxed stability condition provided that the corresponding L–K functional is suitably constructed. Of course, the reduction of conservatism is achieved at the price of the computation burden, which can be clearly seen from Table 5.3.

Table 5.4 The MAUBs h_M for $d_M = -d_m$ in Example 4

d_M	0.1	0.2	0.5	0.8
Theorem 5.5	5.546	3.382	1.237	1.042
Theorem 5.6	6.601	3.691	1.549	1.370
Theorem 5.7	6.614	3.745	1.952	1.595
Theorem 5.8 (N = 2)	6.706	3.976	2.087	1.628
Theorem 5.8 (N = 3)	6.767	4.031	2.113	1.635

Example 4 Consider system (5.1) with

$$A = \begin{bmatrix} 0 & 1 \\ -1 & -2 \end{bmatrix}, \quad A_d = \begin{bmatrix} 0 & 0 \\ -1 & 1 \end{bmatrix}.$$

The comparative results among Theorems 5.5, 5.6, 5.7 and 5.8 with $N = 2, 3$ are listed in Table 5.4. It is also found that Theorem 5.6 is less conservative than Theorem 5.5 and that Theorem 5.7 is less conservative than Theorem 5.6, and so on. Similar conclusions can be also made.

5.3 Stability Analysis for Discrete-Time Systems

Let us consider the following linear discrete-time system with time-varying delay:

$$\begin{cases} x(k+1) = Ax(k) + A_d x(k - h(k)), & k \geq 0 \\ x(k) = \phi(k), & -h_2 \leq k \leq 0 \end{cases} \tag{5.43}$$

where $x(k) \in \mathbb{R}^n$ is the state vector; A and A_d are constant system matrices; $\phi(k)$ is the initial condition; with given integers h_1 and h_2, the time-varying delay $h(k)$ is assumed to satisfy

$$1 \leq h_1 \leq h(k) \leq h_2, \quad k \geq 0. \tag{5.44}$$

In order to obtain delay-dependent conditions for the system (5.43), the double summation term $\sum_{i=k-h_2}^{k-h_1-1} \sum_{j=i}^{k-1} y^T(j)Ry(j)$ is usually contained in L–K functionals, where $y(j) = x(j+1) - x(j)$.

Then, to estimate the summation term

$$\delta_1(k) := \sum_{i=k-h_2}^{k-h_1-1} y^T(i)Ry(i)$$

arising in the forward difference of L–K functionals becomes an essential point in achieving a less conservative stability criterion in the form of LMIs.

Different estimating ways may lead to different types of conditions with different conservatism. To take full advantage of the information of the time-varying delay $h(k)$, $\delta_1(k)$ is usually split into two parts when estimated: $\sum_{i=k-h(k)}^{k-h_1-1} y^T(i)Ry(i)$ and $\sum_{i=k-h_2}^{k-h(k)-1} y^T(i)Ry(i)$.

Here, we will employ four typical ways to estimate $\delta_1(k)$. One way is the combination of the Jensen summation inequality and conventional RCL and meanwhile, a simple L–K functional is used in which no augmented vector is involved. The second and third estimating ways are to combine the Wirtinger-based summation inequality with the conventional and improved RCLs, respectively. The fourth way is to directly employ the Wirtinger-based FMB summation inequality without the cooperation of RCLs.

In the latter three estimating ways the same augmented L–K functional is utilized, in which the augmented vector is deliberated constructed to coordinate with the Wirtinger-based summation inequality no matter with or without free matrices. As a result, four different stability criteria are developed for system (5.43). The relationships among the four conditions are discussed and some conclusions are consequently made.

Before proceeding, the following notations are defined:

$$h_k := h(k), \quad h_{k1} := h_k - h_1, \quad h_{2k} := h_2 - h_k, \quad h_{21} = h_2 - h_1, \quad \alpha = h_{k1}/h_{21},$$
$$x_k(i) := x(k+i), \quad y_k(i) := x_k(i+1) - x_k(i),$$
$$\xi_0(k) := col\{x_k(0), x_k(-h_1), x_k(-h_k), x_k(-h_2)\},$$
$$\upsilon_0(k) := \sum_{i=-h_1}^{0} \frac{x_k(i)}{h_1+1}, \quad \upsilon_1(k) := \sum_{i=-h_k}^{-h_1} \frac{x_k(i)}{h_{k1}+1}, \quad \upsilon_2(k) := \sum_{i=-h_2}^{-h_k} \frac{x_k(i)}{h_{2k}+1},$$
$$\xi(k) := col\,\{\xi_0(k), \upsilon_0(k), \upsilon_1(k), \upsilon_2(k)\}, \quad \delta_0(k) := \sum_{i=-h_1}^{-1} y_k(i)^T R_1 y_k(i),$$
$$V_1(k) := \sum_{i=-h_1}^{-1} x_k^T(i)Q_1x_k(i) + \sum_{i=-h_2}^{-h_1-1} x_k^T(i)Q_2x_k(i) \tag{5.45}$$
$$V_2(k) := h_1 \sum_{i=-h_1}^{-1}\sum_{j=i}^{-1} y_k^T(j)R_1y_k(j), +h_{21} \sum_{i=-h_2}^{-h_1-1}\sum_{j=i}^{-1} y_k^T(j)Ry_k(j). \tag{5.46}$$

5.3.1 Via the Jensen Summation Inequality

Construct the L–K functional candidate as follows:

$$V(k) = x^T(k)Px(k) + V_1(k) + V_2(k) \tag{5.47}$$

where $V_1(k)$ and $V_2(k)$ are, respectively, defined in (5.45) and (5.46). Based on the simple L–K functional candidate (5.47), a stability condition is obtained for system (5.43) via the combination of the Jensen summation inequality and conventional RCL.

Theorem 5.9 *For given integers h_1 and h_2 satisfying (5.44), system (5.43) is asymptotically stable if there exist matrices P, Q_1, Q_2, R_1, $R \in \mathbb{S}_+^n$ and $X \in \mathbb{R}^{n\times n}$ such that the following LMIs hold:*

$$\begin{bmatrix} R & X \\ * & R \end{bmatrix} \geq 0 \tag{5.48}$$

$$\Lambda < 0 \tag{5.49}$$

where

$$\Lambda = \Lambda_0 - e_{12}^T R e_{12} - \begin{bmatrix} e_{23} \\ e_{34} \end{bmatrix}^T \begin{bmatrix} R & X \\ * & R \end{bmatrix} \begin{bmatrix} e_{23} \\ e_{34} \end{bmatrix}, \tag{5.50}$$

$$\begin{aligned} \Lambda_0 &= e_s^T P e_s - e_1^T P e_1 + e_1^T Q_1 e_1 - e_2^T Q_1 e_2 + e_2^T Q_2 e_2 - e_4^T Q_2 e_4 \\ &\quad + (e_s - e_1)^T (h_1^2 R_1 + h_{21}^2 R)(e_s - e_1) \end{aligned} \tag{5.51}$$

with

$$e_{12} = e_1 - e_2, \quad e_{23} = e_2 - e_3, \quad e_{34} = e_3 - e_4, \quad e_s = Ae_1 + A_d e_3,$$
$$e_i = \begin{bmatrix} 0_{n\times(i-1)n} & I_n & 0_{n\times(4-i)n} \end{bmatrix}, \quad i \in \{1, \ldots, 4\}.$$

Proof The forward difference of $V(k)$ along the trajectory of system (5.43) is computed as follows:

$$\begin{aligned} \Delta V(k) &= V(k+1) - V(k) \\ &= x^T(k+1)Px(k+1) - x^T(k)Px(k) + x^T(k)Q_1x(k) \\ &\quad - x^T(k-h_1)Q_1x(k-h_1) + x^T(k-h_1)Q_2x(k-h_1) \\ &\quad - x^T(k-h_2)Q_2x(k-h_2) + h_1^2 y_k^T(0)R_1y_k(0) \\ &\quad + h_{21}^2 y_k^T(0)Ry_k(0) - h_1\delta_1(k) - h_{21}\delta_2(k) \\ &= \xi_0^T(k)\Lambda_0\xi_0(k) - h_1\delta_1(k) - h_{21}\delta_2(k) \\ &\geq \xi_0^T(k)(\Lambda_0 - e_{12}^T R e_{12})\xi_0(k) - h_{21}\delta_2(k) \end{aligned} \tag{5.52}$$

where Λ_0 is defined in (5.51).

With (5.48), applying the Jensen summation inequality and RCL leads to

$$\begin{aligned} h_{21}\delta_2(k) &= h_{21} \sum_{i=-h_k}^{-h_1-1} y_k^T(i)Ry_k(i) + h_{21} \sum_{i=-h_2}^{-h_k-1} y_k^T(i)Ry_k(i) \\ &\geq \frac{1}{\alpha}\xi_0^T(k)e_{23}^T R e_{23}\xi_0(k) + \frac{1}{1-\alpha}\xi_0^T(k)e_{34}^T R e_{34}\xi_0(k) \end{aligned}$$

$$= \xi_0^T(k) \begin{bmatrix} e_{23} \\ e_{34} \end{bmatrix}^T \begin{bmatrix} \frac{1}{\alpha} R & 0 \\ * & \frac{1}{1-\alpha} R \end{bmatrix} \begin{bmatrix} e_{23} \\ e_{34} \end{bmatrix} \xi_0(k)$$

$$\geq \xi_0^T(k) \begin{bmatrix} e_{23} \\ e_{34} \end{bmatrix}^T \begin{bmatrix} R & X \\ * & R \end{bmatrix} \begin{bmatrix} e_{23} \\ e_{34} \end{bmatrix} \xi_0(k)$$

According to the above discussions, we get

$$\Delta V(k) \leq \xi^T(k) \Lambda \xi(k)$$

where Λ is defined in (5.50). From the above inequality, it is seen that the inequality (5.49) ensures that there exists a sufficiently small scalar $\varepsilon > 0$ such that $\Delta V(x_k) \leq \|x(k)\|^2 < 0$ for any $x(k) \neq 0$. This completes the proof. ∎

Remark 5.8 It is seen that Theorem 5.9 is a delay-dependent condition, which is usually more relaxed than delay-independent ones. However, owe to inherent conservatism of the Jensen inequality, Theorem 5.9 is still conservative. More relaxed conditions could be expected if more accurate summation inequalities are employed.

5.3.2 *Via the Wirtinger-Based Summation Inequality*

Construct the L–K functional candidate as follows:

$$V(k) = \chi_0^T(k) P \chi_0(k) + V_1(k) + V_2(k) \tag{5.53}$$

where $V_1(k)$ and $V_2(k)$ are, respectively, defined in (5.45) and (5.46), and

$$\chi_0(k) := col \left\{ x_k(0), \sum_{i=-h_1}^{-1} x_k(i), \sum_{i=-h_2}^{-h_1-1} x_k(i) \right\}.$$

Theorem 5.10 *For given integers h_1 and h_2 satisfying (5.44), system (5.43) is asymptotically stable if there exist matrices $P \in \mathbb{S}_+^{3n}$, $Q_1, Q_2, R_1, R \in \mathbb{S}_+^n$ and $X \in \mathbb{R}^{2n \times 2n}$ such that the following LMIs hold for $h_k \in \{h_1, h_2\}$:*

$$\begin{bmatrix} \tilde{R} & X \\ * & \tilde{R} \end{bmatrix} \geq 0, \tag{5.54}$$

$$\Xi(h_k) < 0, \tag{5.55}$$

where

$$\Xi(h_k) = \Xi_0(h_k) + \Xi_1 + \Xi_{20} - \Xi_{21}, \tag{5.56}$$

$$\Xi_0(h_k) = \Gamma_2^T P\Gamma_2 - \Gamma_1^T P\Gamma_1 + \mathrm{Sym}\{(\Gamma_2 - \Gamma_1)^T P\Gamma_0(h_k)\}, \tag{5.57}$$

$$\Xi_1 = e_1^T Q_1 e_1 - e_2^T Q_1 e_2 + e_2^T Q_2 e_2 - e_4^T Q_2 e_4, \tag{5.58}$$

with

$$\Xi_{20} = e_{s1}^T(h_1^2 R_1 + h_{21}^2 R)e_{s1} - T_0^T \Pi^T \tilde{R}_1 \Pi T_0,$$

$$\Xi_{21} = \begin{bmatrix} \Pi T_1 \\ \Pi T_2 \end{bmatrix}^T \begin{bmatrix} \tilde{R} & X \\ * & \tilde{R} \end{bmatrix} \begin{bmatrix} \Pi T_1 \\ \Pi T_2 \end{bmatrix},$$

$$\Gamma_1 = \begin{bmatrix} e_1 \\ -e_1 \\ -e_2 - e_3 \end{bmatrix}, \quad \Gamma_2 = \begin{bmatrix} e_s \\ -e_2 \\ -e_3 - e_4 \end{bmatrix},$$

$$\Gamma_0(h_k) = \begin{bmatrix} e_0 \\ (h_1 + 1)e_5 \\ (h_{k1} + 1)e_6 + (h_{2k} + 1)e_7 \end{bmatrix},$$

$$\tilde{R}_1 = \mathrm{diag}\{R_1, 3R_1\}, \quad \tilde{R} = \mathrm{diag}\{R, 3R\}, \quad e_s = Ae_1 + A_d e_3, \quad e_{s1} = e_s - e_1,$$

$$\Pi = col\{\Pi_0, \Pi_1\}, \quad \Pi_0 = \begin{bmatrix} I & -I & 0 \end{bmatrix}, \quad \Pi_1 = \begin{bmatrix} I & I & -2I \end{bmatrix},$$

$$T_0 = col\{e_1, e_2, e_5\}, \quad T_1 = col\{e_2, e_3, e_6\}, \quad T_2 = col\{e_3, e_4, e_7\},$$

$$e_i = \begin{bmatrix} 0_{n\times(i-1)n} & I_n & 0_{n\times(7-i)n} \end{bmatrix}, \quad i \in \{1, \ldots, 7\}.$$

Proof The forward difference of $V(k)$ is computed as follows:

$$\begin{aligned} \Delta V(k) &= V(k+1) - V(k) \\ &= \xi^T(k)\Big(\Xi_0(h_k) + \Xi_1\Big)\xi(k) + h_1^2 y_k^T(0) R_1 y_k(0) + h_{21}^2 y_k^T(0) R y_k(0) \\ &\quad - h_1\delta_0(k) - h_{21}\delta_1(k) \end{aligned}$$

where $\Xi_0(h_k)$ and Ξ_1 are, respectively, defined in (5.57) and (5.58).

Applying the Wirtinger-based summation inequality to $h_1\delta_0(k)$ leads to

$$h_1\delta_0(k) \geq \xi^T(k) T_0^T \Pi^T \tilde{R}_1 \Pi T_0 \xi(k)$$

With (5.54), applying the Wirtinger-based summation inequality and conventional RCL to $h_{21}\delta_1(k)$ leads to

$$\begin{aligned} h_{21}\delta_1(k) &= h_{21} \sum_{i=-h_k}^{-h_1-1} y_k^T(i) R y_k(i) + h_{21} \sum_{i=-h_2}^{-h_k-1} y_k^T(i) R y_k(i) \\ &\geq \frac{1}{\alpha}\xi^T(k) T_1^T \Pi^T \tilde{R} \Pi T_1 \xi(k) + \frac{1}{1-\alpha}\xi^T(k) T_2^T \Pi^T \tilde{R} \Pi T_2 \xi(k) \end{aligned}$$

$$
\begin{aligned}
&= \xi^T(k)\begin{bmatrix} \Pi T_1 \\ \Pi T_2 \end{bmatrix}^T \begin{bmatrix} \frac{1}{\alpha}\tilde{R} & 0 \\ * & \frac{1}{1-\alpha}\tilde{R} \end{bmatrix}\begin{bmatrix} \Pi T_1 \\ \Pi T_2 \end{bmatrix}\xi(k) \\
&\geq \xi^T(k)\Xi_{21}\xi(k)
\end{aligned}
$$

Based on the above discussions, we get

$$
\Delta V(x_k) \leq \xi^T(k)\Xi(h_k)\xi(k) \tag{5.59}
$$

where $\Xi(h_k)$ is defined in (5.56). Since $\Xi(h_k)$ is affine with the delay h_k, the inequality $\Xi(h_k) < 0$ for $\forall h_k \in [h_1, h_2]$ is ensured by the two inequalities $\Xi(h_1) < 0$ and $\Xi(h_2) < 0$. Therefore, there exists a sufficiently small scalar $\varepsilon > 0$ such that $\Delta V(x_k) \leq \varepsilon\|x(k)\|^2 < 0$ for any $x(k) \neq 0$, which ensures the asymptotic stability of system (5.43). This completes the proof. ■

Remark 5.9 Based on the augmented L–K functional candidate (5.53), a delay-dependent stability condition is obtained in Theorem 5.10 via the combination of the Wirtinger-based summation inequality and conventional RCL. In order to take full advantage of the Wirtinger-based summation inequality, the vector $\chi_0(k)$ is augmented by adding two summation terms $\sum_{i=-h_1}^{-1} x_k(i)$ and $\sum_{i=-h_2}^{-h_1-1} x_k(i)$. When calculating the forward difference of $\chi_0^T(k)P\chi_0(k)$, cross-product terms for $\sum_{i=-h_1}^{-1} x_k(i)$ and $\sum_{i=-h_2}^{-h_1-1} x_k(i)$ with other state-related vectors will rise. These terms are helpful in reducing the conservatism by cooperating with the use of the Wirtinger-based summation inequality. It can be proved that Theorem 5.10 is less conservative than Theorem 5.9. However, if the simple L–K functional is used, Theorem 5.10 is actually equivalent with Theorem 5.9 even though the Wirtinger-based summation inequality produces a tighter bound than the Jensen summation inequality.

Lemma 5.2 (The conventional RCL [29]) *For matrices $R \in \mathbb{S}_+^n$ and $X \in \mathbb{R}^{n\times n}$ such that*

$$
\begin{bmatrix} R & X \\ * & R \end{bmatrix} \geq 0
$$

the following matrix inequality holds for $\forall \alpha \in (0, 1)$:

$$
\begin{bmatrix} \frac{1}{\alpha}R & 0 \\ * & \frac{1}{1-\alpha}R \end{bmatrix} \geq \begin{bmatrix} R & X \\ * & R \end{bmatrix}
$$

Lemma 5.3 (The improved RCL [19]) *For matrices $R \in \mathbb{S}_+^n$ and $X \in \mathbb{R}^{n\times n}$, the following matrix inequality holds for $\forall \alpha \in (0, 1)$:*

$$
\begin{bmatrix} \frac{1}{\alpha}R & 0 \\ 0 & \frac{1}{1-\alpha}R \end{bmatrix} \geq \begin{bmatrix} R & X \\ X^T & R \end{bmatrix} + \begin{bmatrix} (1-\alpha)Y_1 & 0 \\ 0 & \alpha Y_2 \end{bmatrix} \tag{5.60}
$$

where $Y_1 = R - XR^{-1}X^T$ and $Y_2 = R - X^TR^{-1}X$.

If the improved RCL is considered, a more relaxed stability condition is obtained in the following theorem.

Theorem 5.11 *For given integers h_1 and h_2 satisfying (5.44), system (5.43) is asymptotically stable if there exist matrices $P \in \mathbb{S}_+^{3n}$, $Q_1, Q_2, R_1, R \in \mathbb{S}_+^n$ and $X \in \mathbb{R}^{2n\times 2n}$ such that the following LMIs hold:*

$$\begin{bmatrix} \tilde{\Xi}(h_1) & T_1^T \Pi^T X \\ * & -\tilde{R} \end{bmatrix} < 0, \tag{5.61}$$

$$\begin{bmatrix} \tilde{\Xi}(h_2) & T_2^T \Pi^T X^T \\ * & -\tilde{R} \end{bmatrix} < 0, \tag{5.62}$$

where

$$\tilde{\Xi}(h_k) = \Xi_0(h_k) + \Xi_1 + \Xi_{20} - \tilde{\Xi}_{21}(\alpha),$$
$$\tilde{\Xi}_{21}(\alpha) = \begin{bmatrix} \Pi T_1 \\ \Pi T_2 \end{bmatrix}^T \begin{bmatrix} (2-\alpha)\tilde{R} & X \\ * & (1+\alpha)\tilde{R} \end{bmatrix} \begin{bmatrix} \Pi T_1 \\ \Pi T_2 \end{bmatrix},$$

and other notations are all defined in Theorem 5.10.

Remark 5.10 Based on the same L–K functional, Theorem 5.11 is obtained via the combination of the Wirtinger-based summation inequality and the improved RCL. Since the improved RCL is more accurate than the conventional one, it can be proved in theory that Theorem 5.11 is more relaxed than Theorem 5.10. So, it is predictable that if a more accurate RCL is considered, a less conservative condition is expectable provided that the same L–K functional is utilized.

Remark 5.11 There is a common feature in Theorems 5.9, 5.10 and 5.11, i.e., the RCL is required to achieve the resulting tractable condition. So these three conditions can be grouped into one type: stability conditions via the combination of summation inequalities and RCL. Here are some conclusions:

(i) To achieve a more relaxed stability condition, the augmented L–K functional candidate should be suitably constructed for a more accurate summation inequality.

(ii) Provided that the same L–K functional candidate and the same summation inequality are used, a more accurate RCL leads to a more relaxed stability condition.

5.3.3 Via the Wirtinger-Based FMB Inequality

Theorem 5.12 *For given integers h_1 and h_2 satisfying (5.44), system (5.43) is asymptotically stable if there exist matrices $P \in \mathbb{S}_+^{3n}$, $Q_1, Q_2, R_1, R \in \mathbb{S}_+^n$ and $M_0, M_1, N_0, N_1 \in \mathbb{R}^{3n\times n}$ such that the following LMIs hold:*

$$\begin{bmatrix} \hat{\Xi}(h_1) & h_{21} T_2^T N \\ * & -\tilde{R} \end{bmatrix} < 0, \tag{5.63}$$

$$\begin{bmatrix} \hat{\Xi}(h_2) & h_{21}T_1^T M \\ * & -\tilde{R} \end{bmatrix} < 0, \tag{5.64}$$

where

$$\begin{aligned}
&\hat{\Xi}(h_k) = \Xi_0(h_k) + \Xi_1 + \Xi_{20} + h_{21}\Upsilon_1, \\
&\Upsilon_1 = T_1^T \text{Sym}\{M\Pi\}T_1 + T_2^T \text{Sym}\{N\Pi\}T_2, \\
&M = \begin{bmatrix} M_0 & M_1 \end{bmatrix}, \quad N = \begin{bmatrix} N_0 & N_1 \end{bmatrix},
\end{aligned} \tag{5.65}$$

and other notations are all defined in Theorem 5.10.

Proof Consider the L–K functional (5.53). Its forward difference is computed as follows:

$$\Delta V(k) = \xi^T(k)\Big(\Xi_0(h_k) + \Xi_1 + \Xi_{20}\Big)\xi(k) - h_{21}\delta_1(k).$$

Applying the Wirtinger-based FMB summation inequality to $-\delta_1(k)$ leads to

$$\begin{aligned}
-\delta_1(k) = &- \sum_{i=-h_k}^{-h_1-1} y_k^T(i)Ry_k(i) - \sum_{i=-h_2}^{-h_k-1} y_k^T(i)Ry_k(i) \\
&\le \xi^T(k)T_1^T\Big(h_{k1}M\tilde{R}^{-1}M^T + \text{Sym}\{M\Pi\}\Big)T_1\xi(k) \\
&\quad + \xi^T(k)T_2^T\Big(h_{2k}N\tilde{R}^{-1}N^T + \text{Sym}\{N\Pi\}\Big)T_2\xi(k) \\
&= \xi^T(k)(\Upsilon_1 + \Upsilon_2(h_k))\xi(k)
\end{aligned}$$

where Υ_1 is defined in (5.65), and

$$\Upsilon_2(h_k) = h_{k1}T_1^T M\tilde{R}^{-1}M^T T_1 + h_{2k}T_2^T N\tilde{R}^{-1}N^T T_2.$$

As a result, we have

$$\Delta V(k) \le \xi^T(k)\Big(\Xi_0(h_k) + \Xi_1 + \Xi_{20} + h_{21}\Upsilon_1 + h_{21}\Upsilon_2(h_k)\Big)\xi(k)$$

The remainder is omitted for simplicity. This completes the proof. ■

Remark 5.12 Based on the same augmented L–K functional candidate (5.53), Theorem 5.12 is developed via the Wirtinger-based FMB summation inequality. Since the summation inequality with free matrices is used, the RCL is no longer required. It is seen that unlike the previous three stability conditions, Theorem 5.12 belongs to a different type: stability conditions directly obtained via summation inequalities with free matrices. Then, a question naturally arises: compared with Theorems 5.10 and 5.11, is Theorem 5.12 more or less conservative? Since two types of estimating ways are employed, the two types of stability conditions have two fundamentally different structures. Therefore, the conservatism of the two types of conditions cannot be compared in theory. That is, it cannot be said that Theorem 5.10 or 5.11 is less conservative than Theorem 5.12 or that Theorem 5.12 is less conservative than Theorem 5.10 or 5.11. The results of numerical examples will clearly illustrate this point in the following subsection.

5.3.4 Discrete-Time Numerical Examples

In this subsection, two numerical examples are given, in which the maximum allowable upper bounds (MAUBs) are carefully compared, including the number of variables (NVs).

Example 5 Consider system (5.43) with

$$A = \begin{bmatrix} 0.6480 & 0.0400 \\ 0.1200 & 0.6540 \end{bmatrix}, \quad A_d = \begin{bmatrix} -0.1512 & -0.0518 \\ 0.0259 & -0.1091 \end{bmatrix}.$$

This example is borrowed from [27]. The comparison results among the four conditions are listed in Table 5.5. It is expectable that the MAUBs obtained by Theorem 5.10 are always larger than those obtained by Theorem 5.9. This further verifies the statement that a more accurate summation inequality can lead to a more relaxed condition only if an appropriately augmented L–K functional is constructed. It is also seen that the MAUBs obtained by Theorem 5.11 are always larger than or equal to those obtained by Theorem 5.10. This means that the improved RCL-based condition is less conservative than the conventional RCL-based one.

Particularly, it is found that when the values of h_1 are small, the MAUBs obtained by Theorem 5.12 are larger than those obtained by Theorem 5.11 while when the values of h_1 are large, the MAUBs obtained by Theorem 5.12 become smaller than those obtained by Theorem 5.11. This interesting phenomenon clearly verifies some conclusions stated in Remark 5.12.

From Table 5.5, it is also found that the number of decision variables involved in Theorem 5.12 is obviously larger than those involved in other conditions. This is because the Wirtinger-based FMB summation inequality is used twice in developing this condition.

Table 5.5 The MAUBs h_2 for different h_1 in Example 5

h_1	1	3	5	7	11	13	20	25	NVs
Theorem 5.9	11	13	15	17	21	23	30	35	$3.5n^2 + 2.5n$
Theorem 5.10	16	18	20	22	25	27	34	39	$10.5n^2 + 3.5n$
Theorem 5.11	17	19	21	22	26	28	35	40	$10.5n^2 + 3.5n$
Theorem 5.12	18	19	21	22	26	27	34	39	$18.5n^2 + 3.5n$

Table 5.6 The MAUBs h_2 for different h_1 in Example 6 ($\times$ means infeasibility)

h_1	12	16	20	40	60	100	120	140
Theorem 5.9	×	×	×	×	×	×	×	×
Theorem 5.10	38	65	80	122	140	152	153	152
Theorem 5.11	39	66	82	122	140	153	153	152
Theorem 5.12	40	67	82	121	139	150	151	151

Example 6 Consider system (5.43) with

$$A = \begin{bmatrix} 1 & 0.01 \\ -0.02 & 1.001 \end{bmatrix}, \quad Ad = \begin{bmatrix} 0 & 0 \\ 0.01 & 0 \end{bmatrix}.$$

This example is borrowed from [14]. The MAUBs checked by Theorems 5.10, 5.11 and 5.12 are listed in Table 5.6. It is seen that Theorem 5.9 fails to check the stability of this system. It is found that the MAUBs obtained by Theorem 5.11 are always larger than or equal to those obtained by Theorem 5.10. Additionally, it is also found that when the values of h_1 are small, the MAUBs obtained by Theorem 5.12 are larger than those obtained by Theorem 5.10 or 5.11 while when the values of h_1 are large, the MAUBs obtained by Theorem 5.12 are smaller than those obtained by Theorem 5.10 or 5.11. These above observations further verify the conclusions stated in Remark 5.12.

5.4 Conclusion

This chapter has studied the stability problem of linear time-delay systems with a time-varying delay for both the continuous-time and discrete-time cases. To take full advantage of the interest of various kinds of integral/summation inequalities, different augmented Lyapunov–Krasovskii functionals have been deliberately constructed. As a result, different conservative stability conditions have been obtained for linear time-delay systems with a time-varying delay. The numerical examples have clearly shown that more accurate integral/summation inequalities lead to more relaxed stability conditions if an appropriate L–K functional is chosen.

References

1. Gu K, Kharitonov VL, Chen J (2003) Stability of time-delay systems. Birkhuser, Boston
2. Fridman E (2014) Introduction to time-delay systems: analysis and control. Springer, Berlin
3. Shao H, Han QL (2011) New stability criteria for linear discrete-time systems with interval-like time-varying delays. IEEE Trans Autom Control 56:619–625

4. Fridman E, Shaked U (2002) An improved stabilization method for linear time-delay systems. IEEE Trans Autom Control 47:1931–1937
5. Meng X, Lam J, Du B, Gao H (2010) A delay-partitioning approach to the stability analysis of discrete-time systems. Automatica 46:610–614
6. Han QL (2009) A discrete delay decomposition approach to stability of linear retarded and neutral systems. Automatica 45:517–524
7. He Y, Wang QG, Lin C, Wu M (2007) Delay-range-dependent stability for systems with time-varying delay. Automatica 43:371–376
8. Xu S, Lam J (2005) Improved delay-dependent stability criteria for time-delay systems. IEEE Trans Autom Control 50:384–387
9. Chen J, Lu J, Xu S (2016) Summation inequality and its application to stability analysis for time-delay systems. IEEE Trans Autom Control 10:391–395
10. Chen J, Xu S, Zhang B, Liu G (2017) A note on relationship between two classes of integral inequalities. IEEE Trans Autom Control 62:4044–4049
11. Chen J, Xu S, Jia X, Zhang B (2017) Novel summation inequalities and their applications to stability analysis for systems with time-varying delay. IEEE Trans Autom Control 62:2470–2475
12. Chen J, Xu S, Ma Q, Li Y, Chu Y, Zhang Z (2017) Two novel general summation inequalities to discrete-time systems with time-varying delay. J Frankl Inst 354:5537–5558
13. Gyurkovics E (2015) A note on Wirtinger-type integral inequalities for time-delay systems. Automatica 61:44–46
14. Nam PT, Trinh H, Pathirana PN (2015) Discrete inequalities based on multiple auxiliary functions and their applications to stability analysis of time-delay systems. J Frankl Inst 352:5810–5831
15. Seuret A, Gouaisbaut F (2013) Wirtinger-based integral inequality: application to time-delay systems. Automatica 49:2860–2866
16. Seuret A, Gouaisbaut F (2015) Hierarchy of LMI conditions for the stability analysis of time-delay systems. Syst. Control Lett. 2:1–7
17. Seuret A, Gouaisbaut F, Fridman E (2015) Stability of discrete-time systems with time-varying delays via a novel summation inequality. IEEE Trans Autom Control 60:2740–2745
18. Zhang CK, He Y, Jiang YL, Wu M (2016) An improved summation inequality to discrete-time systems with time-varying delay. Automatica 74:10–15
19. Zhang CK, He Y, Jiang L, Wu M, Zeng HB (2016) Stability analysis of systems with time-varying delay via relaxed integral inequalities. Syst Control Lett 92:52–61
20. Zhang CK, He Y, Jiang L, Wu M (2017) Notes on stability of time-delay systems: bounding inequalities and augmented Lyapunov–Krasovskii functionals. IEEE Trans Autom Control 62:5331–5336
21. Zhang CK, He Y, Jiang L, Wu M, Zeng HB (2017) Summation inequalities to bounded real lemmas of discrete-time systems with time-varying delay. IEEE Trans Autom Control 62:2582–2588
22. Zhang XM, Han QL (2015) Abel lemma-based finite-sum inequality and its application to stability analysis for linear discrete time-delay systems. Automatica 57:199–202
23. Zhang XM, Han QL, Seuret A, Gouaisbaut F (2017) An improved reciprocally convex inequality and an augmented Lyapunov-Krasovskii functional for stability of linear systems with time-varying delay. Automatica 84:221–226
24. Chen J, Xu S, Chen W, Zhang B, Ma Q, Zou Y (2016) Two general integral inequalities and their applications to stability analysis for systems with time-varying delay. Int J Robust Nonlinear Control 26:4088–4103
25. Chen J, Park JH, Xu S (2018) Stability analysis of continuous-time systems with time-varying delay using new Lyapunov–Krasovskii functionals. J Frankl Inst 355:5957–5967
26. Lee TH, Park JH, Xu S (2017) Relaxed conditions for stability of time-varying delay systems. Automatica 75:11–15

27. Xu S, Lam J, Zhang B, Zou Y (2014) A new result on the delay-dependent stability of discrete systems with time-varying delays. Int J Robust Nonlinear Control 24:2512–2521
28. Xu S, Lam J, Zhang B, Zou Y (2015) New insight into delay-dependent stability of time-delay systems. Int J Robust Nonlinear Control 25:961–970
29. Park PG, Ko JW, Jeong C (2011) Reciprocally convex approach to stability of systems with time-varying delays. Automatica 47:235–238

Chapter 6
Stability Analysis for Neural Networks with Time-Varying Delay

6.1 Introduction

Neural networks have found a wide range of applications in engineering and research areas, such as image processing, pattern recognition and optimization problem [1–9]. On the other hand, when implementing artificial neural networks, time delay inevitably arises that may bring undesirable dynamical behaviors. Therefore, the stability problem of neural networks with a time-varying delay has been attracting much attention these years [9–23].

The L–K functional method is a powerful tool to study the stability of time-delay neural networks. For continuous-time neural network with a time-varying delay, the constructed L–K functionals usually contain the quadratic, single- and double-integral functionals [21, 24–30]. In order to take full advantage of the state and delay information, various augmented L–K functionals are now proposed in the literature, in which the involved vectors are augmented by adding state- and delay-related vectors. Additionally, the delay-partitioning method is often used to construct a L–K functional [31–33]. By partitioning the whole delay interval into several subregions, the L–K functional can be constructed to contain more state and delay information. But a L–K functional that contains too many terms may result in a very complicated stability condition with a large number of decision variables. So, constructing a *proper* L–K functional is meaningful and necessary, which should lead to a relaxed stability condition without introducing too many decision variables.

In this chapter, we construct a new L–K functional for continuous-time neural networks with a time-varying delay that includes two complement triple-integral functionals. All of the vectors in the quadratic, single- and double-integral functionals are augmented by adding some state- and delay-related vectors. Three zero equations are added to the time derivative of the L–K functional [26, 34, 35]. The GFMW integral inequality proposed in Chap. 3 is used to estimate the single-integral terms arising in the time derivative of the L–K functional [24, 36]. As a result, a new stability condition is achieved. Several numerical examples show that the new stabil-

J. H. Park et al., *Dynamic Systems with Time Delays: Stability and Control*,
https://doi.org/10.1007/978-981-13-9254-2_6

ity condition is more relaxed than some of the existing ones while with a relatively small number of decision variables.

As for the stability of discrete-time neural networks with a time-varying delay, remarkable results have also been reported in the literature [26, 31, 37–43]. For example, a novel L–K functional was developed [16], in which the information of the mid-delay was fully taken into account. By dividing the delayed interval into multiple subregions, augmented L–K functionals were constructed via the delay-partitioning method [32, 44, 45]. Recently, triple-summation functions were included in the L–K functional [14], apart from the single- and double-summation ones. It is also seen that in order to continually reduce the conservatism, more and more state and delay information of discrete-time neural networks is involved in L–K functionals.

Based on the above discussions, we construct a new L–K functional for discrete-time neural networks with a time-varying delay. In order to further reduce the conservatism, the zero-equation technique and the summation-inequality method are used [35, 46]. Firstly, three zero equations are constructed which are added to the forward difference of the L–K functional. Secondly, the GFMB summation inequality is employed to estimate the single-summation terms arising in the forward difference of the L–K functional. Finally, numerical examples show that the obtained stability condition is less conservative than some of the existing ones.

6.2 Stability Analysis for the Continuous-Time Case

6.2.1 Problem Formation and Preliminaries

Consider the following generalized neural network:

$$\begin{cases} \dot{x}(t) = -Ax(t) + W_0 f(W_2 x(t)) + W_1 f(W_2 x(t - h(t))), \\ x(t) = \phi(t), \quad -h_M \le t \le 0, \end{cases} \tag{6.1}$$

where $x(t) = col\{x_1(t), x_2(t), \ldots, x_n(t)\} \in \mathbb{R}^n$ is the neural state vector with n neurons; $f(W_2 x(t)) = col\{f_1(W_{21}x(t)), f_2(W_{22}x(t)), \ldots, f_n(W_{2n}x(t))\}$ is the neuron activation function with W_{2i} denoting the ith row of W_2; $A = \text{diag}\{a_1, a_2, \ldots, a_n\} > 0$, W_0, W_1 and W_2 are constant real matrices. $h(t)$ is the time-varying delay satisfying the following constraints:

$$0 \le h(t) \le h_M, \quad \dot{h}(t) \le \mu, \tag{6.2}$$

where h_M and μ are real constants.

The activation functions $f_i(W_{2i}x(t))$, $i \in \{1, 2, \ldots, n\}$, satisfy $f_i(0) = 0$ and

$$k_i^- \le \frac{f_i(s_1) - f_i(s_2)}{s_1 - s_2} \le k_i^+, \quad s_1 \ne s_2 \in \mathbb{R}, \tag{6.3}$$

where k_i^- and k_i^+ are known constants that may be positive, negative, or zero.

For convenience, we define

$$h_t = h(t), \quad h_{Mt} = h_M - h_t,$$
$$K_2 = \text{diag}\{k_1^-, k_2^-, \ldots, k_n^-\}, \quad K_1 = \text{diag}\{k_1^+, k_2^+, \ldots, k_n^+\}.$$

The following lemmas will be used to derive the main results.

Lemma 6.1 ([19, 24]) *For matrices $R_1, R_2 \in \mathbb{S}_+^n$ and $X \in \mathbb{R}^{n\times n}$, the following matrix inequalities hold for $\forall p, q \in \mathbb{R}^+$:*

$$\begin{bmatrix} \frac{q}{p}R_1 & 0 \\ 0 & \frac{p}{q}R_2 \end{bmatrix} \geq \begin{bmatrix} \frac{q}{p+q}T_1 & X \\ X^T & \frac{p}{p+q}T_2 \end{bmatrix}$$

where

$$T_1 = R_1 - XR_2^{-1}X^T, \quad T_2 = R_2 - X^TR_1^{-1}X.$$

Lemma 6.2 ([24]) *For a matrix $R \in \mathbb{S}_+^n$, any matrices L, M, and a differentiable vector function $\omega(t)$: $[a, b] \to \mathbb{R}^n$, the following inequalities hold:*

$$-\int_a^b \omega^T(s)R\omega(s)ds \leq \text{Sym}\{\rho_1^T L\chi_1 + \rho_2^T M\chi_2\}$$
$$+ (b-a)\left(\rho_1^T LR^{-1}L^T\rho_1 + \frac{1}{3}\rho_2^T MR^{-1}M^T\rho_2\right),$$

$$-\int_a^b \dot{\omega}^T(s)R\dot{\omega}(s)ds \leq \text{Sym}\{\rho_1^T L\hat{\chi}_1 + \rho_2^T M\hat{\chi}_2\}$$
$$+ (b-a)\left(\rho_1^T LR^{-1}L^T\rho_1 + \frac{1}{3}\rho_2^T MR^{-1}M^T\rho_2\right)$$

where ρ_1, ρ_2 are any vectors, and

$$\chi_1 = \int_a^b \omega(s)ds, \quad \chi_2 = -\chi_1 + \frac{2}{b-a}\int_a^b\int_s^b \omega(u)duds,$$
$$\hat{\chi}_1 = \omega(b) - \omega(a) \quad \hat{\chi}_2 = \omega(b) + \omega(a) - \frac{2}{b-a}\int_a^b \omega(s)ds.$$

Lemma 6.3 ([36]) *For a matrix $R \in \mathbb{S}_+^n$ and a differentiable vector function $\omega(t)$: $[a, b] \to \mathbb{R}^n$, the following two inequalities hold:*

$$\int_a^b \int_u^b \dot{\omega}^T(s) R\dot{\omega}(s) ds du \geq \vartheta^T \Pi_1^T \bar{R} \Pi_1 \vartheta,$$
$$\int_a^b \int_a^u \dot{\omega}^T(s) R\dot{\omega}(s) ds du \geq \vartheta^T \Pi_2^T \bar{R} \Pi_2 \vartheta$$

where $\vartheta = col\left\{\omega(b), \omega(a), \int_a^b \frac{\omega(s)}{b-a} ds, \int_a^b \int_u^b \frac{2\omega(s)}{(b-a)^2} ds du\right\}$, $\bar{R} = \text{diag}\{2R, 4R\}$, *and*

$$\Pi_1 = \begin{bmatrix} I & 0 & -I & 0 \\ I & 0 & 2I & -3I \end{bmatrix}, \quad \Pi_2 = \begin{bmatrix} 0 & I & -I & 0 \\ 0 & I & -4I & 3I \end{bmatrix}.$$

6.2.2 Stability Analysis

We construct the following L–K functional candidate for neural network (6.1):

$$V(t) = \sum_{i=0}^{3} V_i(t), \tag{6.4}$$

where

$$\begin{aligned}
V_0(t) &= \eta_0^T(t) P \eta_0(t), \\
V_1(t) &= \int_{t-h_t}^{t} \eta_1^T(s) Q_1 \eta_1(s) ds + \int_{t-h_M}^{t} \eta_1(s)^T Q_2 \eta_1(s) ds \\
&\quad + \int_{-h_M}^{0} \int_{t+u}^{t} \eta_2^T(s) Z \eta_2(s) ds du \\
V_2(t) &= 2\sum_{i=1}^{n} \int_0^{W_{2i}x(t)} \Big(\lambda_{1i}(k_i^+ s - f(s)) + \lambda_{2i}(f(s) - k_i^- s)\Big) ds, \\
V_3(t) &= \int_{t-h_M}^{t} \int_{u_1}^{t} \int_{u_2}^{t} \dot{x}^T(u_3) R_1 \dot{x}(u_3) du_3 du_2 du_1 \\
&\quad + \int_{t-h_M}^{t} \int_{t-h_M}^{u_1} \int_{u_2}^{t} \dot{x}^T(u_3) R_2 \dot{x}(u_3) du_3 du_2 du_1,
\end{aligned}$$

with

$$\begin{aligned}
\eta_0(t) &= col\left\{x(t), \int_{t-h_M}^{t} x(s) ds\right\}, \quad \eta_1(t) = col\{x(t), f(W_2 x(t))\}, \\
\eta_2(t) &= col\{x(t), \dot{x}(t)\}.
\end{aligned}$$

Remark 6.1 In the L–K functional (6.4), the vectors $\eta_0(t)$, $\eta_1(t)$ and $\eta_2(t)$ are, respectively, augmented by adding state-related vectors $\int_{t-h_M}^{t} x(s)ds$, $f(W_2x(t))$ and $\dot{x}(t)$. These augmentations are very helpful in reducing the conservatism of the obtained stability condition since more state and delay information of the neural network (6.1) is taken into account.

Here is the first theorem for stability of the system (6.1).

Theorem 6.1 *For given scalars h_M, $\mu \in \mathbb{R}$, neural network (6.1) subject to (6.2) and (6.3) is asymptotically stable if there exist matrices $P, Q_1, Q_2, Z \in \mathbb{S}_+^{2n}$, $R_1, R_2 \in \mathbb{S}_+^n$, $\Lambda_1, \Lambda_2, H_1, H_2, H_3, U_1, U_2, U_3 \in \mathbb{D}_+^n$, $Z_a, Z_b \in \mathbb{S}^n$, $X \in \mathbb{R}^{3n\times 3n}$ and any matrices L_1, L_2, M_1, M_2 with appropriate dimensions such that the following LMIs hold:*

$$\begin{bmatrix} \Upsilon(h_M) & h_M g_{11}^T L_1 & h_M g_{12}^T M_1 & F_2^T X^T \\ * & -h_M \bar{Z}_a & 0 & 0 \\ * & * & -3h_M \bar{Z}_a & 0 \\ * & * & * & -\tilde{R}_1 \end{bmatrix} < 0, \tag{6.5}$$

$$\begin{bmatrix} \Upsilon(0) & h_M g_{21}^T L_2 & h_M g_{22}^T M_2 & F_1^T X \\ * & -h_M \bar{Z}_b & 0 & 0 \\ * & * & -3h_M \bar{Z}_b & 0 \\ * & * & * & -\tilde{R}_2 \end{bmatrix} < 0, \tag{6.6}$$

where

$$\Upsilon(h_t) = \Xi_0(h_t) + \Xi_{10} + \Xi_{11}(h_t) + \Xi_2 + \Xi_{30}(\alpha) + \Xi_4, \tag{6.7}$$

$$\Xi_0(h_t) = \text{Sym}\left\{c_3(h_t)^T P c_4\right\}, \tag{6.8}$$

$$\Xi_1 = c_5^T (Q_1 + Q_2)c_5 - (1-u)c_6^T Q_1 c_6 - c_7^T Q_2 c_7 + h_M c_8^T Z c_8, \tag{6.9}$$

$$\Xi_{10} = \Xi_1 + e_1^T Z_a e_1 + e_2^T (Z_b - Z_a)e_2 - e_3^T Z_b e_3,$$

$$\Xi_{11}(h_t) = \text{Sym}\left\{g_{11}^T L_1 c_{11} + g_{12}^T M_1 c_{12}\right\} + \text{Sym}\left\{g_{21}^T L_2 c_{21} + g_{22}^T M_2 c_{22}\right\}, \tag{6.10}$$

$$\Xi_2 = \text{Sym}\{[(K_1W_2e_1 - e_4)^T \Lambda_1 + (e_4 - K_2W_2e_1)^T \Lambda_2]W_2e_s\}, \tag{6.11}$$

$$\begin{aligned} \Xi_{30}(\alpha) = {} & \frac{1}{2}h_M^2 e_s^T (R_1 + R_2)e_s - \begin{bmatrix} F_1 \\ F_2 \end{bmatrix}^T \begin{bmatrix} (1-\alpha)\tilde{R}_1 & X \\ X^T & \alpha\tilde{R}_2 \end{bmatrix} \begin{bmatrix} F_1 \\ F_2 \end{bmatrix} \\ & - E_1^T \bar{R}_1 E_1 - E_2^T \bar{R}_1 E_2 - E_3^T \bar{R}_2 E_3 - E_4^T \bar{R}_2 E_4, \end{aligned} \tag{6.12}$$

$$\begin{aligned} \Xi_4 = {} & \sum_{i=1}^{3} \text{Sym}\{(K_1W_2e_i - e_{i+3})^T H_i (e_{i+3} - K_2W_2e_i)\} \\ & + \sum_{i=1}^{2} \text{Sym}\{(K_1W_2(e_i - e_{i+1}) - e_{i+3} + e_{i+4})^T U_i \\ & \times (e_{i+3} - e_{i+4} - K_2W_2(e_i - e_{i+1}))\} \\ & + \text{Sym}\{(K_1W_2(e_1 - e_3) - e_4 + e_6)^T U_3(e_4 - e_6 - K_2W_2(e_1 - e_3))\}, \end{aligned} \tag{6.13}$$

with

$$
\begin{aligned}
&c_{11} = col\{h_t e_7, e_1 - e_2\}, \quad c_{12} = col\{-h_t e_7 + h_t e_9, e_1 + e_2 - 2e_7\},\\
&c_{21} = col\{h_{Mt} e_8, e_2 - e_3\}, \quad c_{22} = col\{-h_{Mt} e_8 + h_{Mt} e_{10}, e_2 + e_3 - 2e_8\},\\
&c_3(h_t) = col\{e_1, h_t e_7 + h_{Mt} e_8\}, \quad c_4 = col\{e_s, e_1 - e_3\},\\
&c_5 = col\{e_1, e_4\}, \quad c_6 = col\{e_2, e_5\}, \quad c_7 = col\{e_3, e_6\}, \quad c_8 = col\{e_1, e_s\},\\
&e_g = col\{e_1, e_2, \ldots, e_{10}\}, \quad e_s = -Ae_1 + W_0 e_4 + W_1 e_5,\\
&e_i = \left[0_{n\times(i-1)n} \; I_{n\times n} \; 0_{n\times(10-i)n} \right], \; i \in \{1, \ldots, 10\},\\
&\alpha = \frac{h_t}{h_M}, \quad \bar{Z}_a = Z + \begin{bmatrix} 0 & Z_a \\ Z_a & 0 \end{bmatrix}, \quad \bar{Z}_b = Z + \begin{bmatrix} 0 & Z_b \\ Z_b & 0 \end{bmatrix},\\
&F_1 = col\{e_1 - e_2, e_1 + e_2 - 2e_7, e_1 - e_2 + 6e_7 - 6e_9\},\\
&F_2 = col\{e_2 - e_3, e_2 + e_3 - 2e_8, e_2 - e_3 + 6e_8 - 6e_{10}\},\\
&E_1 = col\{e_1 - e_7, e_1 + 2e_7 - 3e_9\}, \quad E_2 = col\{e_2 - e_8, e_2 + 2e_8 - 3e_{10}\},\\
&E_3 = col\{e_2 - e_7, e_2 - 4e_7 + 3e_9\}, \quad E_4 = col\{e_3 - e_8, e_3 - 4e_8 + 3e_{10}\},\\
&\tilde{R}_i = \mathrm{diag}\{R_i, 3R_i, 5R_i\}, \quad \bar{R}_i = \mathrm{diag}\{2R_i, 4R_i\}, \quad i \in \{1, 2\},
\end{aligned}
$$

where g_{11}, g_{12}, g_{21} and g_{22} may be, respectively, chosen as e_g or its part.

Proof Differentiating $V_i(t)$, $i = 0, 1, 2, 3$, along the trajectory of (6.1) leads to:

$$
\begin{aligned}
\dot{V}_0(t) &= 2\eta^T(t) P \dot{\eta}(t)\\
&= \xi^T(t) \Xi_0(h_t) \xi(t),\\
\dot{V}_1(t) &= \eta_1^T(t) Q_1 \eta_1(t) - (1 - \dot{h}_t)\eta_1^T(t - h_t) Q_1 \eta_1(t - h_t) + \eta_1^T(t) Q_2 \eta_1(t)\\
&\quad - \eta_1^T(t - h_M) Q_2 \eta_1(t - h_M) + h_M \eta_2^T(t) Z \eta_2(t) - \int_{t-h_M}^{t} \eta_2^T(s) Z \eta_2(s) ds\\
&\le \xi^T(t) \Xi_1 \xi(t) - \int_{t-h_t}^{t} \eta_2^T(s) Z \eta_2(s) ds - \int_{t-h_M}^{t-h_t} \eta_2^T(s) Z \eta_2(s) ds,\\
\dot{V}_2(t) &= \xi^T(t) \Xi_2 \xi(t),\\
\dot{V}_3(t) &= \frac{h_M^2}{2} \dot{x}^T(t) R_1 \dot{x}(t) - h_{Mt} \int_{t-h_t}^{t} \dot{x}^T(s) R_1 \dot{x}(s) ds\\
&\quad - \int_{t-h_t}^{t} \int_{u_1}^{t} \dot{x}^T(u_2) R_1 \dot{x}(u_2) du_2 du_1 - \int_{t-h_M}^{t-h_t} \int_{u_1}^{t-h_t} \dot{x}^T(u_2) R_1 \dot{x}(u_2) du_2 du_1\\
&\quad + \frac{h_M^2}{2} \dot{x}^T(t) R_2 \dot{x}(t) - h_t \int_{t-h_M}^{t-h_t} \dot{x}^T(s) R_2 \dot{x}(s) ds\\
&\quad - \int_{t-h_t}^{t} \int_{t-h_t}^{u_1} \dot{x}^T(u_2) R_2 \dot{x}(u_2) du_2 du_1 - \int_{t-h_M}^{t-h_t} \int_{t-h_M}^{u_1} \dot{x}^T(u_2) R_2 \dot{x}(u_2) du_2 du_1,
\end{aligned}
$$

where $\Xi_0(h_t)$, Ξ_1 and Ξ_2 are, respectively, defined in (6.8), (6.9) and (6.11), and

$$\xi(t) = col\Big\{x(t), x(t-h_t), x(t-h_M), f(W_2x(t)),$$
$$f(W_2x(t-h_t)), f(W_2x(t-h_M)),$$
$$\frac{1}{h_t}\int_{t-h_t}^{t} x(s)ds, \frac{1}{h_{Mt}}\int_{t-h_M}^{t-h_t} x(s)ds,$$
$$\frac{2}{h_t^2}\int_{t-h_t}^{t}\int_{s}^{t} x(u)duds, \frac{2}{h_{Mt}^2}\int_{t-h_M}^{t-h_t}\int_{s}^{t-h_t} x(u)duds\Big\}.$$

Adding the following two zero equations

$$x^T(t)Z_ax(t) - x^T(t-h_t)Z_ax(t-h_t) - 2\int_{t-h_t}^{t} x^T(s)Z_a\dot{x}(s)ds = 0,$$

$$x^T(t-h_t)Z_bx(t-h_t) - x^T(t-h_M)Z_bx(t-h_M) - 2\int_{t-h_M}^{t-h_t} x^T(s)Z_b\dot{x}(s)ds = 0$$

to $\dot{V}_1(t)$ leads to:

$$\dot{V}_1(t) \le \xi^T(t)\Xi_{10}\xi(t) + \varphi(t), \tag{6.14}$$

where $\varphi(t) = -\int_{t-h_t}^{t} \eta_2^T(s)\bar{Z}_a\eta_2(s)ds - \int_{t-h_M}^{t-h_t} \eta_2^T(s)\bar{Z}_b\eta_2(s)ds$.

Now, applying Lemma 6.2 to $\varphi(t)$ leads to

$$\begin{aligned}\varphi(t) &\le \text{Sym}\{\rho_{11}^T L_1\eta_{11} + \rho_{12}^T M_1\eta_{12}\} + \text{Sym}\{\rho_{21}^T L_2\eta_{21} + \rho_{22}^T M_2\eta_{22}\}\\ &\quad + h_t\left(\rho_{11}^T L_1\bar{Z}_a^{-1}L_1^T\rho_{11} + \frac{1}{3}\rho_{12}^T M_1\bar{Z}_a^{-1}M_1^T\rho_{12}\right)\\ &\quad + h_{Mt}\left(\rho_{21}^T L_2\bar{Z}_b^{-1}L_2^T\rho_{21} + \frac{1}{3}\rho_{22}^T M_2\bar{Z}_b^{-1}M_2^T\rho_{22}\right)\\ &= \xi^T(t)\Big(\Xi_{11}(h_t) + \Xi_{12}(h_t)\Big)\xi(t),\end{aligned} \tag{6.15}$$

where $\Xi_{11}(h_t)$ is defined in (6.10), and

$$\begin{aligned}\Xi_{12}(h_t) = h_t&\left(g_{11}^T L_1\bar{Z}_a^{-1}L_1^T g_{11} + \frac{1}{3}g_{12}^T M_1\bar{Z}_a^{-1}M_1^T g_{12}\right)\\ &+ h_{Mt}\left(g_{21}^T L_2\bar{Z}_b^{-1}L_2^T g_{21} + \frac{1}{3}g_{22}^T M_2\bar{Z}_b^{-1}M_2^T g_{22}\right),\end{aligned}$$

with

$$\rho_{ij} = g_{ij}\xi(t), \quad \eta_{ij} = c_{ij}\xi(t), \quad i, j \in \{1, 2\}.$$

Applying the auxiliary function-based inequality [36] and Lemma 6.1 to the single-integral terms in $\dot{V}_3(t)$ leads to:

$$
\begin{aligned}
& h_{Mt}\int_{t-h_t}^{t}\dot{x}^T(s)R_1\dot{x}_t(s)ds + h_t\int_{t-h_M}^{t-h_t}\dot{x}^T(s)R_2\dot{x}(s)ds \\
\geq & \frac{h_{Mt}}{h_t}\xi^T(t)F_1^T\tilde{R}_1F_1\xi(t) + \frac{h_t}{h_{Mt}}\xi^T(t)F_2^T\tilde{R}_2F_2\xi(t) \\
\geq & \xi^T(t)\begin{bmatrix} F_1 \\ F_2 \end{bmatrix}^T \begin{bmatrix} (1-\alpha)T_1 & X \\ X^T & \alpha T_2 \end{bmatrix}\begin{bmatrix} F_1 \\ F_2 \end{bmatrix}\xi(t),
\end{aligned}
$$

where $\alpha = \frac{h_t}{h_M}$, $T_1 = \tilde{R}_1 - X\tilde{R}_2^{-1}X^T$ and $T_2 = \tilde{R}_2 - X^T\tilde{R}_1^{-1}X$.

Applying Lemma 6.3 to the double-integral terms in $\dot{V}_3(t)$ yields:

$$
\begin{aligned}
\int_{-h_t}^{0}\int_{u_1}^{0}\dot{x}_t^T(u_2)R_1\dot{x}_t(u_2)du_2du_1 &\geq \xi(t)^TE_1^T\bar{R}_1E_1\xi(t), \\
\int_{-h_M}^{-h_t}\int_{u_1}^{-h_t}\dot{x}_t^T(u_2)R_1\dot{x}_t(u_2)du_2du_1 &\geq \xi(t)^TE_2^T\bar{R}_1E_2\xi(t), \\
\int_{-h_t}^{0}\int_{-h_t}^{u_1}\dot{x}_t^T(u_2)R_2\dot{x}_t(u_2)du_2du_1 &\geq \xi(t)^TE_3^T\bar{R}_2E_3\xi(t), \\
\int_{-h_M}^{-h_t}\int_{-h_M}^{u_1}\dot{x}_t^T(u_2)R_2\dot{x}_t(u_2)du_2du_1 &\geq \xi(t)^TE_4^T\bar{R}_2E_4\xi(t).
\end{aligned}
$$

Therefore, we have

$$
\dot{V}_3(t) \leq \xi^T(t)\Big(\Xi_{30}(\alpha) + \Xi_{31}(\alpha)\Big)\xi(t), \tag{6.16}
$$

where $\Xi_{30}(\alpha)$ is defined in (6.12) and

$$
\Xi_{31}(\alpha) = (1-\alpha)F_1^TX\tilde{R}_2^{-1}X^TF_1 + \alpha F_2^TX^T\tilde{R}_1^{-1}XF_2.
$$

The following inequalities can be directly obtained from (6.3) with $H, U \in \mathbb{D}_+^n$, and $s, s_1, s_2 \in \mathbb{R}$:

$$
\begin{aligned}
\rho_1(s, H) &:= 2\rho_{11}^T(s)H\rho_{12}(s) \geq 0, \\
\rho_2(s_1, s_2, U) &:= 2\rho_{21}^T(s_1, s_2)U\rho_{22}(s_1, s_2) \geq 0,
\end{aligned}
$$

where

$$
\begin{aligned}
\rho_{11}(s) &= K_1W_2x(s) - g(W_2x(s)), \\
\rho_{12}(s) &= g(W_2x(s)) - K_2W_2x(s), \\
\rho_{21}(s_1, s_2) &= K_1W_2(x(s_1) - x(s_2)) - g(W_2x(s_1)) + g(W_2x(s_2)), \\
\rho_{22}(s_1, s_2) &= g(W_2x(s_1)) - g(W_2x(s_2)) - K_2W_2(x(s_1) - x(s_2)).
\end{aligned}
$$

Then, it follows that

$$\rho_1(t, H_1) + \rho_1(t - h_t, H_2) + \rho_1(t - h_M, H_3) + \rho_2(t, t - h_t, U_1) \\ +\rho_2(t - h_t, t - h_M, U_2) + \rho_2(t, t - h_M, U_3) \geq 0,$$

which means

$$\xi^T(t)\Xi_4\xi(t) \geq 0, \tag{6.17}$$

where Ξ_4 is defined in (6.13).

As a result, it follows from (6.14), (6.15) to (6.16) and (6.17) that

$$\dot{V}(t) \leq \xi^T(t)\Big(\Upsilon(h_t) + \Xi_{22}(h_t) + \Xi_{31}(\alpha)\Big)\xi(t), \tag{6.18}$$

where $\Upsilon(h_t)$ is defined in (6.7). Define $\Phi(h_t) = \Upsilon(h_t) + \Xi_{22}(h_t) + \Xi_{31}(\alpha)$. Since $\Phi(h_t)$ is affine with h_t, the inequality $\Phi(h_t) < 0$ for $h_t \in [0, h_M]$ is ensured by $\Phi(0) < 0$ and $\Phi(h_M) < 0$.

From the Schur complement, the inequalities (6.5) and (6.6) are, respectively, equivalent to $\Phi(0) < 0$ and $\Phi(h_M) < 0$. So, the conditions in Theorem 6.1 ensure that there exists a sufficiently small positive scalar ε such that $V(t) \leq -\varepsilon\|x(t)\|^2 < 0$ for any $x(t) \neq 0$, which implies asymptotical stability of neural network (6.1). This completes the proof. ■

Remark 6.2 Based on the L–K functional (6.4), a new stability criterion is derived for neural network (6.1). It is worth pointing out that each ρ_{ij}, $i, j \in \{1, 2\}$, can be chosen as the whole $\xi(t)$ or part of it. A different choice may lead to different conservatism and different computation burden of the obtained stability criterion. In general, when we choose the expressions of ρ_{ij} that contain more state and delay information of the neural network (6.1), a more relaxed stability condition is obtained that usually involves a larger number of decision variables. Here we consider the following two cases about expressions of ρ_{ij}, where $\rho_{ij} = g_{ij}\xi(t)$:

CASE I:

$$g_{11} = col\{e_1, e_2, e_7\}, \quad g_{12} = col\{e_1, e_2, e_7, e_9\}, \\ g_{21} = col\{e_2, e_3, e_8\}, \quad g_{22} = col\{e_2, e_3, e_8, e_{10}\};$$

CASE II:

$$g_{11} = g_{12} = g_{21} = g_{22} = e_g.$$

In Case I, the free matrix variables $L_1, L_2 \in \mathbb{R}^{3n\times 2n}$ and $M_1, M_2 \in \mathbb{R}^{4n\times 2n}$ are involved in Theorem 6.1 while in Case II, the free matrix variables $L_1, L_2, M_1, M_2 \in \mathbb{R}^{10n\times 2n}$ are involved. It is found that choosing different expressions of ρ_{ij} can lead to a big gap between the numbers of decision variables involved.

Table 6.1 The MAUBs h_M for different μ in Example 1

μ	0.40	0.45	0.50	0.55	NVs
[42] Theorem 1	5.2420	4.4301	4.1055	3.9231	$87.5n^2 + 11.5n$
[29] Corollary 3	7.4203	6.6190	6.3428	6.2095	$11.5n^2 + 13.5n$
[28] Theorem 1	8.3498	7.3817	7.0219	6.8156	$73n^2 + 13n$
Theorem 6.1 (Case I)	8.2941	7.3971	7.0796	6.8827	$47n^2 + 14n$
Theorem 6.1 (Case II)	8.5669	7.6260	7.2809	7.0683	$99n^2 + 14n$

6.2.3 Numerical Examples

Example 1 Consider neural network (6.1) with

$$W_0 = \begin{bmatrix} 0.0503 & 0.0454 \\ 0.0987 & 0.2075 \end{bmatrix}, \quad W_1 = \begin{bmatrix} 0.2381 & 0.9320 \\ 0.0388 & 0.5062 \end{bmatrix},$$
$$A = \mathrm{diag}\{1.5, 0.7\}, \quad K_1 = \mathrm{diag}\{0.3, 0.8\},$$
$$K2 = \mathrm{diag}\{0, 0\}, \quad W_2 = \mathrm{diag}\{1, 1\}.$$

This example is checked by stability conditions proposed in [28, 29, 42] and Theorem 6.1 proposed in this chapter. The comparison results are listed in Table 6.1, including the numbers of decision variables (NVs). From Table 6.1, it is seen that MAUBs obtained by Case I of Theorem 6.1 are larger than those obtained by other conditions proposed in [28, 29, 42], except for the case of $\mu = 0.4$.

Meanwhile, the number of decision variables involved in Case I is less than those involved in other conditions. Additionally, it is found that MAUBs obtained by Case II of Theorem 6.1 are always larger than those obtained by other conditions at the price of more decision variable involved.

Example 2 Consider neural network (6.1) with

$$A = \mathrm{diag}\{1.2769, 0.6231, 0.9230, 0.4480\},$$
$$W_0 = \begin{bmatrix} -0.0373 & 0.4852 & -0.3351 & 0.2336 \\ -1.6033 & 0.5988 & -0.3224 & 1.2352 \\ 0.3394 & -0.086 & -0.3824 & -0.5785 \\ -0.1311 & 0.3253 & -0.9534 & -0.5015 \end{bmatrix},$$
$$W_1 = \begin{bmatrix} 0.8674 & -1.2405 & -0.5325 & 0.022 \\ 0.0474 & -0.9164 & 0.0360 & 0.9816 \\ 1.8495 & 2.6117 & -0.3788 & 0.8428 \\ -2.0413 & 0.5179 & 1.1734 & -0.2775 \end{bmatrix},$$
$$W_2 = \mathrm{diag}\{1, 1, 1, 1\}, \quad K_2 = \mathrm{diag}\{0, 0, 0, 0\},$$
$$K_1 = \mathrm{diag}\{0.1137, 0.1279, 0.7994, 0.2368\}.$$

Table 6.2 The MAUBs h_M for different μ in Example 2

μ	0.1	0.5	0.9	NVs
[30] Theorem 1	4.1903	3.0779	2.8268	$66.5n^2 + 18.5n$
[33] Corollary 3.1	4.2143	3.1059	2.7494	$13.5n^2 + 6.5n$
[28] Theorem 1	4.2778	3.2152	2.9361	$73n^2 + 13n$
Theorem 6.1 (Case I)	4.3063	3.2305	2.9611	$47n^2 + 14n$
Theorem 6.1 (Case II)	4.3231	3.2592	2.9846	$99n^2 + 14n$

This example is checked by stability conditions proposed in [28, 30, 33] and Theorem 6.1. It is easily seen from Table 6.2 that MAUBs obtained by both Cases I and II of Theorem 6.1 are all larger than those obtained by other conditions proposed in [28, 30, 33], which implies that both Cases I and II are more relaxed than other conditions.

6.3 Stability Analysis for the Discrete-Time Case

6.3.1 Problem Formation and Preliminaries

Consider the following discrete-time neural network with a time-varying delay:

$$x(k+1) = Cx(k) + Af(x(k)) + A_d f(x(k - h(k))), \tag{6.19}$$

where $x(k)$ is the state vector associated with n neurons; $C = \text{diag}\{c_1, c_2, \ldots, c_n\}$ is the state feedback coefficient matrix; A and A_d are the connection weight matrices; $f(x(k))$ represents the neural activation function satisfying $f_i(0) = 0$ and

$$\sigma_i^- \le \frac{f_i(s_1) - f_i(s_2)}{s_1 - s_2} \le \sigma_i^+, \quad s_1 \ne s_2, \quad i \in \{1, 2, \ldots, n\}, \tag{6.20}$$

where $f(x(k)) = col\{f_1(x_1(k)), f_2(x_2(k)), \ldots, f_n(x_n(k))\}$ and σ_i^- and σ_i^+ are known scalars; $h(k)$ is the time-varying delay satisfying

$$1 \le h_1 \le h(k) \le h_2, \tag{6.21}$$

where h_1 and h_2 are known integers.

In what follows, we will analyze the stability for neural network (6.19) via the L–K functional method. Before proceeding, we define the following functions and notations for $k \in \mathbb{N}$:

$$s_1(k) = k+1, \quad s_2(k) = \frac{(k+1)(k+2)}{2}, \quad y(k) = x(k+1) - x(k),$$
$$K_1 = \mathrm{diag}\{\sigma_1^+, \sigma_2^+, \ldots, \sigma_n^+\}, \quad K_2 = \mathrm{diag}\{\sigma_1^-, \sigma_2^-, \ldots, \sigma_n^-\},$$
$$h_k = h(k), \quad h_{k1} = h_k - h_1, \quad h_{2k} = h_2 - h_k, \quad h_{21} = h_2 - h_1.$$

The following summation inequalities developed in Chap. 4 are useful in deriving the main result.

Lemma 6.4 *For matrices $R \in \mathbb{S}_+^n$, N_0, N_1 with appropriate dimensions and a vector function $\{x_a(i)|i \in [0, m]\}$ where $x_a(i) = x(a+i)$, the following inequalities hold:*

$$-\sum_{i=0}^{m-1} x_a^T(i) R x_a(i) \leq \sum_{p=0}^{1} \frac{m}{2p+1} \omega_p^T N_p R^{-1} N_p^T \omega_p + \sum_{p=0}^{1} \mathrm{Sym}\left\{\omega_p^T N_p \chi_p\right\}, \tag{6.22}$$

$$-\sum_{i=0}^{m-1} x_a^T(i) R x_a(i) \leq -\sum_{p=0}^{1} \frac{2p+1}{m} \chi_p^T R \chi_p \tag{6.23}$$

where ω_0 and ω_1 are any vectors, and

$$\chi_0 = \sum_{i=0}^{m-1} x_a(i),$$
$$\chi_1 = -\chi_0 + \frac{2}{m+1} \sum_{i=0}^{m-1} \sum_{j=i}^{m-1} x_a(j).$$

6.3.2 Stability Analysis

For neural network (6.19), we construct the following L–K functional candidate:

$$V(k) = V_0(k) + V_1(k) + V_2(k), \tag{6.24}$$

where

$$V_0(k) = \eta_0^T(k) P \eta_0(k),$$
$$V_1(k) = \sum_{i=k-h_1}^{k-1} \eta_1^T(i) Q_1 \eta_1(i) + \sum_{i=k-h_2}^{k-h_1-1} \eta_1^T(i) Q_2 \eta_1(i),$$
$$V_2(k) = \sum_{i=-h_1}^{-1} \sum_{j=k+i}^{k-1} \eta_2^T(j) R_1 \eta_2(j) + \sum_{i=-h_2}^{-h_1-1} \sum_{j=k+i}^{k-1} \eta_2^T(j) R_2 \eta_2(j),$$

$$\eta_0(k) = col\left\{x(k), \sum_{i=-h_1}^{-1} x_k(i), \sum_{i=-h_2}^{-h_1-1} x_k(i)\right\},$$

$$\eta_1(i) = col\{x(i), f(x(i))\}, \quad \eta_2(i) = col\{x(i), y(i)\},$$

in which P, Q_1, Q_2, R_1 and R_2 are all positive definite matrices that need to be determined.

Remark 6.3 To take full advantage of the state and delay information of the neural network (6.19), the new L–K functional (6.24) is deliberately constructed, in which all of the vectors in the quadratic, single- and double-summation functionals are augmented. Especially, to coordinate with the application of the GFMB summation inequality (6.22), the vector $\eta_2(i)$ is augmented by adding the difference vector $y(i)$. After combined with three zero equations, the R_1- and R_2-dependent summation terms in $\Delta V_2(k)$ are estimated by the GFMB inequality.

The following is the main result for stability of the discrete neural networks (6.19).

Theorem 6.2 *For given h_1 and h_2, neural network (6.19) with the delay h_k satisfying (6.21) is asymptotically stable, if there exist matrices $P \in \mathbb{S}_+^{3n}$, $Q_1, Q_2, R_1, R_2 \in \mathbb{S}_+^{2n}$, $G_1, G_2, G_3 \in \mathbb{S}^n$, $U_i, T_j \in \mathbb{D}_+^n$, $i \in \{1, 2, 3, 4\}$, $j \in \{1, 2, 3\}$, and any matrices $N_{l2}, N_{l3}, l \in \{0, 1\}$, with appropriate dimensions such that the following inequalities hold:*

$$\begin{bmatrix} \Psi(h_1) & h_{21}\Upsilon_3 \\ * & -h_{21}\tilde{\mathscr{R}}_3 \end{bmatrix} < 0, \tag{6.25}$$

$$\begin{bmatrix} \Psi(h_2) & h_{21}\Upsilon_2 \\ * & -h_{21}\tilde{\mathscr{R}}_2 \end{bmatrix} < 0, \tag{6.26}$$

$$\mathscr{R}_1 > 0, \tag{6.27}$$

where

$$\begin{aligned}
&\Upsilon_i = \begin{bmatrix} g_{0i}^T N_{0i} & g_{1i}^T N_{1i} \end{bmatrix}, \quad \tilde{\mathscr{R}}_i = \mathrm{diag}\{\mathscr{R}_i, 3\mathscr{R}_i\}, \quad i \in \{2, 3\}, \\
&\Psi(h_k) = \Xi_0(h_k) + \Xi_1 + \Xi_{21} + \Xi_{22}(h_k) + \Xi_3, \qquad (6.28) \\
&\Pi_0(h_k) = col\{e_0, s_1(h_1)e_9, s_1(h_{k1})e_{10} + s_1(h_{2k})e_{11}\}, \\
&\Pi_2 = col\{e_s, -e_2, -e_3 - e_4\}, \quad \Pi_1 = col\{e_1, -e_1, -e_2 - e_3\}, \\
&\Xi_0(h_k) = \Pi_2^T P \Pi_2 - \Pi_1^T P \Pi_1 + \mathrm{Sym}\{\Pi_0(h_k)^T P(\Pi_2 - \Pi_1)\}, \qquad (6.29) \\
&\Xi_1 = \begin{bmatrix} e_1 \\ e_5 \end{bmatrix}^T Q_1 \begin{bmatrix} e_1 \\ e_5 \end{bmatrix} - \begin{bmatrix} e_2 \\ e_6 \end{bmatrix}^T (Q_1 - Q_2) \begin{bmatrix} e_2 \\ e_6 \end{bmatrix} - \begin{bmatrix} e_4 \\ e_8 \end{bmatrix}^T Q_2 \begin{bmatrix} e_4 \\ e_8 \end{bmatrix}, \qquad (6.30) \\
&\mathscr{R}_i = R_i + \begin{bmatrix} 0 & G_i \\ G_i & G_i \end{bmatrix}, \quad i = 1, 2, \quad \mathscr{R}_3 = R_2 + \begin{bmatrix} 0 & G_3 \\ G_3 & G_3 \end{bmatrix}, \\
&\Xi_{21} = \begin{bmatrix} e_1 \\ e_s - e_1 \end{bmatrix}^T (h_1 R_1 + h_{21} R_2) \begin{bmatrix} e_1 \\ e_s - e_1 \end{bmatrix}
\end{aligned}$$

$$+ e_1^T G_1 e_1 - e_2^T (G_1 - G_2) e_2 - e_3^T (G_2 - G_3) e_3 - e_4^T G_3 e_4, \tag{6.31}$$

$$\Xi_{22}(h_k) = -\frac{1}{h_1}\left(F_{01}^T \mathscr{R}_1 F_{01} + 3F_{11}^T \mathscr{R}_1 F_{11}\right) + \sum_{p=0}^{1} \text{Sym}\left\{g_{p2}^T N_{p2} F_{p2} + g_{p3}^T N_{p3} F_{p3}\right\}, \tag{6.32}$$

with

$$\begin{aligned}
F_{01} &= col\{-e_1 + s_1(h_1)e_9, e_1 - e_2\},\\
F_{11} &= col\{-e_1 - s_1(h_1)e_9 + (h_1 + 2)e_{12}, e_1 + e_2 - 2e_9\},\\
F_{02} &= col\{-e_2 + s_1(h_{k1})e_{10}, e_2 - e_3\},\\
F_{12} &= col\{-e_2 - s_1(h_{k1})e_{10} + (h_{k1} + 2)e_{13}, e_2 + e_3 - 2e_{10}\},\\
F_{03} &= col\{-e_3 + s_1(h_{2k})e_{11}, e_3 - e_4\},\\
F_{13} &= col\{-e_3 - s_1(h_{2k})e_{11} + (h_{2k} + 2)e_{14}, e_3 + e_4 - 2e_{11}\},\\
\Xi_3 &= \text{Sym}\left\{\sum_{i=1}^{4}(K_1 e_i - e_{i+4})^T U_i (e_{i+4} - K_2 e_i)\right\}\\
&\quad + \text{Sym}\left\{\sum_{i=1}^{3}(K_1(e_i - e_{i+1}) - e_{i+4} + e_{i+5})^T T_i\right.\\
&\quad \left.\times (e_{i+4} - e_{i+5} - K_2(e_i - e_{i+1}))\right\},
\end{aligned} \tag{6.33}$$

$$e_i = \left[0_{n\times(i-1)n}\ I_{n\times n}\ 0_{n\times(14-i)n}\right], \quad i \in \{1, \ldots, 14\}, \quad e_0 = 0_{n\times 14n},$$
$$e_s = Ce_1 + Ae_5 + A_d e_7, \quad g_{all} = col\{e_1, e_2, \ldots, e_{14}\},$$

where each g_{ij}, $i \in \{0, 1\}$, $j \in \{2, 3\}$, may be chosen as g_{all} or part of it.

Proof The intermediate augmented state vector is defined as:

$$\xi(k) = col\{x(k), x(k - h_1), x(k - h_k), x(k - h_2), f(x(k)), f(x(k - h_1)), x(k - h_k), x(k - h_2), \mu_1(k), \mu_2(k), \mu_3(k), \mu_4(k), \mu_5(k), \mu_6(k)\},$$

where

$$\mu_1(k) = \sum_{i=-h_1}^{0} \frac{x_k(i)}{s_1(h_1)}, ;\ \mu_2(k) = \sum_{i=-h_k}^{-h_1} \frac{x_k(i)}{s_1(h_{k1})}, \quad \mu_3(k) = \sum_{i=-h_2}^{-h_k} \frac{x_k(i)}{s_1(h_{2k})},$$

$$\mu_4(k) = \sum_{i=-h_1}^{0}\sum_{j=i}^{0} \frac{x_k(j)}{s_2(h_1)}, \quad \mu_5(k) = \sum_{i=-h_k}^{-h_1}\sum_{j=i}^{-h_1} \frac{x_k(j)}{s_2(h_{k1})}, \quad \mu_6(k) = \sum_{i=-h_2}^{-h_k}\sum_{j=i}^{-h_k} \frac{x_k(j)}{s_2(h_{2k})}.$$

The forward differences of $V_i(k)$, $i \in \{0, 1, 2\}$, along the trajectory of (6.19) are, respectively, computed as

$$
\begin{aligned}
\Delta V_0(k) &= V_0(k+1) - V_0(k) \\
&= \eta_0^T(k+1)P\eta_0(k+1) - \eta_0^T(k)P\eta_0(k) \\
&= \xi^T(k)(\Pi_2 + \Pi_0(h_k))^T P(\Pi_2 + \Pi_0(h_k))\xi(k) \\
&\quad - \xi^T(k)(\Pi_1 + \Pi_0(h_k))^T P(\Pi_1 + \Pi_0(h_k))\xi(k) \\
&= \xi^T(k)\Xi_0(h_k)\xi(k), \qquad (6.34) \\
\Delta V_1(k) &= V_1(k+1) - V_1(k) \\
&= \eta_1^T(k)Q_1\eta_1(k) - \eta_1^T(k-h_1)Q_1\eta_1(k-h_1) \\
&\quad + \eta_1^T(k-h_1)Q_2\eta_1(k-h_1) - \eta_1^T(k-h_2)Q_2\eta_1(k-h_2) \\
&= \xi^T(k)\Xi_1\xi(k), \qquad (6.35) \\
\Delta V_2(k) &= V_2(k+1) - V_2(k) \\
&= \eta_2^T(k)(h_1R_1 + h_{21}R_2)\eta_2(k) - \sum_{i=k-h_1}^{k-1} \eta_2^T(i)R_1\eta_2(i) \\
&\quad - \sum_{i=k-h_k}^{k-h_1-1} \eta_2^T(i)R_2\eta_2(i) - \sum_{i=k-h_2}^{k-h_k-1} \eta_2^T(i)R_2\eta_2(i), \qquad (6.36)
\end{aligned}
$$

where $\Xi_0(h_k)$ and Ξ_1 are, respectively, defined in (6.29) and (6.30).

From the results in [26, 34], the following three zero equations

$$
\begin{aligned}
& x^T(k)G_1x(k) - x^T(k-h_1)G_1x(k-h_1) \\
&\quad - \sum_{i=k-h_1}^{k-1} y^T(i)G_1y(i) - 2\sum_{i=k-h_1}^{k-1} y^T(i)G_1x(i) = 0, \\
& x^T(k-h_1)G_2x(k-h_1) - x^T(k-h_k)G_2x(k-h_k) \\
&\quad - \sum_{i=k-h_k}^{k-h_1-1} y^T(i)G_2y(i) - 2\sum_{i=k-h_k}^{k-h_1-1} y^T(i)G_2x(i) = 0, \\
& x^T(k-h_k)G_3x(k-h_k) - x^T(k-h_2)G_3x(k-h_2) \\
&\quad - \sum_{i=k-h_2}^{k-h_k-1} y^T(i)G_3y(i) - 2\sum_{i=k-h_2}^{k-h_k-1} y^T(i)G_3x(i) = 0
\end{aligned}
$$

hold, where G_1, G_2 and G_3 are all symmetric matrices.

Adding the left-hand sides of the above three zero equations to (6.36) leads to

$$\Delta V_2(k) = \xi^T(k)\Xi_{21}\xi(k) - \sum_{i=k-h_1}^{k-1} \eta_2^T(i)\mathscr{R}_1\eta_2(i)$$
$$- \sum_{i=k-h_k}^{k-h_1-1} \eta_2^T(i)\mathscr{R}_2\eta_2(i) - \sum_{i=k-h_2}^{k-h_k-1} \eta_2^T(i)\mathscr{R}_3\eta_2(i), \tag{6.37}$$

where Ξ_{21} is defined in (6.31).

With (6.27), applying (6.23) to the $\mathscr{R}_1$-dependent summation term leads to:

$$- \sum_{i=k-h_1}^{k-1} \eta_2^T(i)\mathscr{R}_1\eta_2(i) \leq -\frac{1}{h_1}\chi_{01}^T\mathscr{R}_1\chi_{01} - \frac{3}{h_1}\chi_{11}^T\mathscr{R}_1\chi_{11}, \tag{6.38}$$

where $\chi_{p1} = F_{p1}\xi(k)$, $p \in \{0, 1\}$.

It follows from (6.25) to (6.26) that $\mathscr{R}_2 > 0$ and $\mathscr{R}_3 > 0$. Then, applying the GFMB inequality (6.22) to the $\mathscr{R}_2$- and $\mathscr{R}_3$-dependent summation terms leads to:

$$- \sum_{i=k-h_k}^{k-h_1-1} \eta_2^T(i)\mathscr{R}_2\eta_2(i)$$
$$\leq \sum_{p=0}^{1} \frac{h_{k1}}{2p+1}\omega_{p2}^T N_{p2}\mathscr{R}_2^{-1}N_{p2}^T\omega_{p2} + \sum_{p=0}^{1} \mathrm{Sym}\{\omega_{p2}^T N_{p2}\chi_{p2}\}, \tag{6.39}$$

$$- \sum_{i=k-h_2}^{k-h_k-1} \eta_2^T(i)\mathscr{R}_3\eta_2(i)$$
$$\leq \sum_{p=0}^{1} \frac{h_{2k}}{2p+1}\omega_{p3}^T N_{p3}\mathscr{R}_3^{-1}N_{p3}^T\omega_{p3} + \sum_{p=0}^{1} \mathrm{Sym}\{\omega_{p3}^T N_{p3}\chi_{p3}\}, \tag{6.40}$$

where $\chi_{pq} = F_{pq}\xi(k)$ and $\omega_{pq} = g_{pq}\xi(k)$, $p \in \{0, 1\}$, $q \in \{2, 3\}$.

Combining (6.38), (6.39) and (6.40) with (6.37) yields:

$$\Delta V_2(k) \leq \xi^T(k)\Big(\Xi_{21} + \Xi_{22}(h_k) + \Xi_{23}(h_k)\Big)\xi(k), \tag{6.41}$$

where $\Xi_{22}(h_k)$ is defined in (6.32), and

$$\Xi_{23}(h_k) = \sum_{p=0}^{1} \frac{h_{k1}}{2p+1} g_{p2}^T N_{p2}\mathscr{R}_2^{-1}N_{p2}^T g_{p2} + \sum_{p=0}^{1} \frac{h_{2k}}{2p+1} g_{p3}^T N_{p3}\mathscr{R}_3^{-1}N_{p3}^T g_{p3}.$$

The following inequalities hold from (6.20):

$$u_i(s) \geq 0, \quad i \in \{1, 2, 3, 4\},$$
$$t_j(s_1, s_2) \geq 0, \quad j \in \{1, 2, 3\},$$

where $u_i(s) = 2[K_1x(s) - f(x(s))]^T U_i[f(x(s)) - K_2x(s)]$ and $t_j(s_1, s_2) = 2[K_1(x(s_1) - x(s_2)) - f(x(s_1)) + f(x(s_2))]^T T_j[f(x(s_1)) - f(x(s_2)) - K_2(x(s_1) - x(s_2))]$ with $U_i = \text{diag}\{u_{1i}, u_{2i}, \ldots, u_{ni}\} \geq 0$ and $T_j = \text{diag}\{t_{1j}, t_{2j}, \ldots, t_{nj}\} \geq 0$.

As a result, the following inequalities

$$\begin{aligned} &u_1(k) + u_2(k - h_1) + u_3(k - h_k) + u_4(k - h_2) \\ &+t_1(k, k - h_1) + t_2(k - h_1, k - h_k) + t_3(k - h_k, k - h_2) \geq 0 \end{aligned}$$

hold, which leads to

$$\xi^T(k)\Xi_3\xi(k) \geq 0,$$

where Ξ_3 is defined in (6.33).

According to the above discussions, the forward difference of $V(k)$ defined in (6.24) is computed as

$$\Delta V(k) \leq \xi^T(k)\Big(\Psi(h_k) + \Xi_{23}(h_k)\Big)\xi(k),$$

where $\Psi(h_k)$ is defined in (6.28).

Define $\Gamma(h_k) = \Psi(h_k) + \Xi_{23}(h_k)$. Since $\Gamma(h_k)$ is affine with the time delay h_k, the endpoint restrictions $\Gamma(h_1) < 0$ and $\Gamma(h_2) < 0$ surely guarantee $\Gamma(h_k) < 0$ with $h_k \in [h_1, h_2]$.

From the Schur complement, it is seen that the two inequalities (6.25) and (6.26) are, respectively, equivalent to $\Gamma(h_1) < 0$ and $\Gamma(h_2) < 0$. Therefore, there exists a sufficient small scalar $\varepsilon > 0$ such that $\Delta V(k) \leq -\varepsilon\|x(k)\|^2 < 0$ for any $x(k) \neq 0$, which means that neural network (6.19) is asymptotically stable. This completes the proof. ■

Remark 6.4 In developing Theorem 6.2, the GFMB summation inequality (6.22) is utilized to estimate the single-summation terms $-\sum_{i=k-h_k}^{k-h_1-1} \eta_2^T(i)\mathscr{R}_2\eta_2(i)$ and $-\sum_{i=k-h_2}^{k-h_k-1} \eta_2^T(i)\mathscr{R}_3\eta_2(i)$, respectively.

In the resulting bounds, each ω_{pq}, $p \in \{0, 1\}$, $q \in \{2, 3\}$, can be chosen as the whole $\xi(k)$ or part of it. Generally speaking, if more part is contained in ω_{pq}, a more relaxed condition can be obtained. By considering the expressions of χ_p, we study two cases for the choices of ω_{pq}, where $\omega_{pq} = g_{pq}\xi(k)$:

CASE I:

$$g_{02} = col\{e_2, e_3, e_{10}\}, \quad g_{12} = col\{e_2, e_3, e_{10}, e_{13}\},$$
$$g_{03} = col\{e_3, e_4, e_{11}\}, \quad g_{13} = col\{e_3, e_4, e_{11}, e_{14}\}.$$

Table 6.3 The MAUBs h_2 for different h_1 in Example 3

h_1	2	4	6	8	10	20	NVs
Corollary 3.3 [38]	30	32	34	36	38	48	$22.5n^2 + 4.5n$
Corollary 3.3 [37]	32	34	36	38	40	52	$20n^2 + 14n$
Theorem 1 [43]	99	101	103	105	107	117	$13.5n^2 + 11.5n$
Theorem 6.2 (Case I)	3119	3121	3123	3125	3127	3137	$42n^2 + 14n$
Theorem 6.2 (Case II)	3120	3122	3124	3126	3128	3138	$126n^2 + 14n$

CASE II: Each g_{pg} is set to g_{all}.

In Case I, free matrix variables $N_{02}, N_{03} \in \mathbb{R}^{3n\times 2n}$ and $N_{12}, N_{13} \in \mathbb{R}^{4n\times 2n}$ are involved in Theorem 6.2 while in Case II, free matrix variables $N_{02}, N_{03}, N_{12}, N_{13} \in \mathbb{R}^{14n\times 2n}$ are involved. It is easily seen that when choosing different expressions of ω_{pq}, different numbers of variables are involved. Furthermore, choosing different expressions of ω_{pq} may lead to different conservative conditions. The following numerical examples will show that the Case II-based stability condition produces larger maximum allowable upper bounds (MAUBs) than the Case I-based one.

6.3.3 Numerical Examples

Example 3 Consider the discrete-time neural network (6.19) with

$$C = \begin{bmatrix} 0.1 & 0 \\ 0 & 0.3 \end{bmatrix}, \quad A = \begin{bmatrix} 0.02 & 0 \\ 0 & 0.004 \end{bmatrix}, \quad A_d = \begin{bmatrix} -0.01 & 0.01 \\ -0.02 & -0.01 \end{bmatrix},$$
$$K_1 = \text{diag}\{1, 1\}, \quad K_2 = \text{diag}\{0, 0\}.$$

In order to show the effectiveness of the proposed approach, the MAUBs h_2 for different h_1 in this example are carefully checked by Theorem 6.2 proposed in this chapter and some of the existing conditions, including the number of decision variables (NVs). From Table 6.3, it is seen that the MAUBs obtained by Case I of Theorem 6.2 are remarkably larger than those obtained by other conditions, which means that Case I is less conservative (although at the price of some more variables). It is noted that compared with Case I, Case II improves the results a little but with a much larger number of decision variables.

Example 4 Consider discrete-time neural network (6.19) with

$$C = \begin{bmatrix} 0.8 & 0 \\ 0 & 0.9 \end{bmatrix}, \quad A = \begin{bmatrix} 0.001 & 0 \\ 0 & 0.005 \end{bmatrix}, \quad A_d = \begin{bmatrix} -0.1 & 0.01 \\ -0.2 & -0.1 \end{bmatrix},$$
$$K_1 = \text{diag}\{1, 1\}, \quad K_2 = \text{diag}\{0, 0\}.$$

Table 6.4 The MAUBs h_2 for different h_1 in Example 4

h_1	4	6	8	10	12	15	NVs
Theorem 1 [26]	20	20	21	21	21	23	$61.5n^2 + 17.5n$
Corollary 1 [31]	20	20	21	21	22	23	$44n^2 + 13n$
Theorem 1 [43]	20	20	21	22	22	24	$13.5n^2 + 11.5n$
Theorem 6.2 (Case I)	20	20	21	22	23	24	$42n^2 + 14n$
Theorem 6.2 (Case II)	20	21	21	22	23	24	$126n^2 + 14n$

Table 6.5 The MAUBs h_2 for different h_1 in Example 5

h_1	2	4	6	8	10	NVs
Theorem 1 [43]	9	11	12	14	16	$13.5n^2 + 11.5n$
Theorem 1 [26]	18	19	21	23	25	$61.5n^2 + 17.5n$
Theorem 6.2 (Case I)	18	19	21	23	25	$42n^2 + 14n$
Theorem 6.2 (Case II)	18	19	21	23	25	$126n^2 + 14n$

This example is widely used in the literature to compare the conservatism of different stability conditions. The relevant comparison results checked by Theorem 6.2 and some of existing conditions are listed in Table 6.4. It is seen that both cases of Theorem 6.2 are less conservative than other conditions since MAUBs obtained by Theorem 6.2 are larger than or at least equal to those obtained by other conditions. It is found that when $h_1 = 6$, the MAUB obtained by Case II is larger than that obtained by Case I, which further verifies some conclusions stated in Remark 6.4. However, this little improvement is achieved at the cost of a much larger number of variables. Therefore, Case I may be a priority choice if the computational burden is especially concerned.

Example 5 Consider discrete-time neural network (6.19) with

$$C = \begin{bmatrix} 0.4 & 0 & 0 \\ 0 & 0.3 & 0 \\ 0 & 0 & 0.1 \end{bmatrix}, \quad A = \begin{bmatrix} -0.2 & -0.2 & 0.1 \\ 0 & -0.3 & 0.2 \\ -0.2 & -0.1 & -0.2 \end{bmatrix}, \quad A_d = \begin{bmatrix} -0.2 & 0.1 & 0 \\ -0.2 & 0.3 & 0.1 \\ 0.1 & -0.2 & 0.3 \end{bmatrix},$$
$$K_1 = \mathrm{diag}\{0.6, 0.3, 0\}, \quad K_2 = \mathrm{diag}\{0, -0.4, -0.2\}.$$

The comparison results are checked by results in [26, 43] and both cases of Theorem 6.2. From Table 6.5, it is seen that all of the MAUBs obtained by Case I are larger than the corresponding ones obtained by the condition proposed in [43]. Meanwhile, it is also seen that although the MAUBs obtained by Case I are equal to the corresponding ones obtained by the condition proposed in [26], the number of decision variables involved in Case I is smaller than that involved in [26]. These findings clearly show the improvement of the proposed method.

6.4 Conclusion

This chapter has studied the stability problem of neural networks with a time-varying delay. For both continuous- and discrete-time cases, new L–K functionals are constructed in which the involved vectors are augmented by adding state- and delay-related vectors. During the process of dealing with the time derivative (or the forward difference) of L–K functionals, the zero-equation technique, and GFMB inequalities are utilized. Therefore, new stability conditions have been obtained, respectively, for continuous- and discrete-time neural networks with a time-varying delay. Several numerical examples have demonstrated the effectiveness of the proposed approach.

References

1. Liu GP (2012) Nonlinear identification and control: a neural network approach. Springer Science & Business Media, Berlin
2. Cao J, Ho DW (2005) A general framework for global asymptotic stability analysis of delayed neural networks based on LMI approach. Chaos Solitons Fractals 24:1317–1329
3. Liu Y, Guo BZ, Park JH, Lee SM (2018) Non-fragile exponential synchronization of delayed complex dynamical networks with memory sampled-data control. IEEE Trans Neural Netw Learn Syst 29:118–128
4. Tang Z, Park JH, Feng J (2018) Impulsive effects on quasi-synchronization of neural networks with parameter mismatches and time-varying delay. IEEE Trans Neural Netw Learn Syst 29:908–919
5. Wei Y, Park JH, Karimi HR, Tian YC, Jung HY (2018) Improved stability and stabilization results for stochastic synchronization of continuous-time semi-Markovian jump neural networks with time-varying delay. IEEE Trans Neural Netw Learn Syst 29:2488–2501
6. Lee TH, Trinh HM, Park JH (2018) Stability analysis of neural networks with time-varying delay by constructing novel Lyapunov functionals. IEEE Trans Neural Netw Learn Syst 29:4238–4247
7. Zhang R, Zeng D, Park JH, Liu Y, Zhong S (2018) Quantized sampled-data control for synchronization of inertial neural networks with heterogeneous time-varying delays. IEEE Trans Neural Netw Learn Syst 29:6385–6395
8. Xiong W, Cao J (2005) Global exponential stability of discrete-time Cohen–Grossberg neural networks. Neurocomputing 64:433–446
9. Zhang B, Xu S, Zou Y (2008) Improved delay-dependent exponential stability criteria for discrete-time recurrent neural networks with time-varying delays. Neurocomputing 72:321–330
10. Li H, Lam J, Cheung KC (2012) Passivity criteria for continuous-time neural networks with mixed time-varying delays. Appl Math Comput 218:11062–11074
11. Mahmoud MS, Xia Y (2011) Improved exponential stability analysis for delayed recurrent neural networks. J Frankl Inst 348:201–211
12. Shi P, Li F, Wu L, Lim CC (2017) Neural network-based passive filtering for delayed neutral-type semi-Markovian jump systems. IEEE Trans Neural Netw Learn Syst 28:2101–2114
13. Wang T, Gao H, Qiu J (2016) A combined adaptive neural network and nonlinear model predictive control for multirate networked industrial process control. IEEE Trans Neural Netw Learn Syst 27:416–425
14. Wang T, Xue M, Fei S, Li T (2013) Triple Lyapunov functional technique on delay-dependent stability for discrete-time dynamical networks. Neurocomputing 122:221–228

15. Wu Z, Su H, Chu J, Zhou W (2009) New results on robust exponential stability for discrete recurrent neural networks with time-varying delays. Neurocomputing 72:3337–3342
16. Wu Z, Su H, Chu J, Zhou W (2010) Improved delay-dependent stability condition of discrete recurrent neural networks with time-varying delays. IEEE Trans Neural Netw 21:692–697
17. Zhang XM, Han QL (2009) New Lyapunov–Krasovskii functionals for global asymptotic stability of delayed neural networks. IEEE Trans Neural Netw 20:533–539
18. Zhang XM, Han QL (2013) Global asymptotic stability for a class of generalized neural networks with interval time-varying delays. IEEE Trans Neural Netw 22:1180–1192
19. Zhang XM, Han QL, Seuret A, Gouaisbaut F (2017) An improved reciprocally convex inequality and an augmented Lyapunov–Krasovskii functional for stability of linear systems with time-varying delay. Automatica 84:221–226
20. Zhang XM, Han QL, Zeng Z (2018) Hierarchical type stability criteria for delayed neural networks via canonical Bessel–Legendre inequalities. IEEE Trans Cybern 48:1660–1671
21. Zhang H, Yang F, Liu X, Zhang Q (2013) Stability analysis for neural networks with time-varying delay based on quadratic convex combination. IEEE Trans Neural Netw Learn Syst 24:513–521
22. Zhang L, Zhu Y, Zheng WX (2017) State estimation of discrete-time switched neural networks with multiple communication channels. IEEE Trans Cybern 47:1028–1040
23. Wang J, Zhang XM, Han QL (2016) Event-triggered generalized dissipativity filtering for neural networks with time-varying delays. IEEE Trans Neural Netw Learn Syst 27:77–88
24. Chen J, Park JH, Xu S (2019) Stability analysis for neural networks with time-varying delay via improved techniques. IEEE Trans Cybern. https://doi.org/10.1109/TCYB.2018.2868136
25. Ding S, Wang Z, Wu Y, Zhang H (2016) Stability criterion for delayed neural networks via Wirtinger-based multiple integral inequality. Neurocomputing 214:53–60
26. Kwon OM, Park MJ, Park JH, Lee SM, Cha EJ (2013) New criteria on delay-dependent stability for discrete-time neural networks with time-varying delays. Neurocomputing 121:185–194
27. Li T, Wang T, Song A, Fei S (2013) Combined convex technique on delay-dependent stability for delayed neural networks. IEEE Trans Neural Netw Learn Syst 24:1459–1466
28. Zhang CK, He Y, Jiang L, Lin WJ, Mu M (2017) Delay-dependent stability analysis of neural networks with time-varying delay: a generalized free-weighting-matrix approach. Appl Math Comput 294:102–120
29. Zhang CK, He Y, Jiang L, Wu M (2016) Stability analysis for delayed neural networks considering both conservativeness and complexity. Appl Math Comput 27:1486–1501
30. Zeng HB, He Y, Wu M, Xiao SP (2015) Stability analysis of generalized neural networks with time-varying delays via a new integral inequality. Neurocomputing 161:148–154
31. Feng Z, Zheng WX (2015) On extended dissipativity of discrete-time neural networks with time delay. IEEE Trans Neural Netw Learn Syst 26:3293–3300
32. Ge C, Hua C, Guan X (2014) New delay-dependent stability criteria for neural networks with time-varying delay using delay-decomposition approach. IEEE Trans Neural Netw Learn Syst 25:1378–1383
33. Liu Y, Lee SM, Kwon OM, Park JH (2015) New approach to stability criteria for generalized neural networks with interval time-varying delays. Neurocomputing 149:1544–1551
34. Kim SH, Park P, Jeong C (2010) Robust $\mathscr{H}_\infty$ stabilisation of networked control systems with packet analyser. IET Control Theory Applications 4:1828–1837
35. Lee SY, Lee WI, Park P (2017) Improved stability criteria for linear systems with interval time-varying delays: generalized zero equalities approach. Appl Math Comput 292:336–348
36. Chen J, Xu S, Chen W, Zhang B, Ma Q, Zou Y (2016) Two general integral inequalities and their applications to stability analysis for systems with time-varying delay. Int J Robust Nonlinear Control 26:4088–4103
37. Banu LJ, Balasubramaniam P (2016) Robust stability analysis for discrete-time neural networks with time-varying leakage delays and random parameter uncertainties. Neurocomputing 179:126–134
38. Banu LJ, Balasubramaniam P, Ratnavelu K (2015) Robust stability analysis for discrete-time uncertain neural networks with leakage time-varying delay. Neurocomputing 151:808–816

39. Mathiyalagan K, Sakthivel R, Anthoni SM (2012) Exponential stability result for discrete-time stochastic fuzzy uncertain neural networks. Phys Lett A 376:901–912
40. Song Q, Wang Z (2007) A delay-dependent LMI approach to dynamics analysis of discrete-time recurrent neural networks with time-varying delays. Phys Lett A 368:134–145
41. Shu Y, Liu X, Liu Y (2016) Stability and passivity analysis for uncertain discrete-time neural networks with time-varying delay. Neurocomputing 173:1706–1714
42. Tian J, Zhong S (2011) Improved delay-dependent stability criterion for neural networks with time-varying delay. Appl Math Comput 217:10278–10288
43. Zhang CK, He Y, Jiang L, Wang QG, Wu M (2017) Stability analysis of discrete-time neural networks with time-varying delay via an extended reciprocally convex matrix inequality. IEEE Trans Cybern 47:3040–3049
44. Wang Z, Liu L, Shan QH, Zhang H (2015) Stability criteria for recurrent neural networks with time-varying delay based on secondary delay partitioning method. IEEE Trans Neural Netw Learn Syst 26:2589–2595
45. Song C, Gao H, Zheng WX (2009) A new approach to stability analysis of discrete-time recurrent neural networks with time-varying delay. Neurocomputing 72:2563–2568
46. Chen J, Xu S, Ma Q, Li Y, Chu Y, Zhang Z (2017) Two novel general summation inequalities to discrete-time systems with time-varying delay. J Frankl Inst 354:5537–5558

Part III
Control Problems of Dynamics Systems with Time-Delays

Chapter 7
$\mathscr{H}_\infty$ Control for the Stabilization of Neural Networks with Time-Varying Delay

7.1 Introduction

Neural networks are essential in engineering because of their wide applications such as image processing, pattern recognition, optimization problem, and learning algorithm [1–13]. To apply the neural networks to these applications, the stability must be guaranteed, so the stabilization problem has become a hot topic [14–17]. In [14], the global exponential stabilization of neural networks was investigated in which various activation functions and time-varying continuously distributed delays were considered. In [15], the sampled-data controller was designed for the exponential stabilization of memristive neural networks under actuator saturation by defining a class of logical switched functions which help to construct a tractable model. In [16], the impulsive controllers were designed for both stabilization and synchronization of discrete-time delayed neural networks and delay-independent but impulse-dependent conditions for designing controller were derived by using the time-varying Lyapunov functional approach and a convex combination technique. In [17], the pinning control method was applied for the stabilization of linearly coupled stochastic neural networks and also a minimum number of controllers was proposed.

In practice, there are a number of factors which harms the stability of the neural networks, for example, time-delay, parametric uncertainty, noise, fault, and so on. Especially, time-delay is unavoidable because neural networks consist of a very large number of neurons and layers and that definitely occurs delays in communication between neurons. To handle time-delay, there are a number of useful techniques have been reported till now [18–23]. By utilizing the developed techniques, the stability and stabilization of neural networks with time-delay are one of important issues and has been deeply taken into account in numerous papers [24–31]. In [24], the delay-dependent stability for continuous neural networks with a time-varying delay in the view-point of the tradeoff between conservativeness and calculation complexity by using a new delay-dependent Lyapunov–Krasovskii functional and free-weighting matrix. Authors in [25] proposed a newly augmented Lyapunov–Krasovskii functional and a new approach namely activation bounding partition approach for the

J. H. Park et al., *Dynamic Systems with Time Delays: Stability and Control*,
https://doi.org/10.1007/978-981-13-9254-2_7

stability of neural networks with time-varying delay. The paper [26] developed an extended reciprocally convex matrix inequality which reduces the conservatism without increase computational burden, and then based on the developed approach and a delay-product-type Lyapunov functionals, a delay-variation-dependent stability criterion was established. In [27], two novel Lyapunov functional which consisted of a positive term and a non-positive term unlike general Lyapunov functional were constructed for the stability of neural networks with time-varying delay.

The external disturbances (or noises) are ubiquitous, destroy stability, and hinder the performance of the system. One solution can be $\mathscr{H}_\infty$ control method which reduces the effects of external disturbances on the system to a certain acceptable value namely $\mathscr{H}_\infty$ performance. $\mathscr{H}_\infty$ control method have adopt on the stabilization of neural networks with external disturbance [32–34].

With above motivations, this chapter considers neural networks with time-varying delay and external disturbance and focus on developing a criterion for designing $\mathscr{H}_\infty$ controller such that the designed controller attenuates the effect of the external disturbance on the system under given level by aid of the Lyapunov stability theory, augmented Lyapunov functional, and delay partitioning approach.

7.2 Problem Formulation

In this chapter, we are interested in the following time-varying delayed neural networks:

$$\begin{aligned} \dot{z}(t) &= -Az(t) + Bg(z(t)) + Cg(z(t-h(t))) + Dw(t) + u(t), \\ z(t) &= \phi(t), \quad \forall t \in [-h\ 0], \end{aligned} \tag{7.1}$$

where $z(t) \in \mathbb{R}^n$ is the neuron state vector, n is the number of neurons in a neural network, $u(t) \in \mathbb{R}^n$ is control inputs, $w(t) \in \mathbb{R}^n$ is external disturbance belonging to $\mathscr{L}_2[0, \infty)$, $g(z(t)) = [g_1(z_1(t)), \ldots, g_n(z_n(t))]^T \in \mathbb{R}^n$ means the neuron activation function, $A, B, C, D \in \mathbb{R}^{n\times n}$ are known constant matrices, $h(t)$ is a time-varying delay satisfying $0 \le h(t) \le h$ and $|\dot{h}(t)| \le \mu < 1$ where h and μ are known positive constants, and $\phi(t)$ is a vector-valued continuous function.

To reduce the effect of the external disturbance rejection to the system, we consider the following state feedback controller

$$u(t) = Kz(t). \tag{7.2}$$

Before deriving our main result, an assumption, definition, and two lemmas are given as follows:

Assumption 7.1 Each activation function $g_i(\cdot)$ is continuous, bounded, and satisfies the following:

$$l_i^- \leq \frac{g_i(\alpha) - g_i(\beta)}{\alpha - \beta} \leq l_i^+, \quad i = 1, 2, \ldots, n$$

where $g(0) = 0$, $\alpha, \beta \in \mathbb{R}$, $\alpha \neq \beta$, and l_i^- and l_i^+ are known real scalars.

Definition 7.1 A system is said to be stable with $\mathscr{H}_\infty$ performance if the following conditions are satisfied:

- With zero disturbance, the system is asymptotically stable.
- With zero initial condition and for a given positive constant γ, the following condition holds:

$$\int_0^\infty z^T(t)z(t)dt < \gamma^2 \int_0^\infty w^T(t)w(t)dt,$$

where $z(t)$ is the state vector of the system and $w(t)$ is disturbance in the system which belongs to $\mathscr{L}_2[0, \infty)$.

Lemma 7.1 (Wirtinger-based integral inequality [35]) *For given a matrix $S \in \mathbb{S}_+^n$, and all continuous function z in $[a, b] \to \mathbb{R}^n$ the following inequality holds:*

$$\int_a^b \dot{z}^T(s)S\dot{z}(s)ds \geq \frac{1}{b-a}\varpi^T(t)\mathscr{S}\varpi(t)$$

where

$$\mathscr{S} = \text{diag}\{S, 3S\},$$
$$\varpi(t) = \begin{bmatrix} z(b) - z(a) \\ z(b) + z(a) - \frac{2}{b-a}\int_a^b z(s)ds \end{bmatrix}.$$

Lemma 7.2 (Relaxed Wirtinger-based Integral Inequality [36]) *For given positive constants $a \leq b \leq c$, any differentiable continuous vector $z(t) \in \mathbb{R}^n$, matrices $S \in \mathbb{S}^n$, and $F \in \mathbb{R}^{2n \times 2n}$, the following inequality holds:*

$$\int_a^c \dot{z}^T(s)S\dot{z}(s)ds \geq \frac{1}{c-a}\chi^T(t)(\mathscr{S} - \mathscr{F})\chi(t),$$

where

$$\chi(t) = \begin{bmatrix} z(c) - z(b) \\ z(c) + z(b) - \frac{2}{c-b}\int_b^c z(s)ds \\ z(b) - z(a) \\ z(b) + z(a) - \frac{2}{b-a}\int_a^b z(s)ds \end{bmatrix},$$
$$\mathscr{S} = \begin{bmatrix} S_d & F \\ * & S_d \end{bmatrix},$$

$$\mathscr{F} = \begin{bmatrix} \frac{b-a}{c-a}(S_d - FS_d^{-1}F^T) & 0 \\ 0 & \frac{c-b}{c-a}(S_d - F^T S_d^{-1} F) \end{bmatrix},$$

$$S_d = \text{diag}\{S, 3S\}.$$

7.3 Main Results

In this section, we propose the existence criterion for a $\mathscr{H}_\infty$ controller for the neural networks (7.1) by use of delay partitioning technique in which the delay range is divided into two unequal subintervals, i.e. $[0, \ \alpha h]$ and $[\alpha h, \ h]$ where $0 < \alpha < 1$.

Before we begin, we define the block entry matrices as $r_i \in \mathbb{R}^{16n \times n}$ $(i = 1, 2, \ldots, 16)$, for example, $r_2^T = [0, \ I_n, \ \underbrace{0, \ \ldots, \ 0}_{14}]$.

Let us consider positive scalars $\lambda_{1i}, \ \lambda_{2i}$ where $i = 1, \ldots, n$, positive definite matrices $P, \ Q_1, \ Q_2, \ Q_3, \ Q_4, \ Q_5, \ R_1, \ R_2, \ R_3$ with appropriate dimensions, and define the following Lyapunov functional candidate:

$$V(t) = V_1(t) + V_2(t) + V_3(t) + V_4(t), \tag{7.3}$$

where

$$\begin{aligned}
V_1(t) &= +2\sum_{i=1}^{n}\left(\lambda_{1i}\int_0^{z_i(t)}(g_i(s) - l_i^- s)ds + \lambda_{2i}\int_0^{z_i(t)}(l_i^+ s - g_i(s))ds\right), \\
V_2(t) &= \xi_1^T(t)P\xi_1(t), \\
V_3(t) &= \int_{t-h(t)}^{t}\xi_2^T(s)Q_1\xi_2(s)ds + \int_{t-h}^{t-h(t)}\xi_2^T(s)Q_2\xi_2(s)ds \\
&\quad + \int_{t-h}^{t}\xi_2^T(s)Q_3\xi_2(s)ds + \int_{t-\alpha h}^{t}\xi_2^T(s)Q_4\xi_2(s)ds \\
&\quad + \int_{t-h}^{t-\alpha h}\xi_2^T(s)Q_5\xi_2(s)ds, \\
V_4(t) &= \int_{t-h}^{t}\int_{v}^{t}\dot{z}^T(s)R_1\dot{z}(s)dsdv + \int_{t-\alpha h}^{t}\int_{v}^{t}\dot{z}^T(s)R_2\dot{z}(s)dsdv \\
&\quad + h\int_{t-h}^{t-\alpha h}\int_{v}^{t}\dot{z}^T(s)R_3\dot{z}(s)dsdv,
\end{aligned}$$

with

$$\begin{aligned}
\xi_1(t) = \Big[& z^T(t), \ z^T(t-h(t)), \ z^T(t-h), \ z^T(t-\alpha h), \\
& \textstyle\int_{t-h(t)}^{t} z^T(s)ds, \ \int_{t-h}^{t-h(t)} z^T(s)ds, \ \int_{t-\alpha h}^{t} z^T(s)ds,
\end{aligned}$$

$$\int_{t-h}^{t-\alpha h} z^T(s)ds \Big]^T,$$

$$\xi_2(t) = \big[\, z^T(t),\ \dot{z}^T(t),\ g^T(z(t)) \,\big]^T.$$

The time derivative of Lyapunov functionals can be calculated as:

$$\dot{V}_1(t) = 2\dot{z}^T(t)\Big(\Lambda_1\big(g(z(t)) - L_m z(t)\big) + \Lambda_2\big(L_p z(t) - g(z(t))\big)\Big), \tag{7.4}$$

$$\dot{V}_2(t) = 2\xi_1^T(t)P\xi_3(t), \tag{7.5}$$

$$\begin{aligned}\dot{V}_3(t) &= \xi_2^T(t)\Big(Q_1 + Q_3 + Q_4\Big)\xi_2(t)\\ &\quad + h_D(t)\xi_2^T(t-h(t))\Big(Q_2 - Q_1\Big)\xi_2(t-h(t))\\ &\quad - \xi_2^T(t-h)\Big(Q_2 + Q_3 + Q_5\Big)\xi_2(t-h)\\ &\quad + \xi_2^T(t-\alpha h)\Big(Q_5 - Q_4\Big)\xi_2(t-\alpha h),\end{aligned} \tag{7.6}$$

$$\begin{aligned}\dot{V}_4(t) &= h\dot{z}^T(t)\Big(R_1 + \alpha R_2 + (1-\alpha)R_3\Big)\dot{z}(t) - \int_{t-h}^{t} \dot{z}^T(s)R_1\dot{z}(s)ds\\ &\quad - \int_{t-\alpha h}^{t} \dot{z}^T(s)R_2\dot{z}(s)ds - \int_{t-h}^{t-\alpha h} \dot{z}^T(s)R_3\dot{z}(s)ds,\end{aligned} \tag{7.7}$$

where

$$\begin{aligned}\xi_3(t) = \Big[\, &\dot{z}^T(t),\ h_D(t)\dot{z}^T(t-h(t)),\ \dot{z}^T(t-h),\ \dot{z}^T(t-\alpha h),\\ &z^T(t) - h_D(t)z(t-h(t)),\ h_D(t)z^T(t-h(t)) - z(t-h),\\ &z^T(t) - z(t-\alpha h),\ z^T(t-\alpha h) - z(t-h)\Big]^T,\end{aligned}$$

$$\begin{aligned}\Lambda_1 &= \mathrm{diag}\{\lambda_{11}, \ldots, \lambda_{1n}\},\\ \Lambda_2 &= \mathrm{diag}\{\lambda_{21}, \ldots, \lambda_{2n}\},\\ L_m &= \mathrm{diag}\{l_1^-, \ldots, l_n^-\},\\ L_p &= \mathrm{diag}\{l_1^+, \ldots, l_n^+\}.\end{aligned}$$

From Assumption 7.1, for $n \times n$-dimensional positive diagonal matrices U_i $(i = 1, 2, 3, 4)$ and W_i $(i = 1, 2, 3, 4, 5)$, the followings are true:

$$\begin{aligned}u_i(a) &= 2(L_m z(a) - g(z(a)))^T U_i(g(z(a)) - L_p z(a)) \geq 0,\\ w_i(a,b) &= 2\big[L_m(z(a) - z(b)) - (g(z(a)) - g(z(b)))\big]^T W_i\\ &\quad \times \big[(g(z(a)) - g(z(b))) - L_p(z(a) - z(b))\big] \geq 0,\end{aligned}$$

where $a, b \in \mathbb{R}$.

Then, it is clear that:

$$\begin{aligned} 0 \le\ & u_1(t) + u_2(t-h(t)) + u_3(t-h) + u_4(t-\alpha h) \\ & + w_1(t, t-h(t)) + w_2(t-h(t), t-h) + w_3(t, t-h) \\ & + w_4(t-\alpha h, t-h) + w_5(t-h(t), t-\alpha h). \end{aligned} \tag{7.8}$$

For a positive constant β and any matrix $H \in \mathbb{R}^{n\times n}$, the following equation (7.9) can be obtained by the system dynamics (7.1)

$$\begin{aligned} 0 =\ & 2(z^T(t) + \beta \dot{z}^T(t))H \\ & \times \Big[-\dot{z}(t) - Az(t) + Bg(z(t)) + Cg(z(t-h(t)) + Dw(t) + Kz(t)\Big]. \end{aligned} \tag{7.9}$$

Now, we are the place to apply the delay partitioning approach. We divide the delay range $[0,\ h]$ into two nonequal subintervals $[0,\ \alpha h]$ and $[\alpha h,\ h]$ where a given constant, $0 < \alpha < 1$, is for partitioning weight. When we consider the delay range as $[0,\ \alpha h]$ and $[\alpha h,\ h]$, two cases are faced regarding to time-varying delay $h(t)$.

First case is $0 \le h(t) \le \alpha h$ and the other case is $\alpha h \le h(t) \le h$. Firstly we consider the first case.

Case 1. $0 \le h(t) \le \alpha h$

In this case, the augmented vector in $V_2(t)$ can be rewritten as

$$\begin{aligned} \xi_1(t) = \Big[& z^T(t),\ z^T(t-h(t)),\ z^T(t-h),\ z^T(t-\alpha h), \\ & \textstyle\int_{t-\alpha h}^{t} z^T(s)ds + \int_{t-h(t)}^{t-\alpha h} z^T(s)ds,\ \int_{t-h}^{t-h(t)} z^T(s)ds, \\ & \textstyle\int_{t-\alpha h}^{t} z^T(s)ds,\ \int_{t-h(t)}^{t-\alpha h} z^T(s)ds + \int_{t-h}^{t-h(t)} z^T(s)ds \Big]^T. \end{aligned}$$

And the integral terms associate with matrices R_1 and R_2 in Eq. (7.7) can be rewritten as:

$$\begin{aligned} \int_{t-h}^{t} \dot{z}^T(s)R_1\dot{z}(s)ds =& \int_{t-h(t)}^{t} \dot{z}^T(s)R_1\dot{z}(s)ds + \int_{t-\alpha h}^{t-h(t)} \dot{z}^T(s)R_1\dot{z}(s)ds \\ & + \int_{t-h}^{t-\alpha h} \dot{z}^T(s)R_1\dot{z}(s)ds, \\ \int_{t-\alpha h}^{t} \dot{z}^T(s)R_2\dot{z}(s)ds =& \int_{t-h(t)}^{t} \dot{z}^T(s)R_2\dot{z}(s)ds + \int_{t-\alpha h}^{t-h(t)} \dot{z}^T(s)R_2\dot{z}(s)ds, \end{aligned}$$

then using above relation, the three integral terms in $\dot{V}_4(t)$ can be

$$
\begin{aligned}
&-\int_{t-h}^{t}\dot{z}^T(s)R_1\dot{z}(s)ds-\int_{t-\alpha h}^{t}\dot{z}^T(s)R_2\dot{z}(s)ds-\int_{t-h}^{t-\alpha h}\dot{z}^T(s)R_3\dot{z}(s)ds\\
&=-\int_{t-h(t)}^{t}\dot{z}^T(s)(R_1+R_2)\dot{z}(s)ds-\int_{t-\alpha h}^{t-h(t)}\dot{z}^T(s)(R_1+R_2)\dot{z}(s)ds\\
&\quad-\int_{t-h}^{t-\alpha h}\dot{z}^T(s)(R_1+R_3)\dot{z}(s)ds.
\end{aligned}
$$

By applying Lemma 7.1 to the integral term associated with $R_1 + R_3$ and Lemma 7.2 with any matrix $S_1 \in \mathbb{R}^{2n\times 2n}$ to the integral terms associated with $R_1 + R_2$, we can obtain the following equations:

$$
\begin{aligned}
&-\int_{t-h(t)}^{t}\dot{z}^T(s)(R_1+R_2)\dot{z}(s)ds-\int_{t-\alpha h}^{t-h(t)}\dot{z}^T(s)(R_1+R_2)\dot{z}(s)ds\\
&\le-\frac{1}{\alpha h}\xi_4^T(t)(\mathscr{R}_{1,[h(t)]}-\mathscr{T}_{1,[h(t)]})\xi_4(t),
\end{aligned}
\tag{7.10}
$$

and

$$
-\int_{t-h}^{t-\alpha h}\dot{z}^T(s)(R_1+R_3)\dot{z}(s)ds\le-\frac{1}{(1-\alpha)h}\xi_5^T(t)R_{13}^d\xi_5(t),
\tag{7.11}
$$

where

$$
\begin{aligned}
\xi_4(t)&=\begin{bmatrix} z(t)-z(t-h(t))\\ z(t)+z(t-h(t))-\frac{2}{h(t)}\int_{t-h(t)}^{t}z(s)ds\\ z(t-h(t))-z(t-\alpha h)\\ z(t-h(t))+z(t-\alpha h)-\frac{2}{\alpha h-h(t)}\int_{t-\alpha h}^{t-h(t)}z(s)ds\end{bmatrix},\\
\xi_5(t)&=\begin{bmatrix} z(t-\alpha h)-z(t-h)\\ z(t-\alpha h)+z(t-h)-\frac{2}{(1-\alpha)h}\int_{t-h}^{t-\alpha h}z(s)ds\end{bmatrix},\\
\mathscr{R}_{1,[h(t)]}&=\begin{bmatrix}\frac{2\alpha h-h(t)}{\alpha h}R_{12}^d & S_1\\ * & \frac{\alpha h+h(t)}{\alpha h}R_{12}^d\end{bmatrix},\\
\mathscr{T}_{1,[h(t)]}&=\begin{bmatrix}\frac{\alpha h-h(t)}{\alpha h}S_1(R_{12}^d)^{-1}S_1^T & 0\\ 0 & \frac{h(t)}{\alpha h}S_1^T(R_{12}^d)^{-1}S_1\end{bmatrix},\\
R_{12}^d&=\mathrm{diag}\{R_1+R_2,\ 3(R_1+R_3)\},\\
R_{13}^d&=\mathrm{diag}\{R_1+R_3,\ 3(R_1+R_3)\}.
\end{aligned}
$$

For deriving LMI condition, we define the following vector

$$\begin{aligned}\zeta_1(t) = \big[\, & z^T(t),\ z^T(t-h(t)),\ z^T(t-h),\ z^T(t-\alpha h),\\ & \dot{z}^T(t),\ \dot{z}^T(t-h(t)),\ \dot{z}^T(t-h),\ \dot{z}(t-\alpha h),\\ & \tfrac{1}{h(t)}\textstyle\int_{t-h(t)}^{t} z^T(s)ds,\ \tfrac{1}{(1-\alpha)h}\int_{t-h}^{t-\alpha h} z^T(s)ds,\\ & \tfrac{1}{\alpha h-h(t)}\textstyle\int_{t-\alpha h}^{t-h(t)} z^T(s)ds,\ g^T(z(t)),\ g^T(z(t-h(t))),\\ & g^T(z(t-h)),\ g^T(z(t-\alpha h)),\ w(t)\,\big]^T.\end{aligned}$$

Then, using block entry matrix defined in the beginning of this section and Eqs. (7.10) and (7.11), time derivative of each Lyapunov functionals (7.4)–(7.7) can be expressed as follows:

$$\dot{V}_1(t) = \zeta_1^T(t)\Omega_1\zeta_1(t), \tag{7.12}$$

$$\dot{V}_2(t) = \zeta_1^T(t)\Omega_{2,[h(t),\dot{h}(t)]}\zeta_1(t), \tag{7.13}$$

$$\dot{V}_3(t) = \zeta_1^T(t)\Omega_{3,[\dot{h}(t)]}\zeta_1(t), \tag{7.14}$$

$$\dot{V}_4(t) \le \zeta_1^T(t)\Big(\Omega_{4,[h(t)]} + \frac{1}{\alpha h}\Gamma_7\mathscr{T}_{1,[h(t)]}\Gamma_7^T\Big)\zeta_1(t), \tag{7.15}$$

where

$$\begin{aligned}\Gamma_{1,[h(t)]} &= \Big[r_1,\ r_2,\ r_3,\ r_4,\ h(t)r_9,\ (\alpha h - h(t))r_{11} + (1-\alpha)hr_{10},\\ &\qquad h(t)r_9 + (\alpha h - h(t))r_{11},\ (1-\alpha)r_{10}\Big],\\ \Gamma_{2,[\dot{h}(t)]} &= \Big[r_5,\ h_D(t)r_6,\ r_7,\ r_8,\ r_1 - h_D(t)r_2,\ h_D(t)r_2 - r_3,\\ &\qquad r_1 - r_4,\ r_4 - r_3\Big],\\ \Gamma_3 &= \big[r_1,\ r_5,\ r_{12}\big],\\ \Gamma_4 &= \big[r_2,\ r_6,\ r_{13}\big],\\ \Gamma_5 &= \big[r_3,\ r_7,\ r_{14}\big],\\ \Gamma_6 &= \big[r_4,\ r_8,\ r_{15}\big],\\ \Gamma_7 &= \big[\Gamma_{71}\ \Gamma_{72}\big],\\ \Gamma_{71} &= \big[r_1 - r_2,\ r_1 + r_2 - 2r_9\big],\\ \Gamma_{72} &= \big[r_2 - r_4,\ r_2 + r_4 - 2r_{11}\big],\\ \Gamma_8 &= \big[r_4 - r_3,\ r_4 + r_3 - 2r_{10}\big],\\ \Omega_1 &= \mathrm{Sym}\Big\{r_5\big(\Lambda_1(r_{12}^T - L_m r_1^T) + \Lambda_2(L_p r_1^T - r_{12}^T)\big)\Big\},\\ \Omega_{2,[h(t),\dot{h}(t)]} &= \mathrm{Sym}\Big\{\Gamma_{1,[h(t)]}P\Gamma_{2,[\dot{h}(t)]}^T\Big\},\end{aligned}$$

$$\Omega_{3,[\dot{h}(t)]} = \Big(\Gamma_3(Q_1 + Q_2 + Q_4)\Gamma_3^T + h_D(t)\Gamma_4(Q_2 - Q_1)\Gamma_4^T$$
$$- \Gamma_5(Q_2 + Q_3 + Q_5)\Gamma_5^T + \Gamma_6(Q_5 - Q_4)\Gamma_6^T\Big),$$
$$\Omega_{4,[h(t)]} = hr_5(R_1 + \alpha R_2 + (1-\alpha)R_3)r_5^T - \frac{1}{\alpha h}\Gamma_7\mathscr{R}_{1,[h(t)]}\Gamma_7^T$$
$$- \frac{1}{(1-\alpha)h}\Gamma_8 R_{13}^d \Gamma_8^T.$$

And, the positive Eq. (7.8) and zero Eq. (7.9) can be expressed as follows:

$$0 \leq \text{Right side of (7.8)} = \zeta_1^T(t)\Upsilon_1\zeta_1(t), \tag{7.16}$$
$$0 = \text{Right side of (7.9)} = \zeta_1^T(t)\Upsilon_2\zeta_1(t), \tag{7.17}$$

where

$$\Upsilon_1 = \text{Sym}\Big\{(r_1 L_m - r_{12})U_1(r_{12}^T - L_p r_1^T) + (r_2 L_m - r_{13})U_2(r_{13}^T - L_p r_2^T)$$
$$+ (r_3 L_m - r_{14})U_3(r_{14}^T - L_p r_3^T) + (r_4 L_m - r_{15})U_4(r_{15}^T - L_p r_4^T)\Big\}$$
$$+ \text{Sym}\Big\{\big((r_1 - r_2)L_m - (r_{12} - r_{13})\big)W_1\big((r_{12}^T - r_{13}^T) - L_p(r_1^T - r_2^T)\big)\Big\}$$
$$+ \text{Sym}\Big\{\big((r_2 - r_3)L_m - (r_{13} - r_{14})\big)W_2\big((r_{13}^T - r_{14}^T) - L_p(r_2^T - r_3^T)\big)\Big\}$$
$$+ \text{Sym}\Big\{\big((r_1 - r_3)L_m - (r_{12} - r_{14})\big)W_3\big((r_{12}^T - r_{14}^T) - L_p(r_1^T - r_3^T)\big)\Big\}$$
$$+ \text{Sym}\Big\{\big((r_4 - r_3)L_m - (r_{15} - r_{14})\big)W_4\big((r_{15}^T - r_{14}^T) - L_p(r_4^T - r_3^T)\big)\Big\}$$
$$+ \text{Sym}\Big\{\big((r_2 - r_4)L_m - (r_{13} - r_{15})\big)W_5\big((r_{13}^T - r_{15}^T) - L_p(r_2^T - r_4^T)\big)\Big\},$$
$$\Upsilon_2 = \text{Sym}\Big\{(r_1 + \beta r_5)(-Hr_5^T - HAr_1^T + HBr_{12}^T + HCr_{13}^T$$
$$+ HDr_{16}^T + Gr_1^T)\Big\},$$
$$G = HK.$$

Finally, by Eqs. (7.12)–(7.17), the upper bound of $\dot{V}(t)$ for Case 1 can be estimated as:

$$\dot{V}(t) \leq \zeta_1^T(t)\left(\Psi_{1,[h(t),\dot{h}(t)]} + \frac{1}{\alpha h}\Gamma_7\mathscr{T}_{1,[h(t)]}\Gamma_7^T\right)\zeta_1(t)$$
$$- z^T(t)z(t) + \gamma^2 w^T(t)w(t), \tag{7.18}$$

where

$$\Psi_{1,[h(t),\dot{h}(t)]} = \Omega_1 + \Omega_{2,[h(t),\dot{h}(t)]} + \Omega_{3,[\dot{h}(t)]} + \Omega_{4,[h(t)]} + \Upsilon_1 + \Upsilon_2 + r_1 r_1^T - \gamma^2 r_{16} r_{16}^T,$$

and γ is a positive constant for $\mathscr{H}_\infty$ performance.

Case 2. $\alpha h \le h(t) \le h$

In this case, the augmented vector in $V_2(t)$ can be rewritten as

$$\xi_1(t) = \Big[z^T(t),\, z^T(t-h(t)),\, z^T(t-h),\, z^T(t-\alpha h), \int_{t-h(t)}^{t} z^T(s)ds,\, \int_{t-\alpha h}^{t-h(t)} z^T(s)ds + \int_{t-h}^{t-\alpha h} z^T(s)ds, \int_{t-h(t)}^{t} z^T(s)ds + \int_{t-\alpha h}^{t-h(t)} z^T(s)ds,\, \int_{t-h}^{t-\alpha h} z^T(s)ds \Big]^T,$$

and let us define the following vector:

$$\zeta_2(t) = \big[z^T(t),\, z^T(t-h(t)),\, z^T(t-h),\, z^T(t-\alpha h), \dot{z}^T(t),\, \dot{z}^T(t-h(t)),\, \dot{z}^T(t-h),\, \dot{z}(t-\alpha h), \tfrac{1}{h-h(t)} \int_{t-h}^{t-h(t)} z^T(s)ds,\, \tfrac{1}{\alpha h} \int_{t-\alpha h}^{t} z^T(s)ds, \tfrac{1}{h(t)-\alpha h} \int_{t-h(t)}^{t-\alpha h} z^T(s)ds,\, g^T(z(t)),\, g^T(z(t-h(t))), g^T(z(t-h)),\, g^T(z(t-\alpha h)),\, w(t) \big]^T.$$

Then, the time derivative of Lyapunov functional $V_2(t)$ can be expressed:

$$\dot{V}_2(t) = \zeta_2^T(t) \Omega_{5,[h(t),\dot{h}(t)]} \zeta_2(t), \tag{7.19}$$

where

$$\Omega_{5,[h(t),\dot{h}(t)]} = \text{Sym}\Big\{ \Gamma_{9,[h(t)]} P \Gamma_{2,[\dot{h}(t)]}^T \Big\},$$

$$\Gamma_{9,[h(t)]} = \Big[r_1,\, r_2,\, r_3,\, r_4,\, \alpha h r_{10} + (h(t) - \alpha h) r_{11},\, (h - h(t)) r_9, \alpha h r_{10},\, (h(t) - \alpha h) r_{11} + (h - h(t)) r_9 \Big].$$

In addition, the integral terms associate with matrices R_1 and R_3 in Eq. (7.7) can be rewritten as:

$$\int_{t-h}^{t} \dot{z}^T(s) R_1 \dot{z}(s)ds = \int_{t-\alpha h}^{t} \dot{z}^T(s) R_1 \dot{z}(s)ds + \int_{t-h(t)}^{t-\alpha h} \dot{z}^T(s) R_1 \dot{z}(s)ds + \int_{t-h}^{t-h(t)} \dot{z}^T(s) R_1 \dot{z}(s)ds,$$

and

$$\int_{t-h}^{t-\alpha h} \dot{z}^T(s)R_3\dot{z}(s)ds = \int_{t-h(t)}^{t-\alpha h} \dot{z}^T(s)R_3\dot{z}(s)ds + \int_{t-h}^{t-h(t)} \dot{z}^T(s)R_3\dot{z}(s)ds,$$

then using above relation, the three integral terms in Eq. (7.7) can be

$$\begin{aligned}
&-\int_{t-h}^{t} \dot{z}^T(s)R_1\dot{z}(s)ds - \int_{t-\alpha h}^{t} \dot{z}^T(s)R_2\dot{z}(s)ds - \int_{t-h}^{t-\alpha h} \dot{z}^T(s)R_3\dot{z}(s)ds \\
&= -\int_{t-\alpha h}^{t} \dot{z}^T(s)(R_1+R_2)\dot{z}(s)ds - \int_{t-h(t)}^{t-\alpha h} \dot{z}^T(s)(R_1+R_3)\dot{z}(s)ds \\
&\quad -\int_{t-h}^{t-h(t)} \dot{z}^T(s)(R_1+R_3)\dot{z}(s)ds.
\end{aligned}$$

By applying Lemma 7.1 to the integral term associated with $R_1 + R_2$ and Lemma 7.2 with any matrix $S_2 \in \mathbb{R}^{2n\times 2n}$ to the integral terms associated with $R_1 + R_3$, we can obtain the following equations:

$$-\int_{t-\alpha h}^{t} \dot{z}^T(s)(R_1+R_2)\dot{z}(s)ds \leq -\frac{1}{\alpha h}\xi_6^T(t)R_{12}^d\xi_6(t), \tag{7.20}$$

and

$$\begin{aligned}
&-\int_{t-h(t)}^{t-\alpha h} \dot{z}^T(s)(R_1+R_3)\dot{z}(s)ds - \int_{t-h}^{t-h(t)} \dot{z}^T(s)(R_1+R_3)\dot{z}(s)ds \\
&\leq -\frac{1}{(1-\alpha)h}\xi_7^T(t)(\mathscr{R}_{2,[h(t)]} - \mathscr{T}_{2,[h(t)]})\xi_7(t),
\end{aligned} \tag{7.21}$$

where

$$\xi_6(t) = \begin{bmatrix} z(t) - z(t-\alpha h) \\ z(t) + z(t-\alpha h) - \frac{2}{\alpha h}\int_{t-\alpha h}^{t} z(s)ds \end{bmatrix},$$

$$\xi_7(t) = \begin{bmatrix} z(t-\alpha h) - z(t-h(t)) \\ z(t-\alpha h) + z(t-h(t)) - \frac{2}{h(t)-\alpha h}\int_{t-h(t)}^{t-\alpha h} z(s)ds \\ z(t-h(t)) - z(t-h) \\ z(t-h(t)) + z(t-h) - \frac{2}{h-h(t)}\int_{t-h}^{t-h(t)} z(s)ds \end{bmatrix},$$

$$\mathscr{R}_{2,[h(t)]} = \begin{bmatrix} \frac{(2-\alpha)h-h(t)}{(1-\alpha)h}R_{13}^d & S_2 \\ * & \frac{(1-2\alpha)h+h(t)}{(1-\alpha)h}R_{13}^d \end{bmatrix},$$

$$\mathscr{T}_{2,[h(t)]} = \begin{bmatrix} \frac{h-h(t)}{(1-\alpha)h}S_2(R_{13}^d)^{-1}S_2^T & 0 \\ 0 & \frac{h(t)-\alpha h}{(1-\alpha)h}S_2^T(R_{13}^d)^{-1}S_2 \end{bmatrix}.$$

Using Eqs. (7.20) and (7.21), the $\dot{V}_4(t)$ can be estimated as:

$$\dot{V}_4(t) \leq \zeta_2^T(t)\left(\Omega_{6,[h(t)]} - \frac{1}{(1-\alpha)h}\Gamma_9\mathscr{T}_{2,[h(t)]}\Gamma_9^T\right)\zeta_2(t), \tag{7.22}$$

where

$$\begin{aligned}
\Omega_{6,[h(t)]} &= hr_5(R_1 + \alpha R_2 + (1-\alpha)R_3)r_5^T - \frac{1}{(1-\alpha)h}\Gamma_9\mathscr{R}_{2,[h(t)]}\Gamma_9^T \\
&\quad - \frac{1}{\alpha h}\Gamma_{10}R_{12}^d\Gamma_{10}^T, \\
\Gamma_9 &= \left[\Gamma_{91}\ \Gamma_{92}\right], \\
\Gamma_{91} &= \left[r_4 - r_2,\ r_4 + r_2 - 2r_{11}\right], \\
\Gamma_{92} &= \left[r_2 - r_3,\ r_2 + r_3 - 2r_9\right], \\
\Gamma_{10} &= \left[r_1 - r_4,\ r_1 + r_4 - 2r_{10}\right].
\end{aligned}$$

Finally, by Eqs. (7.12), (7.14), (7.16), (7.17), (7.19), and (7.22), the upper bound of $\dot{V}(t)$ for Case 2 can be estimated as:

$$\begin{aligned}
\dot{V}(t) &\leq \zeta_2^T(t)\left(\Psi_{2,[h(t),\dot{h}(t)]} + \frac{1}{(1-\alpha)h}\Gamma_9\mathscr{T}_{2,[h(t)]}\Gamma_9^T\right)\zeta_2(t) \\
&\quad - z^T(t)z(t) + \gamma^2 w^T(t)w(t),
\end{aligned} \tag{7.23}$$

where

$$\begin{aligned}
\Psi_{2,[h(t),\dot{h}(t)]} &= \Omega_1 + \Omega_{5,[h(t),\dot{h}(t)]} + \Omega_{3,[\dot{h}(t)]} + \Omega_{6,[h(t)]} + \Upsilon_1 + \Upsilon_2 \\
&\quad + r_1 r_1^T - \gamma^2 r_{16} r_{16}^T.
\end{aligned}$$

Therefore, the following theorem can be obtained.

Theorem 7.1 *For given positive constants h, μ, $0 < \alpha < 1$, β, γ, l_i^-, l_i^+ $(i = 1, \ldots, n)$, the system (7.1) is asymptotically stable with $\mathscr{H}_\infty$ performance level γ if there exist positive scalars λ_{1i}, λ_{2i} $(i = 1, \ldots, n)$ and matrices $P \in \mathbb{S}_+^{8n}$, $Q_i \in \mathbb{S}_+^{3n}$ $(i = 1, \ldots, 5)$, $R_i \in \mathbb{S}_+^n$ $(i = 1, 2, 3)$, $U_i \in \mathbb{D}_+^n$ $(i = 1, \ldots, 4)$, $W_i \in \mathbb{D}_+^n$ $(i = 1, \ldots, 5)$, $S_1, S_2 \in \mathbb{R}^{2n\times 2n}$, $H, G \in \mathbb{R}^{n\times n}$ satisfying the following LMIs:$\forall \dot{h}(t) \in \{-\mu, \mu\}$*

$$\begin{cases}
\begin{bmatrix} \Psi_{1,[h(t)=0,\dot{h}(t)]} & \Gamma_{71}S_1 \\ * & -\alpha h R_{12}^d \end{bmatrix} < 0, \\
\begin{bmatrix} \Psi_{1,[h(t)=\alpha h,\dot{h}(t)]} & \Gamma_{72}S_1^T \\ * & -\alpha h R_{12}^d \end{bmatrix} < 0.
\end{cases} \tag{7.24}$$

$$\begin{cases} \begin{bmatrix} \Psi_{2,[h(t)=\alpha h,\dot{h}(t)]} & \Gamma_{91} S_2 \\ * & -(1-\alpha)hR_{13}^d \end{bmatrix} < 0, \\ \begin{bmatrix} \Psi_{2,[h(t)=h,\dot{h}(t)]} & \Gamma_{92} S_2^T \\ * & -(1-\alpha)hR_{13}^d \end{bmatrix} < 0. \end{cases} \quad (7.25)$$

And, the control gain can be obtained by $K = H^{-1}G$.

Proof We firstly show the neural networks (7.1) for Case 1 are stable with $\mathscr{H}_\infty$ performance γ by Definition 7.1.

- **Condition 1: No disturbance**
  When $w(t) = 0$, it is clear that

$$\begin{aligned} \dot{V}(t) \leq\ & \zeta_1^T(t) \left(\Psi_{1,[h(t),\dot{h}(t)]} + \frac{1}{\alpha h} \Gamma_7 \mathscr{T}_{1,[h(t)]} \Gamma_7^T \right) \zeta_1(t) \\ & - z^T(t)z(t) + \gamma^2 w^T(t)w(t) \\ \leq\ & \zeta_1^T(t) \left(\Psi_{1,[h(t),\dot{h}(t)]} + \frac{1}{\alpha h} \Gamma_7 \mathscr{T}_{1,[h(t)]} \Gamma_7^T \right) \zeta_1(t), \end{aligned}$$

  therefore LMIs (7.24) guarantee the stability of the system (7.1) without disturbance.
- **Condition 2: Zero initial conditions**
  If LMIs (7.24) hold, then it is true that

$$\dot{V}(t) + z^T(t)z(t) \leq \gamma^2 w^T(t)w(t).$$

  then integrating both sides of the above equation from 0 to ∞ leads

$$\int_0^\infty \left(\dot{V}(t) + z^T(t)z(t) \right) dt < \int_0^\infty \gamma^2 w^T(t)w(t)dt.$$

  Since $\dot{V}(0) = 0$ and $\dot{V}(\infty) = 0$, the above equation becomes:

$$\int_0^\infty z^T(t)z(t)dt < \gamma^2 \int_0^\infty w^T(t)w(t)dt,$$

  which is the same as the second condition of Definition 7.1.

Therefore, it can be said that the neural networks (7.1) for Case 1 is asymptotically stable with $\mathscr{H}_\infty$ performance index γ according to Definition 7.1.

To follow the same procedure of Case 1, it can be easily obtained that LMIs (7.25) guarantee the stability of the neural networks (7.1) for Case 2 with $\mathscr{H}_\infty$ performance index γ according to Definition 7.1. This completes the proof. ■

7.4 Numerical Examples

In this section, we consider a chaotic neural networks which can be described as the Eq. (7.1) with the following parameters:

$$A=\begin{bmatrix}1&0\\0&1\end{bmatrix},\quad B=\begin{bmatrix}3&5\\0.1&2\end{bmatrix}.\quad C=\begin{bmatrix}-2.5&0.2\\0.1&-1.5\end{bmatrix},$$
$$D=\begin{bmatrix}10&0\\0&10\end{bmatrix},\quad g(z(t))=\tanh(z(t)), \tag{7.26}$$

then, it is clear the upper and lower bound values of activation function $g(\cdot)$ defined in Assumption 7.1 are $l_i^-=-1$ and $l_i^+=1$ for $i=1,\ldots,n$.

Let the maximum value of time-varying delay is $h=1$ and a tuning parameter is $\beta=1$. Then, by Theorem 7.1, the minimum $\mathscr{H}_\infty$ performance γ according to various μ and α can be obtained as stated in Table 7.1. As seen in Table 7.1, the minimum $\mathscr{H}_\infty$ performance are varying according to the weighting factor α, therefore, the choice of the weighting factor α is an important task for the control performance.

We consider time-varying delay as $h(t)=1.05+0.05\sin 6t$ which implies $h=1.1$ and $\mu=0.3$, initial value as $z(0)=(0.3,0.5)$, external disturbances as $w(t)=\left[e^{-0.01t}\sin(0.01t),\ e^{-0.02t}\cos(0.05t)\right]$, and the other parameters as $\beta=0.01$, $\alpha=0.5$, and $\gamma=0.1$, then, the control gain can be calculated by Theorem 7.1 as follows:

$$K=\begin{bmatrix}-223.7388&-1.0108\\-0.4423&-218.9239\end{bmatrix}. \tag{7.27}$$

We firstly show the original behavior of the neural networks (7.1) with (7.26) without external disturbances and control inputs in Fig. 7.1. In addition, the dynamics of the uncontrolled neural networks (7.1) with (7.26) and external disturbances are shown in Fig. 7.2. As seen in Figs. 7.1 and 7.2, the state of the neural networks are not stable. To stabilize the neural networks, we apply designed controller $u(t)$ with control gain (7.27) to the neural networks, then the results are displayed in Fig. 7.3,

Table 7.1 Minimum $\mathscr{H}_\infty$ performance γ for various μ and α when $h=1$

Unit is $[10^{-4}]$

μ	α								
	0.1	0.2	0.3	0.4	0.5	0.6	0.7	0.8	0.9
0.1	0.3140	0.4950	0.4990	0.5710	0.5100	0.3000	0.5100	0.5870	0.4000
0.3	0.3590	0.5710	0.4990	0.5810	0.5510	0.5300	0.5860	0.5090	0.4990
0.5	0.4210	0.3990	0.5780	0.5300	0.5980	0.6080	0.5750	0.4890	0.3890
0.7	0.3720	0.3810	0.5990	0.5580	0.6660	0.6210	0.5530	0.5100	0.4290
0.9	0.3000	0.6070	0.4000	0.5770	0.6090	0.5770	0.4900	0.3000	0.4080

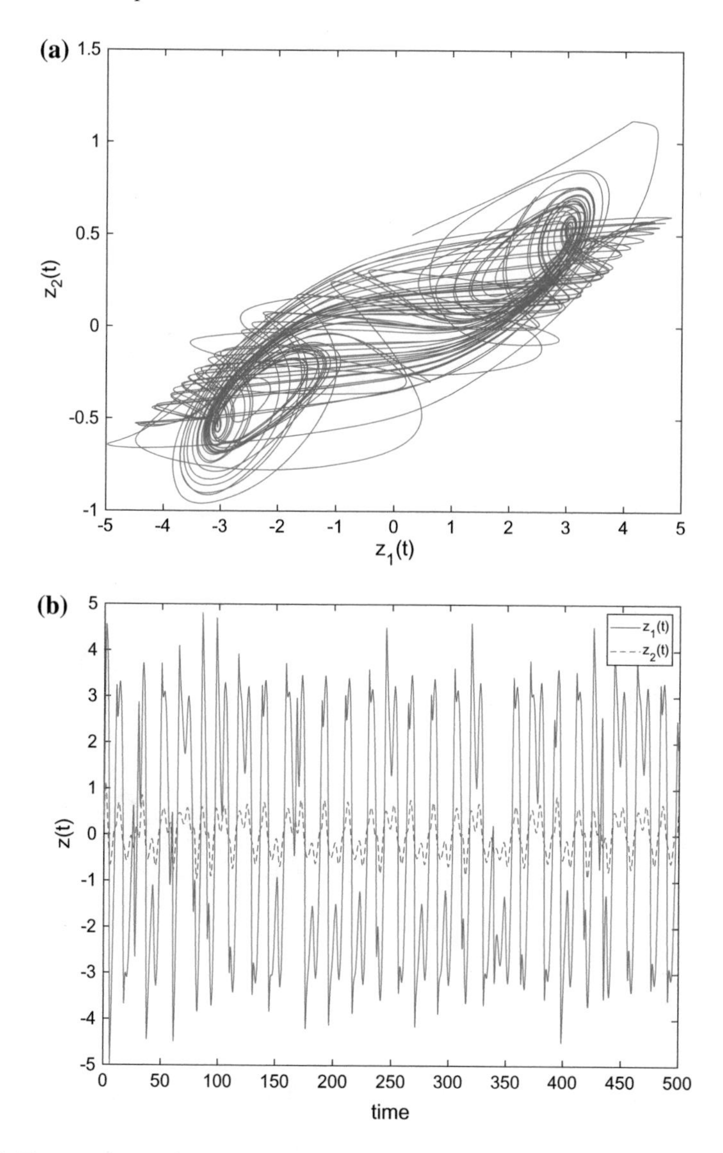

Fig. 7.1 The neural networks (7.1) without external disturbance and controller **a** $z_1(t)$ versus $z_2(t)$ **b** state trajectories of $z(t)$

which implies the designed controller reduces the effect of the disturbance on the state of the neural networks to a prescribed level $\gamma = 0.1$. In addition, the applied control inputs and external disturbances are given in Figs. 7.4 and 7.5, respectively.

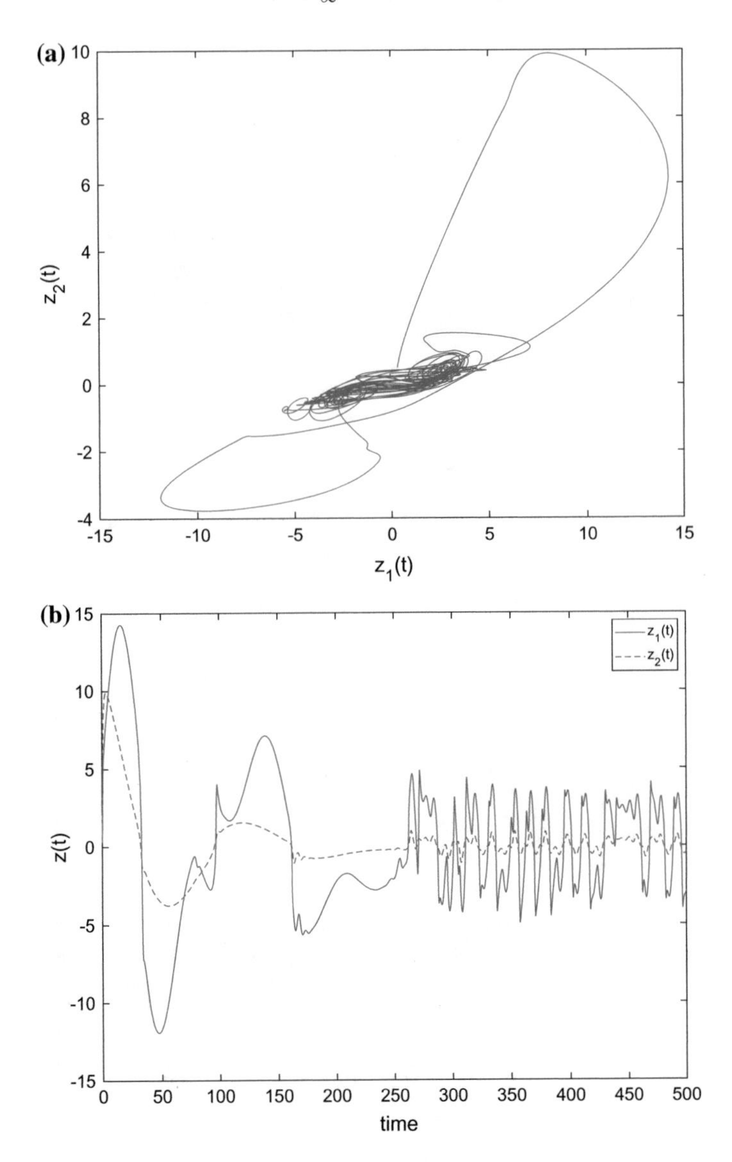

Fig. 7.2 The neural networks (7.1) with disturbance without controller **a** $z_1(t)$ versus $z_2(t)$ **b** state trajectories of $z(t)$

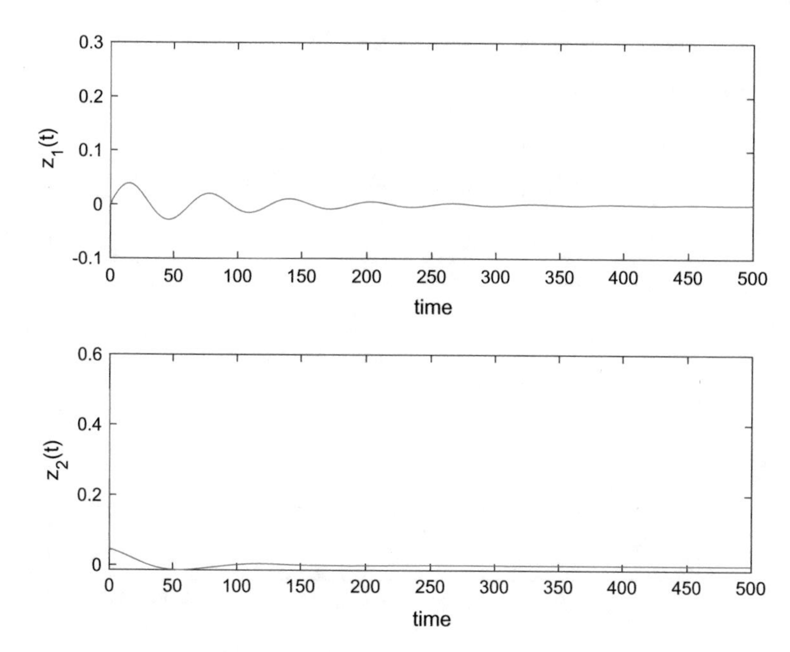

Fig. 7.3 The controlled state trajectories of the neural networks (7.1) with (7.26)

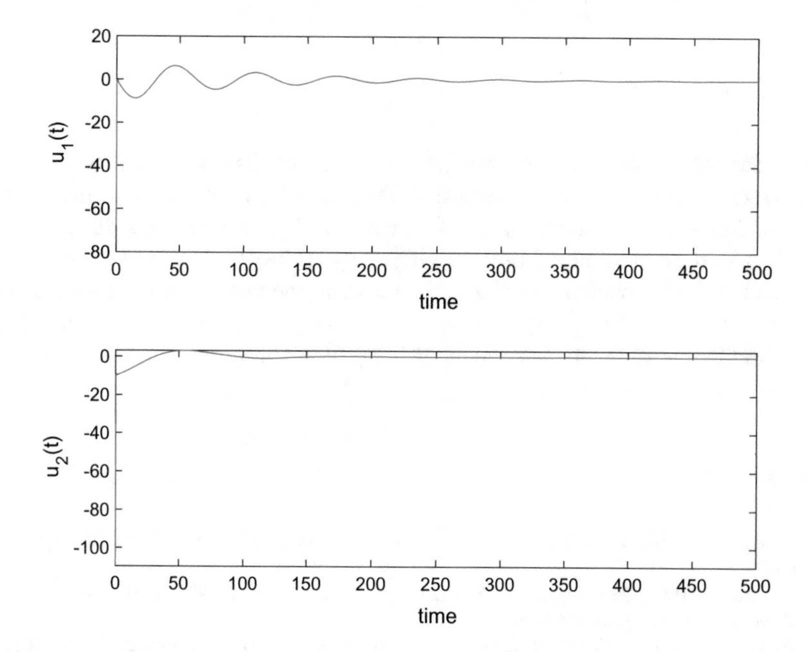

Fig. 7.4 The applied control inputs

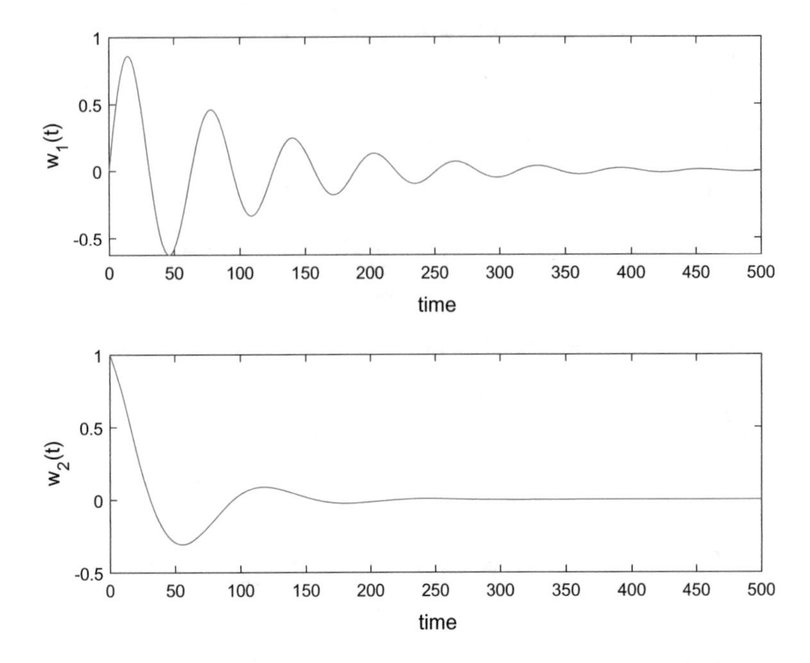

Fig. 7.5 The external disturbances

7.5 Conclusion

In this chapter, the $\mathscr{H}_\infty$ control problem for neural networks with time-varying delays and external disturbance was investigated. Basically, Lyapunov stability theory was used for deriving LMI conditions to design the controller, in which to reduce the conservatism of the condition, the time-varying delay interval divided into two nonequal subintervals by introducing weighting factor α. In the numerical example section, we showed the effect of the weighting factor, α, on $\mathscr{H}_\infty$ performance, γ, and gave some simulation results to describe the validity of the proposed method.

References

1. Haykin S (1998) Neural networks: a comprehensive foundation. Prentice-Hall, Englewood Cliffs
2. Fausett L (1994) Fundamentals of neural networks, Prentice-Hall international editions. Prentice-Hall, Englewood Cliffs
3. Bishop CM (1995) Neural networks for pattern recognition. Oxford University Press, Oxford
4. Haykin S (2009) Neural networks and learning machines. Pearson/Prentice Hall, New York
5. Krizhevsky A, Sutskever I, Hinton GE (2012) Imagenet classification with deep convolutional neural networks. In: Advances in neural information processing systems, pp 1097–1105

6. Hopfield JJ (1982) Neural networks and physical systems with emergent collective computational abilities. Proc Natl Acad Sci USA 79:2554
7. Lee TH, Park JH, Kwon OM, Lee SM (2013) Stochastic sampled-data control for state estimation of time-varying delayed neural networks. Neural Netw 46:99–108
8. Lee TH, Park MJ, Park JH, Kwon OM, Lee SM (2014) Extended dissipative analysis for neural networks with time-varying delays. IEEE Trans Neural Netw Learn Syst 25:1936–1941
9. Chen J, Park JH, Xu S (2019) Stability analysis of discrete-time neural networks with an interval time-varying delay. Neurocomputing 329:248–254
10. Zhang R, Park JH, Zeng D, Liu Y, Zhong S (2018) A new method for exponential synchronization of memristive recurrent neural networks. Inf Sci 466:152–169
11. Zhang R, Zeng D, Park JH, Liu Y, Zhong S (2018) Quantized sampled-data control for synchronization of inertial neural networks with heterogeneous time-varying delays. IEEE Trans Neural Netw Learn Syst 29:6385–6395
12. Park JH, Park CH, Kwon OM, Lee SM (2008) A new stability criterion for bidirectional associative memory neural networks of neutral-type. Appl Math Comput 199:716–722
13. Thuan MV, Trinh H, Hien LV (2016) New inequality-based approach to passivity analysis of neural networks with interval time-varying delay. Neurocomputing 194:301–307
14. Phat VN, Trinh H (2010) Exponential stabilization of neural networks with various activation functions and mixed time-varying delays. IEEE Trans Neural Netw 21:1180–1184
15. Ding S, Wang Z, Rong N, Zhang H (2017) Exponential stabilization of memristive neural networks via saturating sampled-data control. IEEE Trans Cybern 47:3027–3039
16. Chen WH, Lu X, Zheng WX (2015) Impulsive stabilization and impulsive synchronization of discrete-time delayed neural networks. IEEE Trans Neural Netw Learn Syst 26:734–748
17. Lu J, Ho DWC, Wang Z (2009) Pinning stabilization of linearly coupled stochastic neural networks via minimum number of controllers. IEEE Trans Neural Netw 20:1617–1629
18. Nam PT, Trinh H, Pathirana PN, Phat VN (2018) Stability analysis of nonlinear time-delay systems using a novel piecewise positive systems method. IEEE Trans Autom Control 63:291–297
19. Hien LV, Trinh H (2016) New finite-sum inequalities with applications to stability of discrete time-delay systems. Automatica 71:197–201
20. Hien LV, Trinh H (2016) Exponential stability of time-delay systems via new weighted integral inequalities. Appl Math Comput 275:335–344
21. Hien LV, Trinh H (2015) An enhanced stability criterion for time-delay systems via a new bounding technique. J Frankl Inst 352:4407–4422
22. Hien LV, Trinh H (2015) Refined Jensen-based inequality approach to stability analysis of time-delay systems. IET Control Theory Appl 9:2188–2194
23. Nam PT, Pathirana PN, Trinh H (2015) Discrete Wirtinger-based inequality and its application. J Frankl Inst 352:1893–1905
24. Zhang CK, He Y, Jiang L, Wu M (2016) Stability analysis for delayed neural networks considering both conservativeness and complexity. IEEE Trans Neural Netw Learn Syst 27:1486–1501
25. Kwon OM, Park MJ, Lee SM, Park JH, Cha EJ (2013) Stability for neural networks with time-varying delays via some new approaches. IEEE Trans Neural Netw Learn Syst 24:181–193
26. Zhang CK, He Y, Jiang L, Wang QG, Wu M (2017) Stability analysis of discrete-time neural networks with time-varying delay via an extended reciprocally convex matrix inequality. IEEE Trans Cybern 47:3040–3049
27. Lee TH, Trinh HM, Park JH (2018) Stability analysis of neural networks with time-varying delay by constructing novel Lyapunov functionals. IEEE Trans Neural Netw Learn Syst 29:4238–4247
28. Zhang XM, Han QL, Zeng Z (2018) Hierarchical type stability criteria for delayed neural networks via canonical Bessel-Legendre inequalities. IEEE Trans Cybern 48:1660–1671
29. Chen J, Park JH, Xu S (2018) Stability analysis for neural networks with time-varying delay via improved techniques. IEEE Trans Cybern. https://doi.org/10.1109/TCYB.2018.2868136
30. Lakshmanan S, Park JH, Lee TH, Jung HY, Rakkiyappan R (2013) Stability criteria for BAM neural networks with leakage delays and probabilistic time-varying delays. Appl Math Comput 219:9408–9423

31. Lee TH, Park MJ, Park JH, Kwon OM, Jung HY (2015) On stability criteria for neural networks with time-varying delay using Wirtinger-based multiple integral inequality. J Frankl Inst 352:5627–5645
32. Lee TH, Park JH, Jung HY (2018) Design of network-based $\mathscr{H}_\infty$ state estimator for neural networks using network uncertainty compensator. Appl Math Comput 316:205–214
33. Xie Y, Wen J, Peng L (2019) Robust $\mathscr{H}_\infty$ filtering for average dwell time switching systems via a non-monotonic function approach. Int J Control Autom Syst 17:657–666
34. Li X, Wang W, Xu J, Zhang H (2019) Solution to mixed $\mathscr{H}_2/\mathscr{H}_\infty$ control for discrete-time systems with (x, u, v)-dependent noise. Int J Control Autom Syst 17:273–285
35. Seuret A, Gouaisbaut F (2013) Wirtinger-based integral inequality: application to time-delay systems. Automatica 49:2860–2866
36. Zhang CK, He Y, Jiang L, Wu M, Zeng HB (2016) Stability analysis of systems with time-varying delays via relaxed integral inequalities. Syst Control Lett 92:52–61

Chapter 8
Event-Triggered Control for Linear Networked Systems

8.1 Introduction

In the past few decades, networked control systems (NCSs) have received considerable attention due to its increasing demand in networked systems for automation, industrial process control, manufacturing, and many other applications [1]. On the other hand, it is known that the insertion of communication networks in the feedback control loops makes the analysis and synthesis of NCSs much more complex than before [2]. In view of this, the control design problems for NCSs has been studied by many researchers [3–13].

However, the above works are only based on time-triggered control method for system modeling and analysis because of the easy implementation and analysis. In general, the time-triggered scheme leads to inefficient utilization of limited network resources. To mitigate the unnecessary waste of computation and communication resources in conventional time-triggered control, event-triggered control, which advocate the use of actuation only when some function of system state exceeds a threshold, have been proposed. Compared with the conventional time-triggered communication, the scheme of event-triggered control allows a considerable reduction of the network resource occupancy while maintaining control performance. Therefore, many results for event-triggered for linear or nonlinear systems have been published [14–18]. Specifically, the current sampled data will be transmitted once its norm exceeds a fixed value [17]. The problem of control design for event-triggered networked systems with input quantization has been investigated in [19]. Furthermore, the event-triggered $\mathscr{H}_\infty$ control for networked control systems has been studied in [20]. In [21], the problem of an event-triggered scheme and $\mathscr{H}_\infty$ control for NCSs with communication delay and packet loss has been investigated. In [22, 23], self-triggered scheme, where predicting the next triggered time instant depends on the current information, is introduced to guarantee the systems satisfying the required performance. In comparison with the event-triggered scheme, the self-triggered scheme can provide additional energy saving for the sensor and also less complexity in the implementation. However, the average release period, which

J. H. Park et al., *Dynamic Systems with Time Delays: Stability and Control*,
https://doi.org/10.1007/978-981-13-9254-2_8

is based on a self-triggered scheme, is always smaller than the one based on an event-triggered scheme. Furthermore, in the self-triggered scheme, more constraints on the system structure are required for the design or implementation of controllers.

In this chapter, we study the control design problem for event-triggered NCSs. The event-triggered generator is used between the sensor and the controller, which is employed to determine whether the newly sampled data should be sent our or not. Under the event-triggered scheme, a novel model is established for system analysis and control design. In this model, the effect of the networked transmission delay and the properties of the proposed event-triggered scheme are considered. Based on this model, sufficient condition for the stability and controller design are derived, which are described in the form of linear matrix inequality. The criteria are also given the relationship between the parameters of the event-triggered scheme, the transmission delay, and the feedback gain matrix. Therefore, the controller and parameters of event-trigger can be derived by using our approach. In order to demonstrate the effectiveness of the proposed method, a simulation example is given.

8.2 Problem Statement

Consider a class of linear systems described by:

$$\begin{aligned} \dot{x}(t) &= Ax(t) + Bu(t) + F\omega(t), \\ z(t) &= Cx(t) + Du(t), \end{aligned} \tag{8.1}$$

where $x(k) \in \mathbb{R}^n$ is the state vector, $u(t) \in \mathbb{R}^m$ is the control input vector, $z(t) \in \mathbb{R}^p$ is the controlled output vector, $\omega(t) \in \mathbb{R}^p$ denotes a disturbance input belonging to $\mathscr{L}_2[0, \infty)$, and A, B, F, C, and D are system matrices with appropriate dimensions.

In this chapter, it is supposed that the system (8.1) is controlled over a communication networked via a state feedback controller $u(t) = Kx(t)$, where K is a gain matrix to be determined, such that $\mathscr{H}_\infty$ performance of the closed-loop system is satisfied. A digital communication network medium is used to the connected controller and the system (8.1) instead of the traditional point-to-point structure control system. Moreover, in order to reduce the communication load or to reduce the frequency of exchange information, the event-triggered scheme is employed to determine whether the newly sampled signal is sent out or not.

The following event-triggered condition is given to judge whether the current signal needs to be transmitted or not:

$$\begin{aligned} &[x((k+j)h) - x(kh)]^T \Omega [x((k+j)h) - x(kh)] \\ &\leq \sigma x^T((k+j)h) \Omega x((k+j)h), \end{aligned} \tag{8.2}$$

where $\Omega > 0, j = 1, 2, \ldots, \sigma \in [0, 1)$, and h is the sampling period.

Remark 8.1 The above event-triggered scheme is involved by the parameter σ, Ω, h, which determine the burden of network communication. It should be noted that the event-triggered mechanism with the algorithm (8.2) only has a relationship with the difference between the states sampled in discrete instants, and not need to consider what happened in between updates.

Under the algorithm, suppose that the release times are t_0h, t_1h, $\ldots$, where t_0 denotes the initial time. $s_ih = t_{i+1}h - t_ih$ is the release period which corresponds to the sampling period given by the event-triggered scheme (8.2).

For network uncertainties, the effect of the transmission delay on the system is considered here. It is assumed that the time-varying delay in the network communication is d_k and $d_k \in [0, d]$, where d is a positive real constant. Therefore, the states $x(t_0h), x(t_1h), \ldots$ will achieve at the controller side at the instants $t_0h + d_1, t_1h + d_1, t_2h + d_2, \ldots$, respectively.

Based on the analysis above and taking the effect of the transmission delay into account, the system model under the event triggered condition (8.2) can be expressed as

$$\dot{x}(t) = Ax(t) + Bu(t_kh) + B_w\omega(t) \tag{8.3}$$

$$z(t) = Cx(t) + Du(t_kh), \ \ t \in [t_kh + d_k, t_{k+1}h + d_{k+1}). \tag{8.4}$$

Under the following state-feedback control $u(t) = Kx(t)$, for $t \in [t_kh + d_k, t_{k+1}h + d_{k+1})$, Eqs. (8.1)–(8.2) can be rewritten as

$$\dot{x}(t) = Ax(t) + BKx(t_kh) + B_w\omega(t) \tag{8.5}$$

$$z(t) = Cx(t) + DKx(t_kh). \tag{8.6}$$

For the problem discussed above, here the following two cases will be considered: (1) **Case A**: If $t_kh + h + \bar{d} \geq t_{k+1}h + d_{k+1}$, where $\bar{d}$ =max$\{d_k\}$, define a function $d(t)$ as

$$d(t) = t - t_kh, t \in [t_kh + d_k, t_{k+1}h + d_{k+1}). \tag{8.7}$$

Obviously,

$$d_k \leq d(t) \leq (t_{k+1} - t_k)h + d_{k+1} \leq h + \bar{d}. \tag{8.8}$$

(2) **Case B**: If $t_kh + h + \bar{d} < t_{k+1}h + d_{k+1}$, consider the following intervals

$$[t_kh + d_k, t_kh + h + \bar{d}), [t_kh + ih + \bar{d}, t_kh + ih + h + \bar{d}).$$

Since $d_k < \bar{d}$, it can be easily known that τ_M exists such that

$$t_kh + \tau_M h + \bar{d} < t_{k+1}h + d_{k+1} \leq t_kh + \tau_M h + h + \bar{d}$$

and $x(t_kh)$ and $x(t_kh + ih)$ with $i = 1, 2, \ldots, \tau_M$ satisfy (8.2).

Let

$$\begin{cases} I_0 = [t_k h + d_k, t_k h + h + \bar{d}) \\ I_i = [t_k h + ih + \bar{d}, t_k h + ih + h + \bar{d}) \\ I_{\tau_M} = [t_k h + \tau_M h + \bar{d}, t_{k+1} h + d_{k+1}) \end{cases} \tag{8.9}$$

where $i = 1, 2, \ldots, \tau_M - 1$.

One has

$$[t_k h + d_k, t_{k+1} h + d_{k+1}) = \cup_{i=0}^{i=\tau_M} I_i. \tag{8.10}$$

In order to understand the partition of the interval in (8.10), define

$$d(t) = \begin{cases} t - t_k h, & t \in I_0 \\ t - t_k h - ih, & t \in I_i, i = 1, \ldots, \tau_M - 1. \\ t - t_k h - \tau_M h, & t \in I_{\tau_M} \end{cases} \tag{8.11}$$

Then, the following equation can be easily obtained

$$\begin{cases} d_k \leq d(t) \leq h + \bar{d}, & t \in I_0 \\ d_k \leq \bar{d} \leq d(t) \leq h + \bar{d}, & t \in I_i, i = 1, 2, \ldots, \tau_M - 1 \\ d_k \leq \bar{d} \leq d(t) \leq h + \bar{d}, & t \in I_{\tau_M} \end{cases} \tag{8.12}$$

where the third row in (8.14) holds because $t_{k+1}h + d_{k+1} \leq t_k h + (\tau_M + 1)h + \bar{d}$.

Thus, one has

$$0 \leq d_k \leq d(t) \leq h + \bar{d} \triangleq d_M, t \in [t_k h + d_k, t_{k+1} h + d_{k+1}). \tag{8.13}$$

In Case A, for $t \in [t_k h + d_k, t_{k+1} h + d_{k+1})$, define $e_k(t) = 0$.

In Case B, define

$$e_k(t) = \begin{cases} 0, & t \in I_0 \\ x(t_k h) - x(t_k h + ih), & t \in I_i, i = 1, 2, \ldots, \tau_M - 1. \\ x(t_k h) - x(t_k h + \tau_M h), & t \in I_{\tau_M} \end{cases} \tag{8.14}$$

Using the $d(t)$ and $e_k(t)$, (8.6) can be rewritten as

$$\begin{aligned} \dot{x}(t) &= Ax(t) + BKx(t - d(t)) + BKe_k(t) + B_w \omega(t), \\ z(t) &= Cx(t) + DKx(t - d(t)) + DKe_k(t), \end{aligned} \tag{8.15}$$

where $t \in [t_k h + d_k, t_{k+1} h + d_{k+1})$.

Remark 8.2 It is easy to see that if $\sigma = 0$, then $e_k(t) = 0$, the model represented by (8.15) can be reduced to the one in [7]. One can see that if $\sigma \to 0^+$, the dynamics of the systems with the event-triggered scheme will approach to the one with a time-triggered scheme.

The following lemma will be used in next section for deriving main results.

Lemma 8.1 ([24]) *For any matrix* $\begin{bmatrix} M & S \\ * & M \end{bmatrix} \geq 0$, *scalars* $\eta_1(t) > 0$ *and* $\eta_2(t) > 0$ *satisfying* $\eta_1(t) + \eta_2(t) = 1$, *vector functions* $m_1(t)$ *and* $m_2(t) : \mathscr{N} \rightarrow \mathbb{R}^n$, *the following inequality holds*

$$
\begin{aligned}
& -\frac{1}{\eta_1(t)} m_1^T(t) M m_1(t) - \frac{1}{\eta_2(t)} m_2^T(t) M m_2(t) \\
\leq & \begin{bmatrix} m_1(t) \\ m_2(t) \end{bmatrix}^T \begin{bmatrix} M & S \\ * & M \end{bmatrix} \begin{bmatrix} m_1(t) \\ m_2(t) \end{bmatrix}^T .
\end{aligned}
$$

8.3 Main Results

8.3.1 Stabilization Analysis of the Event-Triggered Networked Control Systems

In this subsection, an asymptotic stability and stabilization condition of the event-triggered networked control systems are derived in terms of linear matrix inequalities (LMIs).

First, a stability criterion of the event-triggered networked control systems is given in the following theorem.

Theorem 8.1 *For given scalars* $\gamma > 0, d_M > 0, \sigma > 0$ *and the control gain* K, *the system (8.6) is asymptotically stable with an* $\mathscr{H}_\infty$ *norm bound* γ *if there exist matrices* $P > 0, Q > 0, X > 0, R > 0$ *and* $S_1, S_2, S_3, S_4, \Omega$ *with appropriate dimensions, such that the following LMIs hold:*

$$
\Gamma = \begin{bmatrix}
\Gamma_{11} & \Gamma_{12} & 6R & \Gamma_{14} & \Gamma_{15} & PBK \\
* & -12R & -4S_4 & 6R & \Gamma_{25} & 0 \\
* & * & -12R & 2(S_4^T - S_2^T) & \Gamma_{35} & 0 \\
* & * & * & -4R - Q & \Gamma_{45} & 0 \\
* & * & * & * & \Gamma_{55} & 0 \\
* & * & * & * & * & -\Omega \\
* & * & * & * & * & * \\
* & * & * & * & * & * \\
* & * & * & * & * & *
\end{bmatrix}
$$

$$\begin{matrix} PB_\omega & d_M A^T R & C^T \\ 0 & 0 & 0 \\ 0 & 0 & 0 \\ 0 & 0 & 0 \\ 0 & d_M (RBK)^T & K^T D^T \\ 0 & d_M (RBK)^T & K^T D^T \\ -\gamma^2 I & d_M B_\omega^T R^T & 0 \\ * & -R & 0 \\ * & * & -I \end{bmatrix} < 0, \tag{8.16}$$

where

$$\begin{aligned} \Gamma_{11} &= PA + A^T P + Q + X - 4R, \\ \Gamma_{12} &= 2(S_3^T + S_4^T), \\ \Gamma_{14} &= (S_1^T + S_2^T) - (S_3^T + S_4^T), \\ \Gamma_{15} &= -2R - S_1^T - S_2^T - S_3^T - S_4^T + PBK, \\ \Gamma_{25} &= 6R - 2(S_3 - S_4), \\ \Gamma_{35} &= 6R + 2(S_2^T + S_4^T), \\ \Gamma_{45} &= -2R - S_1 + S_2 + S_3 - S_4, \\ \Gamma_{55} &= -8R + S_1 + S_1^T - S_2 - S_2^T + S_3 + S_3^T - S_4 - S_4^T + \delta\Omega. \end{aligned}$$

Proof Construct the following Lyapunov functional candidate as:

$$\begin{aligned} V(t) = x^T(t)Px(t) &+ \int_{t-d_M}^{t} x^T(s)Qx(s)ds + \int_{t-d(t)}^{t} x^T(s)Xx(s)ds \\ &+ d_M \int_{t-d_M}^{t} \int_{s}^{t} \dot{x}^T(s)R\dot{x}(s)ds. \end{aligned} \tag{8.17}$$

Taking the time-derivative of (8.17) yields

$$\begin{aligned} \dot{V}(t) \le 2x^T(t)P\dot{x}(t) &+ x^T(t)Qx(t) - x^T(t-d_M)Qx(t-d_M) \\ &+ x^T(t)Xx(t) + d_M\dot{x}^T(t)R\dot{x}(t) - d_M \int_{t-d_M}^{t} \dot{x}^T(s)Rx(s)ds \\ &+ \sigma x^T(t-d(t))\Omega x(t-d(t)) - e_k^T(t)\Omega e_k(t). \end{aligned} \tag{8.18}$$

By using Lemma 8.1 and Jensen's inequality, the integral term $-d_M \int_{t-d_M}^{t} \dot{x}(s)Rx(s)$ ds is transformed as

$$\begin{aligned} -d_M \int_{t-d_M}^{t} \dot{x}^T(s)Rx(s)ds &= -d_M \int_{t-d_M}^{t-d(t)} \dot{x}^T(s)Rx(s)ds - d_M \int_{t-d(t)}^{t} \dot{x}^T(s)Rx(s)ds \\ &\quad - \frac{d_M}{d_M - d(t)} \xi^T(t)[e_1^T \; e_2^T] \begin{bmatrix} R & 0 \\ 0 & 3R \end{bmatrix} \begin{bmatrix} e_1 \\ e_2 \end{bmatrix} \xi(t) \end{aligned}$$

$$-\frac{d_M}{d(t)}\xi^T(t)[e_3^T \ \ e_4^T]\begin{bmatrix} R & 0 \\ 0 & 3R \end{bmatrix}\begin{bmatrix} e_3 \\ e_4 \end{bmatrix}\xi(t), \tag{8.19}$$

where

$$\begin{aligned}
\xi^T(t) &= \left[x^T(t) \ \ \frac{\int_{t-d_M}^{t-d(t)} x^T(s)ds}{d_M - d(t)} \ \ \frac{\int_{t-d(t)}^{t} x^T(s)ds}{d(t)} \ \ x^T(t-d_M) \ \ x(t-d(t)) \right], \\
e_1 &= [0 \ \ 0 \ \ 0 \ \ -I \ \ I], \\
e_2 &= [0 \ \ -2I \ \ 0 \ \ I \ \ I], \\
e_3 &= [I \ \ 0 \ \ 0 \ \ 0 \ \ -I], \\
e_4 &= [I \ \ 0 \ \ -2I \ \ 0 \ \ I].
\end{aligned}$$

Then, one has

$$-d_M \int_{t-d_M}^{t} \dot{x}^T(s)Rx(s)ds \le -\xi^T(t)\Sigma^T\Psi\Sigma\xi(t), \tag{8.20}$$

where

$$\Sigma^T = [e_1^T \ \ e_2^T \ \ e_3^T \ \ e_4^T], \ \ \Psi = \begin{bmatrix} R & 0 & S_1 & S_2 \\ * & 3R & S_3 & S_4 \\ * & * & R & 0 \\ * & * & * & 3R \end{bmatrix}.$$

Now, the $\mathscr{H}_\infty$ performance for the closed-loop system (8.15) will be established. If the derivative of $V(t)$ is negative, then $z(t) \to 0$ as $t \to \infty$.

Then, substituting (8.19) into (8.18) gives

$$\dot{V}(t) + z^T(t)z(t) - \gamma^2\omega^T(t)\omega(t) \le \eta^T(t)(\Gamma + \Psi^T R^{-1}\Psi + \Phi^T\Phi)\eta(t), \tag{8.21}$$

where

$$\begin{aligned}
\eta^T(t) &= [\xi^T(t) \ \ e_k^T(t) \ \ w^T(t)], \\
\Gamma &= \begin{bmatrix}
\Gamma_{11} & \Gamma_{12} & 6R & \Gamma_{14} & \Gamma_{15} & PBK & PB_\omega \\
* & -12R & -4S_4 & 6R & \Gamma_{25} & 0 & 0 \\
* & * & -12R & 2(S_4^T - S_2^T) & \Gamma_{35} & 0 & 0 \\
* & * & * & -4R - Q & \Gamma_{45} & 0 & 0 \\
* & * & * & * & \Gamma_{55} & 0 & 0 \\
* & * & * & * & * & -\Omega & 0 \\
* & * & * & * & * & * & -\gamma^2 I
\end{bmatrix}, \\
\Psi &= [d_M RA \ \ 0 \ \ 0 \ \ 0 \ \ d_M RBK \ \ d_M RBK \ \ d_M RB_w], \\
\Phi &= [C \ \ 0 \ \ 0 \ \ 0 \ \ DK \ \ DK \ \ 0].
\end{aligned}$$

If $\dot{V}(t) \leq 0$, then

$$\Gamma + \Psi^T R^{-1} \Psi + \Phi^T \Phi < 0, \tag{8.22}$$

By using Schur complement, (8.22) is equivalent to (8.16). One can see that the system (8.6) with $\omega = 0$ is asymptotically stable and $||z(t)||_2 \leq \gamma ||\omega(t)||_2$ under the zero initial condition. This completes the proof of the theorem. ■

8.3.2 Event-Triggered Controller Design of Networked Control Systems

In this subsection, we devote our attention to designing an event-triggered controller such that system (8.6) is asymptotically stable.

Based on Theorem 8.1, the event-triggered controller design method for system (8.6) is provided in the following theorem.

Theorem 8.2 *For given scalars $\gamma > 0, d_M > 0, \sigma > 0$, the system (8.6) is asymptotically stable with an $\mathscr{H}_\infty$ norm bound γ if there exist matrices $G > 0, \bar{Q} > 0, \bar{X} > 0, \bar{R} > 0, \bar{\Omega} > 0$ and $\bar{S}_1, \bar{S}_2, \bar{S}_3, \bar{S}_4, T$ with appropriate dimensions, such that the following LMIs hold:*

$$\begin{bmatrix} \Gamma_{11} & \Gamma_{12} & 6\bar{R} & \Gamma_{14} & \Gamma_{15} & BT & B_\omega & d_M GA^T & GC^T \\ * & -12\bar{R} & -4\bar{S}_4 & 6\bar{R} & \Gamma_{25} & 0 & 0 & 0 & 0 \\ * & * & -12\bar{R} & 2(\bar{S}_4^T - \bar{S}_2^T) & \Gamma_{35} & 0 & 0 & 0 & 0 \\ * & * & * & -4\bar{R} - \bar{Q} & \Gamma_{45} & 0 & 0 & 0 & 0 \\ * & * & * & * & \Gamma_{55} & 0 & 0 & d_M T^T B^T & T^T D^T \\ * & * & * & * & * & -\bar{\Omega} & 0 & d_M T^T B^T & T^T D^T \\ * & * & * & * & * & * & -\gamma^2 I & d_M B_w^T & 0 \\ * & * & * & * & * & * & * & \delta^2 \bar{R} - 2\delta G & 0 \\ * & * & * & * & * & * & * & * & -I \end{bmatrix} < 0, \tag{8.23}$$

where

$$\begin{aligned} \Gamma_{11} &= AG + GA^T + \bar{Q} + \bar{X} - 4\bar{R}, \\ \Gamma_{12} &= 2(\bar{S}_3^T + \bar{S}_4^T), \\ \Gamma_{14} &= (\bar{S}_1^T + \bar{S}_2^T) - (\bar{S}_3^T + \bar{S}_4^T), \\ \Gamma_{15} &= -2\bar{R} - \bar{S}_1^T - \bar{S}_2^T - \bar{S}_3^T - \bar{S}_4^T + BT, \\ \Gamma_{25} &= 6\bar{R} - 2(\bar{S}_3 - \bar{S}_4), \\ \Gamma_{35} &= 6\bar{R} + 2(\bar{S}_2^T + \bar{S}_4^T), \\ \Gamma_{45} &= -2\bar{R} - \bar{S}_1 + \bar{S}_2 + \bar{S}_3 - \bar{S}_4, \end{aligned}$$

$$\Gamma_{55} = -8\bar{R} + \bar{S}_1 + \bar{S}_1^T - \bar{S}_2 - \bar{S}_2^T + \bar{S}_3 + \bar{S}_3^T - \bar{S}_4 - \bar{S}_4^T + \sigma\bar{\Omega}.$$

Then, the feedback gain matrix is given as $K = TG^{-1}$.

Proof Define

$$\begin{aligned} G &= P^{-1}, T = KG, \bar{Q} = GQG, \bar{R} = GRG, \\ \bar{X} &= GXG, \bar{S}_1 = GS_1G, \bar{S}_2 = GS_2G, \\ \bar{S}_3 &= GS_3G, \bar{S}_4 = GS_4G, \bar{\Omega} = G\Omega G, \end{aligned}$$

Pre- and post-multiplying (8.16) by $\text{diag}\{G, G, G, G, G, G, G, G, G\}^T$ and $\text{diag}\{G, G, G, G, G, G, I, R^{-1}, I\}$, we obtain (8.23). This completes the proof of the theorem. ■

Remark 8.3 It should be noted that $\sigma\Omega$ in (8.16) and (8.23) result from the term $e_k^T(t)\Omega e_k(t)$ by using the trigger condition, which renders the effects of the trigger parameters σ and Ω, the sampling period h has relation with the proposed stability condition.

8.4 Numerical Example and Simulation

In this section, an example is given to show the effectiveness of the proposed method on the design of the $\mathscr{H}_\infty$ controller.

Example. Consider the following inverted pendulum system [19]. The plant's state-space model is described by

$$\dot{x}(t) = \begin{bmatrix} 0 & 1 & 0 & 0 \\ 0 & 0 & \frac{-mg}{M} & 0 \\ 0 & 0 & 0 & 1 \\ 0 & 0 & \frac{g}{l} & 0 \end{bmatrix} x(t) + \begin{bmatrix} 0 \\ \frac{1}{M} \\ 0 \\ \frac{-1}{Ml} \end{bmatrix} u(t) \tag{8.24}$$

where $M = 10$ is the cart mass and $m = 1$ denotes the mass of the pendulum bob, $l = 3$ is the length of the pendulum arm and $g = 10$ denotes the gravitational acceleration.

The initial state is chosen as $x_0 = [0.98 \;\; 0 \;\; 0.2 \;\; 0]^T$. It is clear that the system is unstable when no control input because the eigenvalues of the system matrix have positive values.

The state $x(t) = [x_1^T(t) \; x_2^T(t) \; x_3^T(t) \; x_4^T(t)]^T = [y \; \dot{y} \; \theta \; \dot{\theta}]^T$, where $x_i (i = 1, 2, 3, 4)$ are the cart's position, velocity, the pendulum bob's angular and pendulum bob's angular velocity, respectively.

Choose $C = B_w^T = [1 \quad 1 \quad 1 \quad 1]$, $D = 0.1$, and $\omega(t) = e^{-0.1t}$. By applying Theorem 8.2 with $\sigma = 0.2$, $\gamma = 5$, we can obtain the upper bound for d_M is 0.21, and the corresponding gain matrix and trigger matrix are

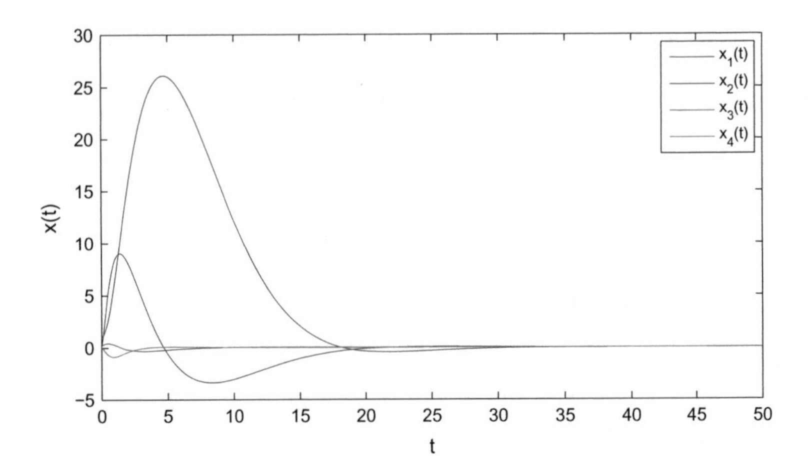

Fig. 8.1 Responses of $x(t)$

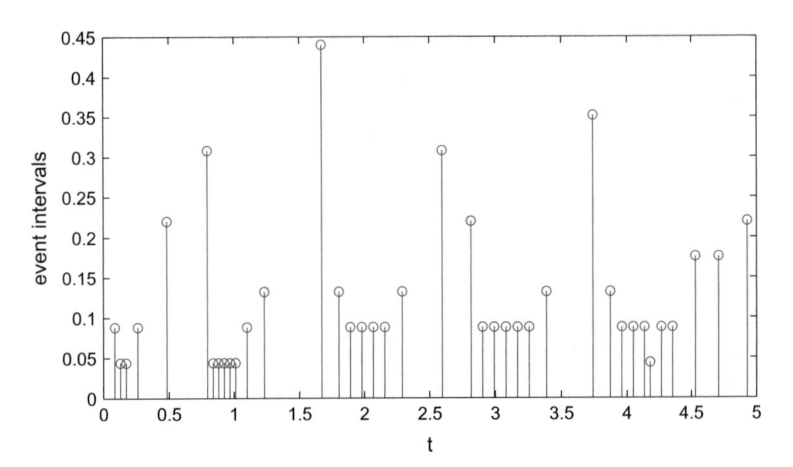

Fig. 8.2 The release instants and release interval

$$K = \begin{bmatrix} 0.8124 & 6.1347 & 286.2458 & 148.3563 \end{bmatrix},$$

$$\Omega = \begin{bmatrix} 0.0089 & 0.0632 & 2.5834 & 1.3568 \\ 0.0632 & 0.5431 & 17.8230 & 9.1543 \\ 3.5436 & 9.9832 & 890.5346 & 489.3274 \\ 1.3842 & 9.4528 & 486.2467 & 272.4084 \end{bmatrix}.$$

Based on the above gain matrix, the state response $x(t)$ is shown in Fig. 8.1. And the release instants and release interval are given in Fig. 8.2. From these figures, it can be seen that the proposed event-trigger scheme can make the considered system be stable with a specific norm bound.

8.5 Conclusion

The event-triggered control for networked control systems is discussed in this chapter. By transforming the event-triggered mechanism intro a time delay model and establishing effectiveness method to deal with the state feedback control, a novel method for event-triggered controller design is proposed and a stability condition with an $\mathscr{H}_\infty$ norm bound is established in terms of LMIs. A simulation example is given to show the validity of the proposed method.

References

1. Tian Y, Levy D (2008) Compensation for control packet dropout in networked control systems. Inf Sci 178:1263–1278
2. Zhang W, Branicky M, Phillips S (2001) Stability of networked control systems. IEEE Control Syst Mag 21:84–99
3. Gao H, Meng X, Chen T (2008) Stabilization of networked control systems with a new delay characterization. IEEE Trans Autom Control 53:2142–2148
4. Han Q (2009) A discrete delay decomposition approach to stability of linear retarded and neutral systems. Automatica 45:517–524
5. Wang Z, Yang F, Ho DWC, Liu X (2007) Robust $\mathscr{H}_\infty$ control for networked systems with random packet losses. IEEE Trans Syst Man Cybern Part B 37:916–924
6. Yue D, Han Q, Peng C (2004) State feedback controller design of networked control systems. IEEE Trans Circuits Syst II: Express Briefs 51:640–644
7. Yue D, Han Q, Lam J (2005) Networked-based robust $\mathscr{H}_\infty$ control with systems with uncertainty. Automatica 41:999–1007
8. Zhang H, Yang D, Chai T (2007) Guaranteed cost networked control for T-S fuzzy systems with time delays. IEEE Trans Syst Man Cybern Part C 37:160–172
9. Wang J, Chen M, Shen H, Park JH, Wu ZG (2017) A Markov jump model approach to reliable event-triggered retarded dynamic output feedback $\mathscr{H}_\infty$ control for networked systems. Nonlinear Anal: Hybrid Syst 26:137–150
10. Long Y, Park JH, Ye D (2017) Finite frequency fault detection for networked systems with access constraint. Int J Robust Nonlinear Control 27:2410–2427
11. Lee TH, Xia J, Park JH (2017) Networked control system with asynchronous samplings and quantizations in both transmission and receiving channels. Neurocomputing 237:25–38
12. Mathiyalagan K, Park JH, Sakthivel R (2015) New results on passivity-based $\mathscr{H}_\infty$ control for networked cascade control systems with application to power plant boiler-turbine system. Nonlinear Anal: Hybrid Syst 17:56–69
13. Mathiyalagan K, Lee TH, Park JH, Sakthivel R (2016) Robust passivity based resilient $\mathscr{H}_\infty$ control for networked control systems with random gain fluctuations. Int J Robust Nonlinear Control 26:426–444
14. Wang X (2009) Event-trigging in cyber-physical systems. PhD dissertation, The University Notre Dame, Notre Dame
15. Shen M, Yan S, Zhang G (2016) A new approach to event-triggered static output feedback control of networked control systems. ISA Trans 65:468–474
16. Wang X, Lemmon MD (2011) Event-triggering in distributed networked control systems. IEEE Trans Autom Control 56:586–601
17. Heemels WPM, Sandee JH, van Den Bosch PPJ (2008) Analysis of event-driven controllers for linear systems. Int J Control 81:571–590

18. Lunze J, Lehmann DA (2010) A state-feedback approach to event-based control. Automatica 46:211–215
19. Hu S, Yue D (2012) Event-triggered control design of linear networked systems with quantizations. ISA Trans 51:153–162
20. Yue D, Tian E, Han Q (2013) A delay systems method for designing event-triggered controllers of networked control systems. IEEE Trans Autom Control 58:475–481
21. Peng C, Yang T (2013) Event-triggered communication and $\mathscr{H}_\infty$ control co-design for networked control systems. Automatica 49:1326–1332
22. Gommans T, Antunes D, Donkers T, Tabuada P, Heemels WPM (2014) Self-triggered linear quadratic control. Automatica 50:1279–1287
23. Peng C, Han Q (2012) A novel self-triggered sampling scheme in networked control systems. In: Proceedings of IEEE 51st annual conference on decision and control, pp 3904–3909
24. Park PG, Ko JW, Jeong C (2011) Reciprocally convex approach to stability of systems with time-varying delays. Automatica 47:235–238

Chapter 9
Design of Dynamic Controller for the Synchronization of Complex Dynamical Networks with a Coupling Delay

9.1 Introduction

It is a natural phenomenon that systems are connected and affect/be affected each other when communication technology is grown. This is normally dealt with the model of complex dynamical networks (CDNs) which is a set of interconnected nodes, so undoubtedly CDNs have build up a sturdy research area and have received a great attention from many fields such as social networks, World Wide Web, epidemic spreading networks, and neural networks [1–17].

Synchronization is one of important issues in research on CDNs applications in chemical reactions, secure communication, power grids, and biological systems. Hence, the investigation of synchronization for CDNs is meaningful in both basic theory and technological practice.

There are mainly two types of synchronization, one is inner synchronization which is about the synchronization between all nodes in a CDN [13, 14]; another is outer synchronization which is about the synchronization between two or more CDNs [15–17].

To achieve the synchronization of CDNs, much effort has made by using various control methods such as adaptive control, pinning control, impulsive control, and sampled-date control, and synchronization schemes including complete synchronization, cluster synchronization, impulsive synchronization, phase synchronization, lag synchronization, and quasi synchronization.

On the other hand, feedback controllers can be classified into static controller and dynamic controller according to the dynamic structure of stabilizing controllers. In contrast with static controllers which is limited to reflect the situation in the closed-loop system after its' installation, the dynamic controllers can react well the change of the system because they have their own dynamic behaviors, generally its' system order is the same with the plant. Thus the dynamic control method has higher flexibility than the static one and provides the wide selection of control parameters. Of course, since there are more controller variables to be designed, the stabilization analysis of the closed-loop system with dynamic controller becomes more complex.

J. H. Park et al., *Dynamic Systems with Time Delays: Stability and Control*,
https://doi.org/10.1007/978-981-13-9254-2_9

As these reasons, we are concerning to design a dynamic feedback controller in this chapter [18–21].

When designing stabilizing controllers for large systems or CDNs, one approach is to design a centralized controller for a fully integrated closed-loop system, and the other is to design individual controllers called decentralized controllers for each subsystem in large-scale systems or each node in CDNs. The centralized control method is to design a controller in the frameworks of a stupendously integrated closed-loop system. It is common sense that the designing process of the centralized control method is simple and the effect of them is outstanding because the controller designed by considering whole behavior of the system. But it sometimes unrealistic for a CDN of a huge number of systems. On the other hand, the decentralized control method designs controllers within the framework of nodes (or subsystems) of a CDN and apply the designed controllers individually to each nodes. So, the decentralized control method is preferred for CDNs than centralized one because it is easier to design the controllers. In this chapter, we will look at controller design in these two cases.

In addition, to reduce the effect of coupling delay on the controller performance, we divide the delay period into two nonequal subintervals [22, 23].

9.2 Problem Formulation

For the investigation of the synchronization of CDNs with a coupling delay, we consider the following a CDN:

$$\dot{x}_i(t) = Ax_i(t) + Bf(x_i(t)) + \sum_{j=1}^{N} c_{ij}x_j(t-h) + u_i(t), \quad i = 1, \ldots, N \tag{9.1}$$

where $x_i = (x_{i1}, x_{i2}, \ldots, x_{in})^T \in \mathbb{R}^n$ is the state vector of the ith node, $f : \mathbb{R}^n \to \mathbb{R}^n$ is a smooth nonlinear vector field, $u_i(t)$ is the control input of ith node, $C = (c_{ij})_{N\times N}$ is the coupling matrix of the network satisfying: $c_{ij} = 1$ when there is a connection between i and $j(i \neq j)$; otherwise $c_{ij} = 0$ $(i \neq j)$; $c_{ii} = -\sum_{j=1, j\neq i}^{N} c_{ij}$, $(i = 1, \ldots, N)$, h is a positive constant for time-delay, and A and B are known constant matrices.

Assumption 9.1 The smooth nonlinear function $f(\cdot)$ is satisfied following Lipschitz condition:

$$\|f(a) - f(b)\| \leq l\|a - b\|.$$

The aim of this chapter is to achieve

$$x_1(t) = x_2(t) = \cdots = x_N(t) = s(t) \quad \text{as } t \to \infty,$$

where $s(t) \in \mathbb{R}^n$ is a solution of a target node, satisfying

$$\dot{s}(t) = As(t) + Bf(s(t)). \tag{9.2}$$

Let us define error vectors as $e_i(t) = s(t) - x_i(t)$, then we can obtain the error dynamics as

$$\dot{e}_i(t) = Ae_i(t) + B\bar{f}_i(t) - \sum_{j=1}^{N} c_{ij} e_j(t-h) - u_i(t), \quad i = 1, \ldots, N \tag{9.3}$$

where $\bar{f}_i(t) = f(s(t)) - f(x_i(t))$.

And, it can be rewritten as

$$\dot{e}(t) = A_N e(t) + B_N \bar{f}(t) - C_N e(t-h) - u(t), \tag{9.4}$$

where $e(t) = [e_1^T(t),\ e_2^T(t),\ \ldots,\ e_N^T(t)]^T$, $\bar{f}(t) = [\bar{f}_1^T(t),\ \bar{f}_2^T(t),\ \ldots,\ \bar{f}_N^T(t)]^T$, $u(t) = [u_1^T(t),\ u_2^T(t),\ \ldots,\ u_N^T(t)]^T$, $A_N = I_N \otimes A$, $B_N = I_N \otimes B$, $C_N = C \otimes I_n$ ($\otimes$ is the Kronecker product).

In this chapter, we consider the following dynamic feedback controller:

$$\begin{aligned} \dot{\zeta}(t) &= A_c \zeta(t) + B_c e(t), \\ u(t) &= C_c \zeta(t), \quad \zeta(0) = 0, \end{aligned} \tag{9.5}$$

where $\zeta(t) \in \mathbb{R}^{Nn}$ is the state vector of controller, and A_c, B_c and C_c are $Nn \times Nn$ dimensional constant gain matrices to be designed.

So, the closed-loop system can be

$$\dot{v}(t) = \bar{A}v(t) - \bar{C}v(t-h) + \bar{B}\bar{f}(t), \tag{9.6}$$

where

$$v(t) = \begin{bmatrix} e(t) \\ \zeta(t) \end{bmatrix}, \quad \bar{A} = \begin{bmatrix} A_N & -C_c \\ B_c & A_c \end{bmatrix}, \quad \bar{C} = \begin{bmatrix} C_N & 0 \\ 0 & 0 \end{bmatrix}, \quad \bar{B} = \begin{bmatrix} B_N \\ 0 \end{bmatrix}. \tag{9.7}$$

Before deriving main theorem, the following lemma is essential.

Lemma 9.1 (Wirtinger-based integral inequality [24]) *For a given positive definite matrix $R \in \mathbb{R}^{n \times n}$, positive scalars a_1, a_2 satisfying $a_1 \le a_2$, and all continuously differentiable function x in $[a_1, a_1] \to \mathbb{R}^n$, the following inequality holds:*

$$\int_{a_1}^{a_2} \dot{x}^T(s)R\dot{x}(s)ds$$
$$\geq \frac{1}{a_2 - a_1} \begin{bmatrix} x(a_2) \\ x(a_1) \\ \frac{1}{a_2-a_1}\int_{a_1}^{a_2} x(s)ds \end{bmatrix}^T \begin{bmatrix} 4R & 2R & -6R \\ * & 4R & -6R \\ * & * & 12R \end{bmatrix} \begin{bmatrix} x(a_2) \\ x^T(a_1) \\ \frac{1}{a_2-a_1}\int_{a_1}^{a_2} x(s)ds \end{bmatrix}.$$

9.3 Centralized Dynamic Controller for Synchronization of a CDN with Coupling Delay

In this section, we consider centralized dynamics controller for the CDN (9.1). The coupling delay is handled by dividing two weighted subintervals.

9.3.1 Synchronization Analysis Under Centralized Dynamic Controller

Firstly, we introduce the analysis of synchronization under centralized dynamic controller.

Theorem 9.1 *For given positive constants h, l, $0 \leq \alpha \leq 1$, β, and known control gain matrices A_c, B_c, C_c, the dynamic feedback controller (9.5) guarantees synchronization of the CDN (9.1), if there exist matrices $P \in \mathbb{S}_{+}^{6Nn}$, $Q, R, W \in S_{+}^{2Nn}$, $H \in \mathbb{R}^{2Nn\times 2Nn}$ satisfying the following LMI:*

$$\Omega = \begin{bmatrix} \Sigma_1 - P_2 + P_3 - \frac{2}{h_1}R & -P_3 & h_1 P_4^T + \frac{6}{h_1}R & h_d P_5 & P_1 + \bar{A}^T H^T & 0 \\ * & \Sigma_2 & -\frac{2}{h_d}R & \Sigma_3 & h_d(P_6^T - P_5) + \frac{6}{h_d}R & 0 & 0 \\ * & * & -W - \frac{4}{h_d}R & -h_1 P_5^T & -h_d P_6^T + \frac{6}{h_d}R & \bar{C}^T H^T & 0 \\ * & * & * & -\frac{12}{h_1}R & 0 & P_2^T & 0 \\ * & * & * & * & -\frac{12}{h_d}R & P_3^T & 0 \\ * & * & * & * & * & hR - H - H^T & H\bar{B} \\ * & * & * & * & * & * & -\beta I \end{bmatrix} < 0, \tag{9.8}$$

where

$$\Sigma_1 = P_2 + P_2^T + Q - \frac{4}{h_1}R + \beta L,$$
$$\Sigma_2 = -Q + W - \frac{4}{h_1}R - \frac{4}{h_d}R,$$
$$\Sigma_3 = -h_1 P_4^T + h_1 P_5^T + \frac{6}{h_1}R,$$
$$h_1 = \alpha h,$$
$$h_d = h - h_1,$$
$$L = \begin{bmatrix} l^2 I & 0 \\ * & 0 \end{bmatrix}.$$

Proof Let us define the following Lyapunov functional candidate:

$$V(t) = \xi_1^T(t)P\xi_1(t) + \int_{t-h_1}^{t} v^T(s)Qv(s)ds + \int_{t-h}^{t-h_1} v^T(s)Wv(s)ds + \int_{t-h}^{t}\int_{r}^{t} v^T(s)Rv(s)ds, \tag{9.9}$$

where

$$P = \begin{bmatrix} P_1 & P_2 & P_3 \\ * & P_4 & P_5 \\ * & * & P_6 \end{bmatrix}$$
$$\xi_1^T(t) = \Big[v^T(t),\ \textstyle\int_{t-h_1}^{t} v^T(s)ds,\ \int_{t-h}^{t-h_1} v^T(s)ds \Big].$$

Time derivative of Lyapunov functional (9.9) can be

$$\dot{V}(t) = 2\xi_1^T(t)P\xi_2(t) + v^T(t)Qv(t) + v^T(t-h_1)(W-Q)v(t-h_1) - v^T(t-h)Wv(t-h) + h\dot{v}^T(t)R\dot{v}(t) - \int_{t-h}^{t} \dot{v}^T(s)R\dot{v}(s)ds, \tag{9.10}$$

where

$$\xi_2^T(t) = \Big[\dot{v}^T(t),\ v^T(t) - v^T(t-h_1),\ v^T(t-h_1) - v^T(t-h) \Big].$$

By Lemma 9.1, the integral term in Eq. (9.10) can be estimated as

$$\begin{aligned}
&-\int_{t-h}^{t} v^T(s)Rv(s)ds \\
&= -\int_{t-h_1}^{t} v^T(s)Rv(s)ds - \int_{t-h}^{t-h_1} v^T(s)Rv(s)ds \\
&\leq \frac{1}{h_1}\begin{bmatrix} v(t) \\ v(t-h_1) \\ \frac{1}{h_1}\int_{t-h_1}^{t} v(s)ds \end{bmatrix}^T \begin{bmatrix} -4R & -2R & 6R \\ * & -4R & 6R \\ * & * & -12R \end{bmatrix} \begin{bmatrix} v(t) \\ v(t-h_1) \\ \frac{1}{h_1}\int_{t-h_1}^{t} v(s)ds \end{bmatrix} \\
&\quad + \frac{1}{h_d}\begin{bmatrix} v(t-h_1) \\ v(t-h) \\ \frac{1}{h_d}\int_{t-h}^{t-h_1} v(s)ds \end{bmatrix}^T \begin{bmatrix} -4R & -2R & 6R \\ * & -4R & 6R \\ * & * & -12R \end{bmatrix} \begin{bmatrix} v(t-h_1) \\ v(t-h) \\ \frac{1}{h_d}\int_{t-h}^{t-h_1} v(s)ds \end{bmatrix}.
\end{aligned} \tag{9.11}$$

From Assumption 9.1 and Eq. (9.6), for a positive constant β, and a positive matrix $H \in \mathbb{R}^{2Nn\times 2Nn}$, the followings are obtained

$$0 \leq \beta(v^T(t)Lv(t) - \bar{f}^T(e(t))\bar{f}(e(t))), \tag{9.12}$$

$$0 = 2\dot{v}^T(t)H[-\dot{v}(t) + \bar{A}v(t) + \bar{B}\bar{f}(e(t)) + \bar{C}v(t-h)]. \tag{9.13}$$

Then, a new upper bound of $\dot{V}(t)$ can be

$$\dot{V}(t) \leq \eta^T(t)\Omega\eta(t), \tag{9.14}$$

where

$$\begin{aligned}
\eta^T(t) =& \Big[v^T(t),\ v^T(t-h_1),\ v^T(t-h),\ \tfrac{1}{h_1}\textstyle\int_{t-h_1}^{t} v^T(s)ds,\ \tfrac{1}{h_d}\int_{t-h}^{t-h_1} v^T(s)ds, \\
& \dot{v}^T(t),\ \bar{f}^T(e(t)) \Big].
\end{aligned}$$

Finally, if LMI (9.8) holds then the error system is asymptotically stable. This completes the proof. ■

9.3.2 Design of Centralized Dynamic Controller

To design control gains A_c, B_c, C_c, we present the following theorem because the inequality (9.8) is not LMI.

Theorem 9.2 *For given positive constants h, l, $0 \leq \alpha \leq 1$, β, the dynamic feedback controller (9.5) guarantees synchronization of the CDN (9.1), if there exist matrices $H_Y, H_S \in \mathbb{S}_+^n$, $\bar{P}_{ij}, \bar{Q}_j, \bar{R}_j, \bar{W}_j, Z_j \in \mathbb{R}^{n\times n}$ for $i = 1, 4, 6$; $j = 1, 2, 3$, $\bar{P}_{ij} \in \mathbb{R}^{n\times n}$ for $i = 2, 3, 5$; $j = 1, \ldots, 4$, satisfying the following LMIs:*

$$\begin{cases} \bar{P} = \begin{bmatrix} \bar{P}_{11} & \bar{P}_{12} & \bar{P}_{21} & \bar{P}_{22} & \bar{P}_{31} & \bar{P}_{32} \\ * & \bar{P}_{13} & \bar{P}_{23} & \bar{P}_{24} & \bar{P}_{33} & \bar{P}_{34} \\ * & * & \bar{P}_{41} & \bar{P}_{42} & \bar{P}_{51} & \bar{P}_{52} \\ * & * & * & \bar{P}_{43} & \bar{P}_{53} & \bar{P}_{54} \\ * & * & * & * & \bar{P}_{61} & \bar{P}_{62} \\ * & * & * & * & * & \bar{P}_{63} \end{bmatrix} > 0, \\ \bar{Q} = \begin{bmatrix} \bar{Q}_1 & \bar{Q}_2 \\ * & \bar{Q}_3 \end{bmatrix} > 0, \\ \bar{R} = \begin{bmatrix} \bar{R}_1 & \bar{R}_2 \\ * & \bar{R}_3 \end{bmatrix} > 0, \\ \bar{W} = \begin{bmatrix} \bar{W}_1 & \bar{W}_2 \\ * & \bar{W}_3 \end{bmatrix} > 0, \\ \begin{bmatrix} H_Y & I \\ * & H_S \end{bmatrix} > 0, \end{cases} \tag{9.15}$$

$$\left[\begin{array}{cccccccccc} \Delta_1 & \beta l H_Y^T & \Delta_2 & \Delta_3 & \Delta_4 & \bar{P}_{31} & 0 & 0 & \bar{P}_{32} & \Delta_5 \\ * & -\beta I & 0 & 0 & 0 & 0 & 0 & 0 & 0 & 0 \\ * & * & \Delta_8 & \Delta_9 & \Delta_{10} & \bar{P}_{32}^T & 0 & 0 & \bar{P}_{33} & \Delta_{11} \\ * & * & * & \Delta_{14} & \Delta_{15} & -\frac{2}{h_d}\bar{R}_1 & 0 & 0 & -\frac{2}{h_d}\bar{R}_2 & \Delta_{16} \\ * & * & * & * & \Delta_{20} & -\frac{2}{h_d}\bar{R}_2^T & 0 & 0 & -\frac{2}{h_d}\bar{R}_3 & \Delta_{21} \\ * & * & * & * & * & -\bar{W}_1 - \frac{4}{h_d}\bar{R}_1 & H_Y^T & H_M & -\bar{W}_2 - \frac{4}{h_d}\bar{R}_2 & -h_1\bar{P}_{51}^T \\ * & * & * & * & * & * & -I & 0 & 0 & 0 \\ * & * & * & * & * & * & * & -I & 0 & 0 \\ * & * & * & * & * & * & * & * & -\bar{W}_3 - \frac{4}{h_d}\bar{R}_3 & -h_1\bar{P}_{52}^T \\ * & * & * & * & * & * & * & * & * & -\frac{12}{h_1}\bar{R}_1 \\ * & * & * & * & * & * & * & * & * & * \\ * & * & * & * & * & * & * & * & * & * \\ * & * & * & * & * & * & * & * & * & * \\ * & * & * & * & * & * & * & * & * & * \\ * & * & * & * & * & * & * & * & * & * \\ * & * & * & * & * & * & * & * & * & * \\ * & * & * & * & * & * & * & * & * & * \\ * & * & * & * & * & * & * & * & * & * \end{array}\right.$$

$$
\left.\begin{array}{cccccccc}
\Delta_6 & h_d\bar{P}_{51} & h_d\bar{P}_{52} & \Delta_7 & \bar{P}_{12}+Z_3 & 0 & 0 & 0 \\
0 & 0 & 0 & 0 & 0 & 0 & 0 & 0 \\
\Delta_{12} & h_d\bar{P}_{53} & h_d\bar{P}_{54} & \bar{P}_{12}^T+A_N^T & \Delta_{13} & 0 & 0 & 0 \\
\Delta_{17} & \Delta_{18} & \Delta_{19} & 0 & 0 & 0 & 0 & 0 \\
\Delta_{22} & \Delta_{23} & \Delta_{24} & 0 & 0 & 0 & 0 & 0 \\
-h_1\bar{P}_{53}^T & \Delta_{25} & \Delta_{26} & 0 & 0 & 0 & 0 & 0 \\
0 & 0 & 0 & 0 & 0 & 0 & 0 & 0 \\
0 & 0 & 0 & 0 & 0 & 0 & 0 & 0 \\
-h_1\bar{P}_{54}^T & \Delta_{27} & \Delta_{28} & 0 & 0 & 0 & 0 & 0 \\
-\frac{12}{h_1}\bar{R}_2 & 0 & 0 & \bar{P}_{21}^T & \bar{P}_{23}^T & 0 & 0 & 0 \\
-\frac{12}{h_1}\bar{R}_3 & 0 & 0 & \bar{P}_{22}^T & \bar{P}_{24}^T & 0 & 0 & 0 \\
* & -\frac{12}{h_d}\bar{R}_1 & -\frac{12}{h_d}\bar{R}_2 & \bar{P}_{31}^T & \bar{P}_{33}^T & 0 & 0 & 0 \\
* & * & -\frac{12}{h_d}\bar{R}_3 & \bar{P}_{32}^T & \bar{P}_{34}^T & 0 & 0 & 0 \\
* & * & * & h\bar{R}_1-2H_Y & h\bar{R}_2-2I & C_N & 0 & B_N \\
* & * & * & * & h\bar{R}_3-2H_S & H_SC_N & 0 & H_S^T \\
* & * & * & * & * & -I & 0 & 0 \\
* & * & * & * & * & * & -I & 0 \\
* & * & * & * & * & * & * & -\beta I
\end{array}\right] < 0, \quad (9.16)
$$

where

$$\Delta_1 = \bar{P}_{21} + \bar{P}_{21}^T + \bar{Q}_1 - \frac{4}{h_1}\bar{R}_1, \quad \Delta_2 = \bar{P}_{22} + \bar{P}_{23}^T + \bar{Q}_2 - \frac{4}{h_1}\bar{R}_2 + l^2 H_Y^T,$$

$$\Delta_3 = -\bar{P}_{21} + \bar{P}_{31} - \frac{2}{h_2}\bar{R}_1, \quad \Delta_4 = -\bar{P}_{22} + \bar{P}_{32} - \frac{2}{h_1}\bar{R}_2,$$

$$\Delta_5 = h_1\bar{P}_{41} + \frac{6}{h_1}\bar{R}_1, \quad \Delta_6 = h_1\bar{P}_{42} + \frac{6}{h_1}\bar{R}_2,$$

$$\Delta_7 = \bar{P}_{11} + H_Y^T A_N^T + Z_1, \quad \Delta_8 = \bar{P}_{24} + \bar{P}_{24}^T + \bar{W}_3 - \frac{4}{h_1}\bar{R}_3 + \beta l^2 I,$$

$$\Delta_9 = -\bar{P}_{23} + \bar{P}_{32}^T - \frac{2}{h_2}\bar{R}_2^T, \quad \Delta_{10} = -\bar{P}_{24} + \bar{P}_{33} - \frac{2}{h_1}\bar{R}_3,$$

$$\Delta_{11} = h_1\bar{P}_{42}^T + \frac{6}{h_1}\bar{R}_2^T, \quad \Delta_{12} = h_1\bar{P}_{43} + \frac{6}{h_1}\bar{R}_3,$$

$$\Delta_{13} = \bar{P}_{13} + A_N^T H_S + Z_2^T, \quad \Delta_{14} = -\bar{Q}_1 + \bar{W}_1 - \frac{4}{h_1}\bar{R}_1 - \frac{4}{h_d}\bar{R}_1,$$

$$\Delta_{15} = -\bar{Q}_2 + \bar{W}_2 - \frac{4}{h_1}\bar{R}_2 - \frac{4}{h_d}\bar{R}_2, \quad \Delta_{16} = -h_1\bar{P}_{41} + h_1\bar{P}_{51}^T + \frac{6}{h_1}\bar{R}_1,$$

$$\Delta_{17} = -h_1\bar{P}_{42} + h_1\bar{P}_{53}^T + \frac{6}{h_1}\bar{R}_2, \quad \Delta_{18} = h_d(\bar{P}_{61} - \bar{P}_{51}) + \frac{6}{h_d}\bar{R}_1,$$

$$\Delta_{19} = h_d(\bar{P}_{62} - \bar{P}_{52}) + \frac{6}{h_d}\bar{R}_2, \quad \Delta_{20} = -\bar{Q}_3 + \bar{W}_3 - \frac{4}{h_1}\bar{R}_3 - \frac{4}{h_d}\bar{R}_3,$$

$$\Delta_{21} = -h_1\bar{P}_{42}^T + h_1\bar{P}_{52}^T + \frac{6}{h_1}\bar{R}_2^T, \quad \Delta_{22} = -h_1\bar{P}_{43} + h_1\bar{P}_{54}^T + \frac{6}{h_1}\bar{R}_3,$$

$$\Delta_{23} = h_d(\bar{P}_{62}^T - \bar{P}_{53}) + \frac{6}{h_d}\bar{R}_2^T, \quad \Delta_{24} = h_d(\bar{P}_{63} - \bar{P}_{54}) + \frac{6}{h_d}\bar{R}_3,$$
$$\Delta_{25} = -h_d\bar{P}_{61} + \frac{6}{h_d}\bar{R}_1, \quad \Delta_{26} = -h_d\bar{P}_{62} + \frac{6}{h_d}\bar{R}_2,$$
$$\Delta_{27} = -h_d\bar{P}_{62}^T + \frac{6}{h_d}\bar{R}_2^T, \quad \Delta_{28} = -h_d\bar{P}_{63} + \frac{6}{h_d}\bar{R}_3.$$

Proof Firstly, by a fact that $2a^Tb \le a^Ta + b^Tb$, the following is true

$$2\dot{v}^T(t)H\bar{C}v(t-h) \le \dot{v}^T(t)H\bar{C}\bar{C}^TH^T\dot{v}(t) + v^T(t-h)v(t-h).$$

By using the above equation, Eq. (9.14) can be changed to

$$\dot{V}(t) \le \eta^T(t)\bar{\Omega}\eta(t), \tag{9.17}$$

where

$$\bar{\Omega} = \begin{bmatrix} \Sigma_1 - P_2 + P_3 - \frac{2}{h_1}R & -P_3 & h_1P_4^T + \frac{6}{h_1}R & h_dP_5 & P_1 + \bar{A}^TH^T & 0 \\ * & \Sigma_2 & -\frac{2}{h_d}R & h_d(P_6^T - P_5) + \frac{6}{h_d}R & 0 & 0 \\ * & * & -W - \frac{4}{h_d}R + I & -h_1P_5^T & -h_dP_6^T + \frac{6}{h_d}R & 0 & 0 \\ * & * & * & -\frac{12}{h_1}R & 0 & P_2^T & 0 \\ * & * & * & * & -\frac{12}{h_d}R & P_3^T & 0 \\ * & * & * & * & * & \Sigma_4 & H\bar{B} \\ * & * & * & * & * & * & -\beta I \end{bmatrix},$$

with

$$\Sigma_4 = hR - H - H^T + H\bar{C}\bar{C}^TH^T,$$

and Σ_1, Σ_2, Σ_3 are defined in Theorem 9.1.

Let us assume that the matrix H is positive definite and define it and its inverse as

$$H = \begin{bmatrix} H_S & H_D \\ * & H_T \end{bmatrix}, \quad H^{-1} = \begin{bmatrix} H_Y & H_M \\ * & H_W \end{bmatrix},$$

where H_S and H_Y are positive definite matrices and H_D and H_M are invertible matrices.

Then, the following is true

$$H^{-1}H = \begin{bmatrix} H_Y H_S + H_M H_D^T & H_Y H_D + H_M H_T \\ H_M^T H_S + H_W H_D^T & H_M^T H_D + H_W H_T \end{bmatrix} = \begin{bmatrix} I & 0 \\ 0 & I \end{bmatrix}.$$

When we define the following two matrices

$$E_a = \begin{bmatrix} H_Y & I \\ H_M^T & 0 \end{bmatrix}, \quad E_b = \begin{bmatrix} I & H_S \\ 0 & H_D \end{bmatrix},$$

we can get the followings:

$$HE_a = E_b, \quad E_a^T H E_a = E_a^T E_b = \begin{bmatrix} H_Y & I \\ I & H_S \end{bmatrix} > 0.$$

Now, we Pre- and Post-multiply $\mathrm{diag}\Big\{\underbrace{E_a^T, \ldots, E_a^T}_{6}, I\Big\}$ and $\mathrm{diag}\Big\{\underbrace{E_a, \ldots, E_a}_{6}, I\Big\}$ to $\bar{\Omega}$, respectively, then it can be obtained

$$\begin{bmatrix} \bar{\Sigma}_1 & -\bar{P}_2 + \bar{P}_3 - \frac{2}{h_1}\bar{R} & -\bar{P}_3 & h_1\bar{P}_4^T + \frac{6}{h_1}\bar{R} & h_d\bar{P}_5 & \bar{P}_1 + E_a^T\bar{A}^T E_b & 0 \\ * & \bar{\Sigma}_2 & -\frac{2}{h_d}\bar{R} & \bar{\Sigma}_3 & h_d(\bar{P}_6^T - \bar{P}_5) + \frac{6}{h_d}\bar{R} & 0 & 0 \\ * & * & -\bar{W} - \frac{4}{h_d}\bar{R} + E_a^T E_a & -h_1\bar{P}_5^T & -h_d\bar{P}_6^T + \frac{6}{h_d}\bar{R} & 0 & 0 \\ * & * & * & -\frac{12}{h_1}\bar{R} & 0 & \bar{P}_2^T & 0 \\ * & * & * & * & -\frac{12}{h_d}\bar{R} & \bar{P}_3^T & 0 \\ * & * & * & * & * & \bar{\Sigma}_4 & E_b^T\bar{B} \\ * & * & * & * & * & * & -\beta I \end{bmatrix},$$

where

$$\bar{\Sigma}_1 = \bar{P}_2 + \bar{P}_2^T + \bar{Q} - \frac{4}{h_1}\bar{R} + \beta E_a^T L E_a,$$

$$\bar{\Sigma}_2 = -\bar{Q} + \bar{W} - \frac{4}{h_1}\bar{R} - \frac{4}{h_d}\bar{R},$$

$$\bar{\Sigma}_3 = -h_1\bar{P}_4^T + h_1\bar{P}_5^T + \frac{6}{h_1}\bar{R},$$

$$\bar{\Sigma}_4 = h\bar{R} - E_a^T E_b - E_b^T E_a + E_b^T \bar{C}\bar{C}^T E_b,$$
$$\bar{P}_i = E_a^T P_i E_a, \quad \text{for } i = 1, \ldots, 6,$$
$$\bar{Q} = E_a^T Q E_a, \quad \bar{W} = E_a^T W E_a, \quad \bar{R} = E_a^T R E_a.$$

Defining LMI variables as

$$P_i = \begin{bmatrix} P_{i1} & P_{i2} \\ * & P_{i3} \end{bmatrix} \text{ for } i = 1, 4, 6, \quad P_i = \begin{bmatrix} P_{i1} & P_{i2} \\ P_{i3} & P_{i4} \end{bmatrix} \text{ for } i = 2, 3, 5,$$
$$Q = \begin{bmatrix} Q_1 & Q_2 \\ * & Q_3 \end{bmatrix} \; W = \begin{bmatrix} W_1 & W_2 \\ * & W_3 \end{bmatrix} \; R = \begin{bmatrix} R_1 & R_2 \\ * & R_3 \end{bmatrix},$$

leads

$$\bar{P}_i = \begin{bmatrix} H_Y^T P_{i1} H_Y + H_Y^T P_{i2} H_M^T + H_M P_{i2}^T H_Y + H_M P_{i3} H_M^T & H_Y^T P_{i1} + H_M P_{i2}^T \\ * & P_{i1} \end{bmatrix},$$
$$; \text{ for } i = 1, 4, 6,$$
$$\bar{P}_i = \begin{bmatrix} H_Y^T P_{i1} H_Y + H_Y^T P_{i2} H_M^T + H_M P_{i3} H_Y + H_M P_{i4} H_M^T & H_Y^T P_{i1} + H_M P_{i3} \\ P_{i1} H_Y + P_{i2} H_M^T & P_{i1} \end{bmatrix},$$
$$; \text{ for } i = 2, 3, 5,$$
$$\bar{Q} = \begin{bmatrix} H_Y^T Q_1 H_Y + H_Y^T Q_2 H_M^T + H_M Q_2^T H_Y + H_M Q_3 H_M^T & H_Y^T Q_1 + H_M Q_2^T \\ * & Q_1 \end{bmatrix},$$
$$\bar{W} = \begin{bmatrix} H_Y^T W_1 H_Y + H_Y^T W_2 H_M^T + H_M W_2^T H_Y + H_M W_3 H_M^T & H_Y^T W_1 + H_M W_2^T \\ * & W_1 \end{bmatrix},$$
$$\bar{R} = \begin{bmatrix} H_Y^T R_1 H_Y + H_Y^T R_2 H_M^T + H_M R_2^T H_Y + H_M R_3 H_M^T & H_Y^T R_1 + H_M R_2^T \\ * & R_1 \end{bmatrix},$$
$$E_a^T L E_a = \begin{bmatrix} H_Y^T & H_M \\ I & 0 \end{bmatrix} \begin{bmatrix} l^2 I & 0 \\ 0 & 0 \end{bmatrix} \begin{bmatrix} H_Y & I \\ H_M^T & 0 \end{bmatrix} = \begin{bmatrix} l^2 H_Y^T H_Y & l^2 H_Y^T \\ * & l^2 I \end{bmatrix},$$
$$E_a^T E_a = \begin{bmatrix} H_Y^T & H_M \\ I & 0 \end{bmatrix} \begin{bmatrix} H_Y & I \\ H_M^T & 0 \end{bmatrix} = \begin{bmatrix} H_Y^T H_Y + H_M H_M^T & H_Y^T \\ * & I \end{bmatrix},$$
$$E_a^T \bar{A}^T H^T E_a = \begin{bmatrix} H_Y^T A_N^T + H_M C_c^T & H_Y^T A_N^T H_S + H_Y^T B_c^T H_D^T + H_M C_c^T H_S + H_M A_c^T H_D^T \\ A_N^T & A_N^T H_S + B_c^T H_D^T \end{bmatrix},$$
$$E_a^T H \bar{B} = \begin{bmatrix} B_N \\ H_S^T \end{bmatrix},$$
$$E_a^T H \bar{C} = \begin{bmatrix} C_N & 0 \\ H_S C_N & 0 \end{bmatrix}.$$

Let us define the following variables

$$Z_1 = H_M C_c^T,$$
$$Z_2 = H_D B_c,$$
$$Z_3 = H_Y^T A_N^T H_S + H_Y^T Z_2^T + Z_1 H_S + H_M A_c^T H_D^T.$$

Then, by Schur complement, it is clear that LMI (9.16) is equivalent to $\bar{\Omega} < 0$. This completes proof. ■

9.4 Decentralized Dynamic Controller for Synchronization of a CDN with Coupling Delay

In this section, we consider the following decentralized dynamic feedback controllers for each node:

$$\begin{aligned} \dot{\zeta}_i(t) &= A_{ci}\zeta(t) + B_{ci}e_i(t), \\ u_i(t) &= C_{ci}\zeta_i(t), \quad \zeta_i(0) = 0, \quad i = 1, \ldots, N, \end{aligned} \tag{9.18}$$

where $\zeta_i(t) \in \mathbb{R}^n$ is the state vector of controller at ith node, and A_{ci}, B_{ci} and C_{ci} are $n \times n$ dimensional constant gain matrices to be designed.

Then, the closed-loop system can be

$$\dot{v}_i(t) = \bar{A}_i v_i(t) - \sum_{j=1}^{N} \bar{C}_{ij} v_j(t-h) + \bar{B}\bar{f}_i(t), \tag{9.19}$$

where

$$v_i(t) = \begin{bmatrix} e_i(t) \\ \zeta_i(t) \end{bmatrix}, \quad \bar{A}_i = \begin{bmatrix} A & -C_{ci} \\ B_{ci} & A_{ci} \end{bmatrix}, \quad \bar{C}_{ij} = \begin{bmatrix} c_{ij}I & 0 \\ 0 & 0 \end{bmatrix}, \quad \bar{B} = \begin{bmatrix} B \\ 0 \end{bmatrix}. \tag{9.20}$$

9.4.1 *Synchronization Analysis Under Decentralized Dynamic Controllers*

Under decentralized dynamic controller (9.18), we present the following theorem to analyze the synchronization of the CDN (9.1).

Theorem 9.3 *For given positive constants h, l, $0 \le \alpha \le 1$, β_i, and known control gain matrices A_{ci}, B_{ci}, C_{ci}, the dynamic feedback controllers (9.18) guarantees synchronization of the CDN (9.1), if there exist matrices $P_i \in \mathbb{S}_+^{6n}$, Q_i, R_i, $W_i \in \mathbb{S}_+^{2n}$, $H_i \in \mathbb{R}^{2n\times 2n}$ satisfying the following LMIs: $\forall i = 1, 2, \ldots, N$*

$$\Upsilon_i = \begin{bmatrix} \Gamma_{i1} & -P_{i2}+P_{i3}-\frac{2}{h_1}R_i & -P_{i3} & h_1P_{i4}^T+\frac{6}{h_1}R_i & h_dP_{i5} & P_{i1}+\bar{A}^TH_i^T & 0 \\ * & \Gamma_{i2} & -\frac{2}{h_d}R_i & \Gamma_{i3} & h_d(P_{i6}^T-P_{i5})+\frac{6}{h_d}R_i & 0 & 0 \\ * & * & -W_i-\frac{4}{h_d}R_i & -h_1P_{i5}^T & -h_dP_{i6}^T+\frac{6}{h_d}R_i & 0 & 0 \\ * & * & * & -\frac{12}{h_1}R_i & 0 & P_{i2}^T & 0 \\ * & * & * & * & -\frac{12}{h_d}R_i & P_{i3}^T & 0 \\ * & * & * & * & * & \Gamma_{i4} & H_i\bar{B} \\ * & * & * & * & * & * & -\beta I \end{bmatrix} < 0, \quad (9.21)$$

where

$$\Gamma_{i1} = P_{i2} + P_{i2}^T + Q_i - \frac{4}{h_1}R_i + \beta L + NI,$$

$$\Gamma_{i2} = -Q_i + W_i - \frac{4}{h_1}R_i - \frac{4}{h_d}R_i,$$

$$\Gamma_{i3} = -h_1P_{i4}^T + h_1P_{i5}^T + \frac{6}{h_1}R_i.$$

$$\Gamma_{i4} = hR_i - H_i - H_i^T + H_iC_{di}C_{di}H_i^T.$$

Proof Let us define the following Lyapunov functional candidate:

$$V(t) = \sum_{i=1}^{N}\left(\xi_{i1}^T(t)P_i\xi_{i1}(t) + \int_{t-h_1}^{t} v_i^T(s)Q_iv_i(s)ds + \int_{t-h}^{t-h_1} v_i^T(s)W_iv_i(s)ds + \int_{t-h}^{t}\int_{r}^{t} v_i^T(s)R_iv_i(s)ds\right), \quad (9.22)$$

where

$$P_i = \begin{bmatrix} P_{i1} & P_{i2} & P_{i3} \\ * & P_{i4} & P_{i5} \\ * & * & P_{i6} \end{bmatrix},$$

$$\xi_{i1}^T(t) = \Big[v_i^T(t),\ \int_{t-h_1}^{t} v_i^T(s)ds,\ \int_{t-h}^{t-h_1} v_i^T(s)ds \Big].$$

Time derivative of Lyapunov functional (9.22) can be

$$\begin{aligned} \dot{V}(t) = \sum_{i=1}^{N} \Big(& 2\xi_{i1}^T(t)P_i\xi_{i2}(t) + v_i^T(t)Q_i v_i(t) \\ & + v_i^T(t-h_1)(W_i - Q_i)v_i(t-h_1) - v_i^T(t-h)W_i v_i(t-h) \\ & + h\dot{v}_i^T(t)R_i\dot{v}_i(t) - \int_{t-h}^{t} \dot{v}_i^T(s)R_i\dot{v}_i(s)ds \Big), \end{aligned} \tag{9.23}$$

where

$$\xi_{i2}^T(t) = \Big[\dot{v}_i^T(t),\ v_i^T(t) - v_i^T(t-h_1),\ v_i^T(t-h_1) - v_i^T(t-h) \Big].$$

By Lemma 9.1, the integral term in Eq. (9.23) can be estimated as

$$\begin{aligned} & -\int_{t-h}^{t} v_i^T(s)R_i v_i(s)ds \\ & = -\int_{t-h_1}^{t} v_i^T(s)R_i v_i(s)ds - \int_{t-h}^{t-h_1} v_i^T(s)R_i v_i(s)ds \\ & \le \frac{1}{h_1}\begin{bmatrix} v_i(t) \\ v_i(t-h_1) \\ \frac{1}{h_1}\int_{t-h_1}^{t} v_i(s)ds \end{bmatrix}^T \begin{bmatrix} -4R_i & -2R_i & 6R_i \\ * & -4R_i & 6R_i \\ * & * & -12R_i \end{bmatrix} \begin{bmatrix} v_i(t) \\ v_i(t-h_1) \\ \frac{1}{h_1}\int_{t-h_1}^{t} v_i(s)ds \end{bmatrix} \\ & + \frac{1}{h_d}\begin{bmatrix} v_i(t-h_1) \\ v_i(t-h) \\ \frac{1}{h_d}\int_{t-h}^{t-h_1} v_i(s)ds \end{bmatrix}^T \begin{bmatrix} -4R_i & -2R_i & 6R_i \\ * & -4R_i & 6R_i \\ * & * & -12R_i \end{bmatrix} \begin{bmatrix} v_i(t-h_1) \\ v_i(t-h) \\ \frac{1}{h_d}\int_{t-h}^{t-h_1} v_i(s)ds \end{bmatrix}. \end{aligned} \tag{9.24}$$

And by Assumption 9.1, for positive constants β_i the following are obtained

$$0 \le \beta_i(v_i^T(t)Lv_i(t) - \bar{f}_i^T(e_i(t))\bar{f}_i(e_i(t))), \tag{9.25}$$

For matrices $H_i \in \mathbb{S}_+^{2n}$, the following is obtained

$$0 = 2\dot{v}^T(t)H[-\dot{v}(t) + \bar{A}v(t) + \bar{B}\bar{f}(e_i(t)) + \bar{C}v(t-h)]. \tag{9.26}$$

Here, it can be easily obtained the following relation

$$\begin{aligned}
&\sum_{i=1}^{N}\left(-2\dot{v}_i^T(t)H_i\sum_{j=1}^{N}\bar{C}_{ij}v_j(t-h)\right)\\
&\le \sum_{i=1}^{N}\left(\dot{v}_i^T(t)H_iC_{di}C_{di}^TH_i^T\dot{v}_i(t) + \sum_{j=1}^{N}v_j^T(t-h)v_j(t-h)\right)\\
&= \sum_{i=1}^{N}\left(\dot{v}_i^T(t)H_iC_{di}C_{di}H_i^T\dot{v}_i(t) + Nv_i^T(t-h)v_i(t-h)\right),
\end{aligned} \tag{9.27}$$

where $C_{di} = \begin{bmatrix} \sum_{j=1}^{N} c_{ij}^2 I & 0 \\ * & 0 \end{bmatrix}$.

Therefore, by using Eqs. (9.23)–(9.27), a new upper bound of $\dot{V}(t)$ can be

$$\dot{V}(t) \le \sum_{i=1}^{N}\eta_i^T(t)\Upsilon_i\eta_i(t), \tag{9.28}$$

where

$$\begin{aligned}
\eta_i^T(t) = \Big[&v_i^T(t),\ v_i^T(t-h_1),\ v_i^T(t-h),\ \tfrac{1}{h_1}\textstyle\int_{t-h_1}^{t}v_i^T(s)ds,\ \tfrac{1}{h_d}\int_{t-h}^{t-h_1}v_i^T(s)ds,\\
&\dot{v}_i^T(t),\ \bar{f}_i^T(e_i(t))\Big].
\end{aligned}$$

Here, it is clear that if LMI (9.8) holds then the closed-loop system (9.19) is asymptotically stable. This completes the proof. ■

9.4.2 *Design of Decentralized Dynamic Controllers*

Now, we are in the position to design control gains A_c, B_c, C_c by the following theorem.

Theorem 9.4 *For given positive constants h, l, $0 \leq \alpha \leq 1$, β_i, the dynamic feedback controller (9.18) guarantees synchronization of the CDN (9.1), if there exist matrices $H_{iY}, H_{iS} \in \mathbb{S}^n_+$, $\bar{P}_{ijk}, \bar{Q}_{ik}, \bar{R}_{ik}, \bar{W}_{ik}, Z_{ik} \in \mathbb{R}^{n\times n}$ for $j = 1, 4, 6$; $k = 1, 2, 3$, $\bar{P}_{ijk} \in \mathbb{R}^{n\times n}$ for $j = 2, 3, 5$; $k = 1, \ldots, 4$, satisfying the following LMIs: $\forall i = 1, \ldots, N$*

$$\begin{cases} \bar{P}_i = \begin{bmatrix} \bar{P}_{i11} & \bar{P}_{i12} & \bar{P}_{i21} & \bar{P}_{i22} & \bar{P}_{i31} & \bar{P}_{i32} \\ * & \bar{P}_{i13} & \bar{P}_{i23} & \bar{P}_{i24} & \bar{P}_{i33} & \bar{P}_{i34} \\ * & * & \bar{P}_{i41} & \bar{P}_{i42} & \bar{P}_{i51} & \bar{P}_{i52} \\ * & * & * & \bar{P}_{i43} & \bar{P}_{i53} & \bar{P}_{i54} \\ * & * & * & * & \bar{P}_{i61} & \bar{P}_{i62} \\ * & * & * & * & * & \bar{P}_{i63} \end{bmatrix} > 0, \\ \bar{Q}_i = \begin{bmatrix} \bar{Q}_{i1} & \bar{Q}_{i2} \\ * & \bar{Q}_{i3} \end{bmatrix} > 0, \\ \bar{R}_i = \begin{bmatrix} \bar{R}_{i1} & \bar{R}_{i2} \\ * & \bar{R}_{i3} \end{bmatrix} > 0, \\ \bar{W}_i = \begin{bmatrix} \bar{W}_{i1} & \bar{W}_{i2} \\ * & \bar{W}_{i3} \end{bmatrix} > 0, \\ \begin{bmatrix} H_{iY} & I \\ * & H_{iS} \end{bmatrix} > 0, \end{cases} \tag{9.29}$$

$$\left[\begin{array}{ccccccccccc}
\Delta_{i,1} & \rho_i H_{iY}^T & N H_{iM} & \Delta_{i,2} & \Delta_{i,3} & \Delta_{i,4} & \bar{P}_{i31} & 0 & 0 & \bar{P}_{i32} & \Delta_{i,5} \\
* & -I & 0 & 0 & 0 & 0 & 0 & 0 & 0 & 0 & 0 \\
* & * & -NI & 0 & 0 & 0 & 0 & 0 & 0 & 0 & 0 \\
* & * & * & \Delta_{i,8} & \Delta_{i,9} & \Delta_{i,10} & \bar{P}_{i32}^T & 0 & 0 & \bar{P}_{i33} & \Delta_{i,11} \\
* & * & * & * & \Delta_{i,14} & \Delta_{i,15} & -\frac{2}{h_d}\bar{R}_{i1} & 0 & 0 & -\frac{2}{h_d}\bar{R}_{i2} & \Delta_{i,16} \\
* & * & * & * & * & \Delta_{i,20} & -\frac{2}{h_d}\bar{R}_{i2}^T & 0 & 0 & -\frac{2}{h_d}\bar{R}_{i3} & \Delta_{i,21} \\
* & * & * & * & * & * & -\bar{W}_{i1} - \frac{4}{h_d}\bar{R}_{i1} & H_{iY}^T & H_{iM} & -\bar{W}_{i2} - \frac{4}{h_d}\bar{R}_{i2} & -h_1 \bar{P}_{i51}^T \\
* & * & * & * & * & * & * & -I & 0 & 0 & 0 \\
* & * & * & * & * & * & * & * & -I & 0 & 0 \\
* & * & * & * & * & * & * & * & * & -\bar{W}_{i3} - \frac{4}{h_d}\bar{R}_{i3} & -h_1 \bar{P}_{i52}^T \\
* & * & * & * & * & * & * & * & * & * & -\frac{12}{h_1}\bar{R}_{i1} \\
* & * & * & * & * & * & * & * & * & * & * \\
* & * & * & * & * & * & * & * & * & * & * \\
* & * & * & * & * & * & * & * & * & * & * \\
* & * & * & * & * & * & * & * & * & * & * \\
* & * & * & * & * & * & * & * & * & * & * \\
* & * & * & * & * & * & * & * & * & * & * \\
* & * & * & * & * & * & * & * & * & * & * \\
* & * & * & * & * & * & * & * & * & * & *
\end{array}\right.$$

$$
\left.\begin{matrix}
\Delta_{i,6} & h_d\bar{P}_{i51} & h_d\bar{P}_{i52} & \Delta_{i,7} & \bar{P}_{i12}+Z_{i3} & 0 & 0 & 0 \\
0 & 0 & 0 & 0 & 0 & 0 & 0 & 0 \\
0 & 0 & 0 & 0 & 0 & 0 & 0 & 0 \\
\Delta_{i,12} & h_d\bar{P}_{i53} & h_d\bar{P}_{i54} & \bar{P}_{i12}^T+A^T & \Delta_{i,13} & 0 & 0 & 0 \\
\Delta_{i,17} & \Delta_{i,18} & \Delta_{i,19} & 0 & 0 & 0 & 0 & 0 \\
\Delta_{i,22} & \Delta_{i,23} & \Delta_{i,24} & 0 & 0 & 0 & 0 & 0 \\
-h_1\bar{P}_{i53}^T & \Delta_{i,25} & \Delta_{i,26} & 0 & 0 & 0 & 0 & 0 \\
0 & 0 & 0 & 0 & 0 & 0 & 0 & 0 \\
0 & 0 & 0 & 0 & 0 & 0 & 0 & 0 \\
-h_1\bar{P}_{i54}^T & \Delta_{i,27} & \Delta_{i,28} & 0 & 0 & 0 & 0 & 0 \\
-\frac{12}{h_1}\bar{R}_{i2} & 0 & 0 & \bar{P}_{i21}^T & \bar{P}_{i23}^T & 0 & 0 & 0 \\
-\frac{12}{h_1}\bar{R}_{i3} & 0 & 0 & \bar{P}_{i22}^T & \bar{P}_{i24}^T & 0 & 0 & 0 \\
* & -\frac{12}{h_d}\bar{R}_{i1} & -\frac{12}{h_d}\bar{R}_{i2} & \bar{P}_{i31}^T & \bar{P}_{i33}^T & 0 & 0 & 0 \\
* & * & -\frac{12}{h_d}\bar{R}_{i3} & \bar{P}_{i32}^T & \bar{P}_{i34}^T & 0 & 0 & 0 \\
* & * & * & h\bar{R}_{i1}-2H_{iY} & h\bar{R}_{i2}-2I & C_{id} & 0 & B \\
* & * & * & * & h\bar{R}_{i3}-2H_{iS} & H_{iS}C_{id} & 0 & H_{iS}^T \\
* & * & * & * & * & -I & 0 & 0 \\
* & * & * & * & * & * & -I & 0 \\
* & * & * & * & * & * & * & -\beta_i I
\end{matrix}\right] < 0, \tag{9.30}
$$

where

$$\Delta_{i,1} = \bar{P}_{i21} + \bar{P}_{i21}^T + \bar{Q}_{i1} - \frac{4}{h_1}\bar{R}_{i1}, \quad \Delta_{i,2} = \bar{P}_{i22} + \bar{P}_{i23}^T + \bar{Q}_{i2} - \frac{4}{h_1}\bar{R}_{i2} + \rho_i^2 H_{iY}^T,$$

$$\Delta_{i,3} = -\bar{P}_{i21} + \bar{P}_{i31} - \frac{2}{h_2}\bar{R}_{i1}, \quad \Delta_{i,4} = -\bar{P}_{i22} + \bar{P}_{i32} - \frac{2}{h_1}\bar{R}_{i2},$$

$$\Delta_{i,5} = h_1\bar{P}_{i41} + \frac{6}{h_1}\bar{R}_{i1}, \quad \Delta_{i,6} = h_1\bar{P}_{i42} + \frac{6}{h_1}\bar{R}_{i2},$$

$$\Delta_{i,7} = \bar{P}_{i11} + H_{iY}^T A^T + Z_{i1}, \quad \Delta_{i,8} = \bar{P}_{i24} + \bar{P}_{i24}^T + \bar{W}_{i3} - \frac{4}{h_1}\bar{R}_{i3} + \rho_i^2 I,$$

$$\Delta_{i,9} = -\bar{P}_{i23} + \bar{P}_{i32}^T - \frac{2}{h_2}\bar{R}_{i2}^T, \quad \Delta_{i,10} = -\bar{P}_{i24} + \bar{P}_{i33} - \frac{2}{h_1}\bar{R}_{i3},$$

$$\Delta_{i,11} = h_1\bar{P}_{i42}^T + \frac{6}{h_1}\bar{R}_{i2}^T, \quad \Delta_{i,12} = h_1\bar{P}_{i43} + \frac{6}{h_1}\bar{R}_{i3},$$

$$\Delta_{i,13} = \bar{P}_{i13} + A^T H_{iS} + Z_{i2}^T, \quad \Delta_{i,14} = -\bar{Q}_{i1} + \bar{W}_{i1} - \frac{4}{h_1}\bar{R}_{i1} - \frac{4}{h_d}\bar{R}_{i1},$$

$$\Delta_{i,15} = -\bar{Q}_{i2} + \bar{W}_{i2} - \frac{4}{h_1}\bar{R}_{i2} - \frac{4}{h_d}\bar{R}_{i2}, \quad \Delta_{i,16} = -h_1\bar{P}_{i41} + h_1\bar{P}_{i51}^T + \frac{6}{h_1}\bar{R}_{i1},$$

$$\Delta_{i,17} = -h_1\bar{P}_{i42} + h_1\bar{P}_{i53}^T + \frac{6}{h_1}\bar{R}_{i2}, \quad \Delta_{i,18} = h_d(\bar{P}_{i61} - \bar{P}_{i51}) + \frac{6}{h_d}\bar{R}_{i1},$$

$$\Delta_{i,19} = h_d(\bar{P}_{i62} - \bar{P}_{i52}) + \frac{6}{h_d}\bar{R}_{i2}, \quad \Delta_{i,20} = -\bar{Q}_{i3} + \bar{W}_{i3} - \frac{4}{h_1}\bar{R}_{i3} - \frac{4}{h_d}\bar{R}_{i3},$$

$$\Delta_{i,21} = -h_1\bar{P}_{i42}^T + h_1\bar{P}_{i52}^T + \frac{6}{h_1}\bar{R}_{i2}^T, \quad \Delta_{i,22} = -h_1\bar{P}_{i43} + h_1\bar{P}_{i54}^T + \frac{6}{h_1}\bar{R}_{i3},$$

$$\Delta_{i,23} = h_d(\bar{P}_{i62}^T - \bar{P}_{i53}) + \frac{6}{h_d}\bar{R}_{i2}^T, \quad \Delta_{i,24} = h_d(\bar{P}_{i63} - \bar{P}_{i54}) + \frac{6}{h_d}\bar{R}_{i3},$$

$$\Delta_{i,25} = -h_d\bar{P}_{i61} + \frac{6}{h_d}\bar{R}_{i1}, \quad \Delta_{i,26} = -h_d\bar{P}_{i62} + \frac{6}{h_d}\bar{R}_{i2},$$

$$\Delta_{i,27} = -h_d\bar{P}_{i62}^T + \frac{6}{h_d}\bar{R}_{i2}^T, \quad \Delta_{i,28} = -h_d\bar{P}_{i63} + \frac{6}{h_d}\bar{R}_{i3},$$

$$\rho_i = \sqrt{N + \beta_i l^2}.$$

Proof To follow the same procedure to Theorem 9.2, let us define the following matrices:

$$H_i = \begin{bmatrix} H_{iS} & H_{iD} \\ * & H_{iT} \end{bmatrix}, \quad H_i^{-1} = \begin{bmatrix} H_{iY} & H_{iM} \\ * & H_{iW} \end{bmatrix},$$

$$E_{ia} = \begin{bmatrix} H_{iY} & I \\ H_{iM}^T & 0 \end{bmatrix}, \quad E_{ib} = \begin{bmatrix} I & H_{iS} \\ 0 & H_{iD} \end{bmatrix},$$

$$P_{ij} = \begin{bmatrix} P_{ij1} & P_{ij2} \\ * & P_{ij3} \end{bmatrix} \text{ for } j = 1, 4, 6, \quad P_{ij} = \begin{bmatrix} P_{ij1} & P_{ij2} \\ P_{ij3} & P_{ij4} \end{bmatrix} \text{ for } j = 2, 3, 5,$$

$$Q_i = \begin{bmatrix} Q_{i1} & Q_{i2} \\ * & Q_{i3} \end{bmatrix} \quad W_i = \begin{bmatrix} W_{i1} & W_{i2} \\ * & W_{i3} \end{bmatrix} \quad R_i = \begin{bmatrix} R_{i1} & R_{i2} \\ * & R_{i3} \end{bmatrix},$$

$$\bar{P}_{ij} = E_{ia}^T P_{ij} E_{ia}, \text{ for } j = 1, \ldots, 6,$$

$$\bar{Q}_i = E_{ia}^T Q_i E_{ia}, \quad \bar{W}_i = E_{ia}^T W_i E_{ia}, \quad \bar{R}_i = E_{ia}^T R_i E_{ia},$$

$$Z_{i1} = H_{iM} C_{ic}^T,$$

$$Z_{i2} = H_{iD} B_{ic},$$

$$Z_{i3} = H_{iY}^T A^T H_{iS} + H_{iY}^T Z_{i2}^T + Z_{i1} H_{iS} + H_{iM} A_{ic}^T H_{iD}^T.$$

Then, by Schur complement, $\mathrm{diag}\Big\{\underbrace{E_{ia}^T, \ldots, E_{ia}^T}_{6}, I\Big\}\Upsilon\,\mathrm{diag}\Big\{\underbrace{E_{ia}, \ldots, E_{ia}}_{6}, I\Big\} < 0$ is equivalent to LMIs (9.30). This completes proof. ■

Remark 9.1 It should noted that the decentralized control method can cover the case of different coupling delays for each nodes because it designs the controllers in the framework of each nodes. Apart from this situation, in this chapter, we consider the coupling delay of each nodes are the same for simplicity and unification with centralized concept.

9.5 Numerical Example

For a numerical example, we consider a CDN consisted of three nodes with the following Lorenz chaotic system [25]:

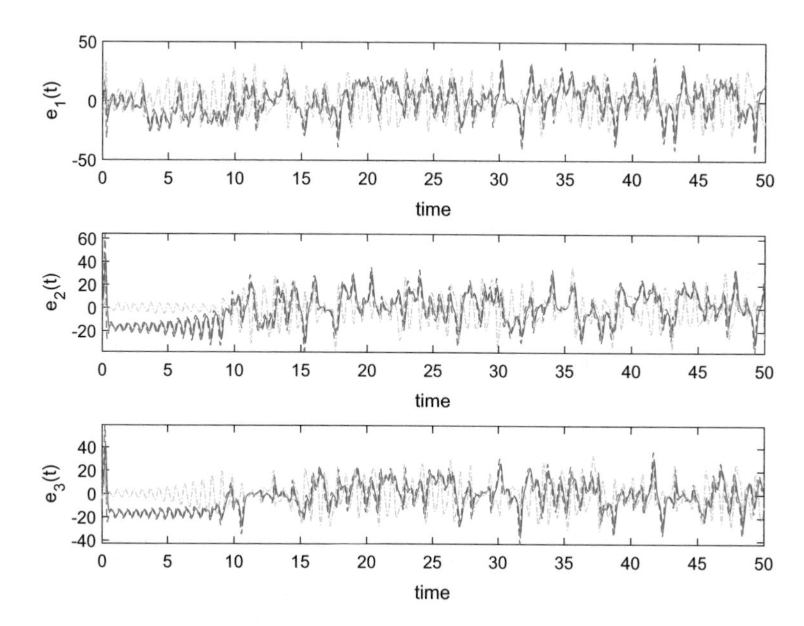

Fig. 9.1 The uncontrolled error signals

$$A = \begin{bmatrix} -10 & 10 & 0 \\ 28 & -1 & 0 \\ 0 & 0 & -8/3 \end{bmatrix}, \quad B = \begin{bmatrix} I & 0 & 0 \\ 0 & -I & 0 \\ 0 & 0 & I \end{bmatrix}, \quad C = 0.2 \begin{bmatrix} -2 & 0 & 1 \\ 0 & -2 & 1 \\ 1 & 1 & -3 \end{bmatrix},$$

$$f(x(t)) = \begin{bmatrix} 0 \\ x_1(t)x_3(t) \\ x_1(t)x_2(t) \end{bmatrix},$$

where the function $f(\cdot)$ satisfies Lipschitz constants $l = 80$.

And the other parameters and initial conditions are set as $\alpha = 0.5$, $\beta = 0.1$, $h = 0.5$, $x_1(0) = (1, 2, 3)$, $x_2(0) = (-5, 2, -7)$, $x_3(0) = (-10, 10, -3)$, $s(0) = (1, 5, -7)$, respectively.

First, the dynamic behavior of the CDN (9.1) without a controller is depicted in Fig. 9.1 which can be seen that the CDN is not synchronized without control.

With the above parameters, the centralized and decentralized dynamic control gains can be obtained by Theorems 9.1 and 9.2, respectively. See Appendices 1 and 2. Figures 9.2 and 9.4 shows the error signals under centralized dynamic controller (9.5) and decentralized dynamic controller (9.18) which implies the synchronization is achieved by designed controller. In addition, the state trajectories of the designed controller by Theorems 9.1 and 9.2 are presented in Figs. 9.3 and 9.5, respectively.

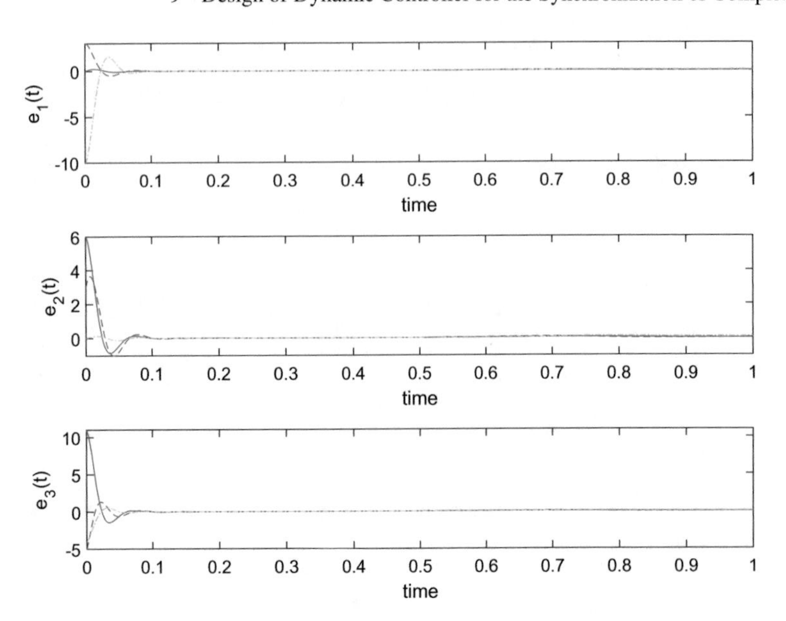

Fig. 9.2 The error signals under the centralized dynamic controller (9.5)

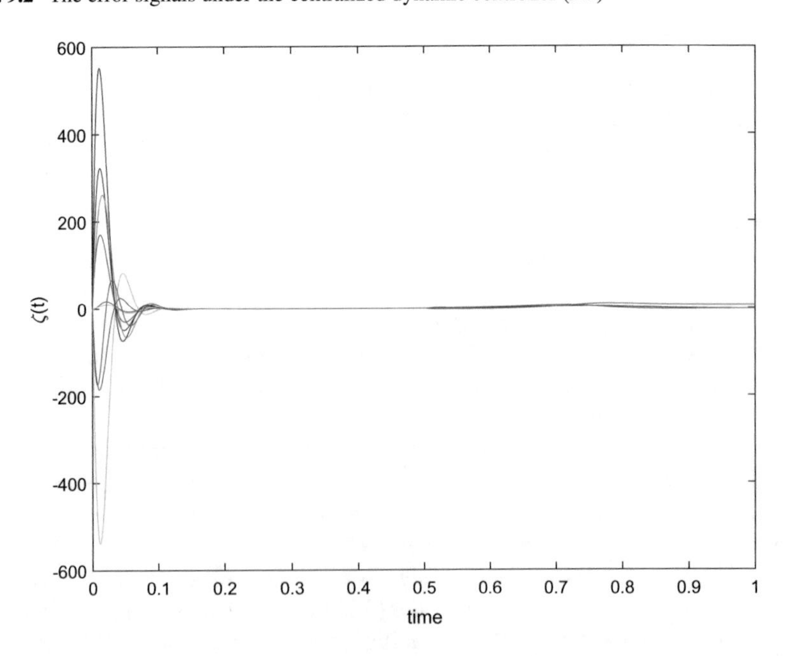

Fig. 9.3 The state trajectories of the centralized dynamic controller (9.5)

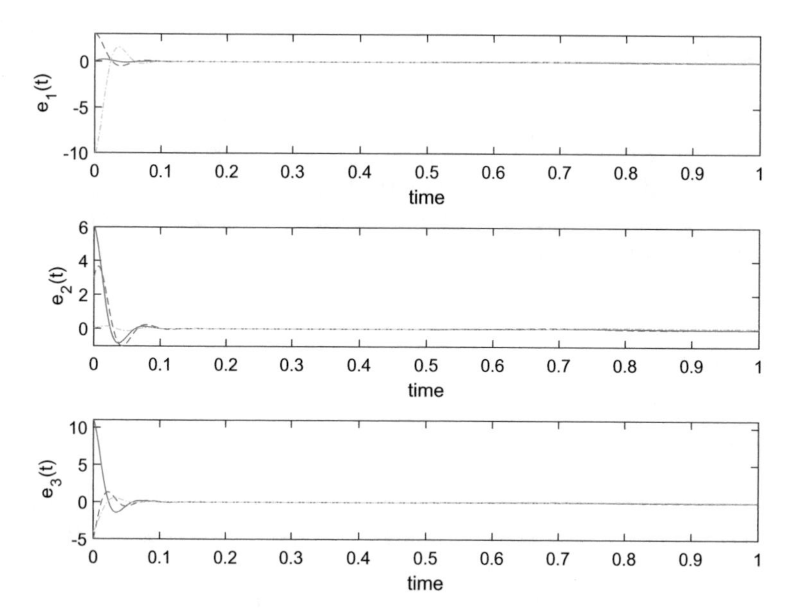

Fig. 9.4 The error signals under the decentralized dynamic controller (9.18)

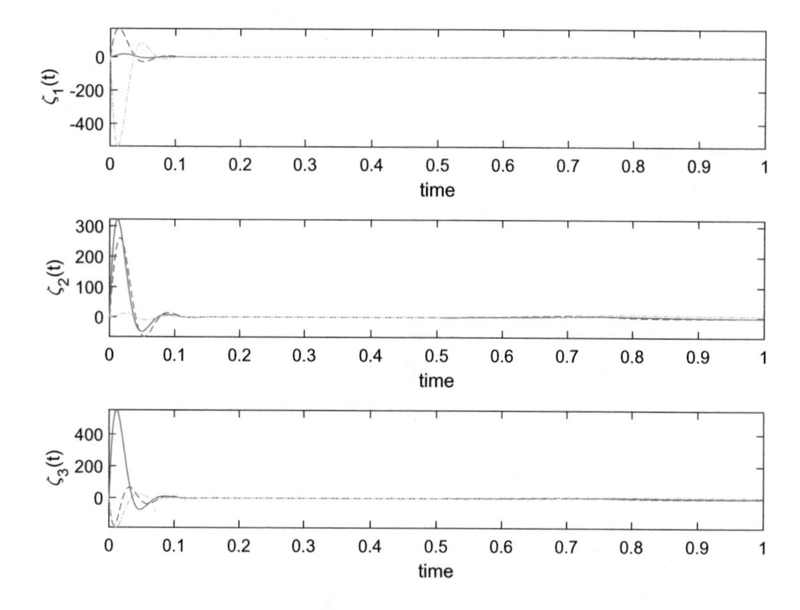

Fig. 9.5 The state trajectories of the decentralized dynamic controller (9.18)

9.6 Conclusion

The synchronization of a CDN with coupling delay was investigated in this chapter. To this end, two types of the controller were designed: centralized dynamics controller and decentralized dynamic controller. Variable delay fragmentation approach was employed to derive the sufficient conditions for the controllers. A numerical example was given to illustrate the validity of the proposed theorems.

Appendix 1

The control parameters of the designed centralized controller:

$$A_c = \left[\begin{matrix} -102.1920 & -0.0001 & -0.0000 & -0.0099 & -0.2000 & 0.0007 \\ -0.0001 & -102.1909 & 0.0000 & 0.0002 & 0.0006 & 0.2001 \\ 0.0000 & 0.0000 & -102.1944 & -0.2016 & 0.0099 & 0.0000 \\ -0.0098 & 0.0001 & -0.2014 & -102.3880 & -0.0001 & 0.0001 \\ -0.2001 & 0.0006 & 0.0099 & -0.0001 & -102.3908 & -0.0001 \\ 0.0007 & 0.1999 & 0.0000 & 0.0000 & -0.0001 & -102.3915 \\ 0.0000 & 0.0001 & 0.0001 & 0.0000 & 0.0011 & 0.1998 \\ -0.0000 & -0.0000 & 0.0017 & 0.1945 & -0.0459 & 0.0003 \\ 0.0002 & 0.0000 & 0.0003 & 0.0459 & 0.1945 & -0.0011 \end{matrix}\right.$$
$$\left.\begin{matrix} -0.0005 & 0.0363 & -0.1968 \\ 0.2000 & 0.0002 & -0.0004 \\ 0.0002 & -0.1966 & -0.0363 \\ 0.0001 & 0.0014 & 0.0003 \\ -0.0001 & -0.0002 & 0.0002 \\ -0.0003 & -0.0000 & -0.0000 \\ -102.1915 & 0.0000 & 0.0000 \\ -0.0001 & -102.1916 & -0.0001 \\ -0.0000 & -0.0000 & -102.1913 \end{matrix}\right],$$

$$B_c = \begin{bmatrix} -0.0005 & -0.0108 & -0.1769 & 0.0001 & 0.0183 & -0.5646 \\ 0.1003 & 0.0185 & -0.0021 & 0.3480 & -0.0171 & -0.0269 \\ -0.0040 & 0.0922 & -0.0081 & -0.0016 & 1.2253 & 0.0158 \\ -0.0439 & -0.1715 & 0.0053 & 0.0199 & 100.9738 & -4.9414 \\ -0.0005 & 0.0034 & 0.2462 & -0.3277 & -4.9409 & -100.9791 \\ 0.1980 & -0.0035 & 0.0003 & -101.1009 & 0.0358 & 0.3262 \\ -100.9011 & -0.0345 & -0.2326 & 0.0016 & -0.0439 & 0.0031 \\ -0.0762 & 99.2220 & 18.3324 & -0.0033 & -0.0272 & 0.0034 \\ 0.2225 & 18.3326 & -99.2215 & -0.0017 & -0.0119 & -0.0462 \end{bmatrix}$$

$$
\left.\begin{matrix}
-0.0020 & 0.0001 & 100.9011 \\
-100.9011 & 0.0401 & -0.0020 \\
-0.0403 & -100.8978 & -0.0001 \\
-0.0152 & 0.8219 & -0.0262 \\
0.0278 & -0.0560 & -0.1632 \\
0.0518 & 0.0021 & 0.0006 \\
0.0998 & 0.0037 & -0.0004 \\
0.0178 & -0.1061 & 0.0065 \\
0.0053 & -0.0115 & 0.0243
\end{matrix}\right],
$$

$$
Cc = \left[\begin{matrix}
-0.0005 & 0.0995 & -0.0040 & -0.0442 & -0.0005 & 0.1980 \\
-0.0109 & 0.0186 & 0.0914 & -0.1713 & 0.0033 & -0.0035 \\
-0.1767 & -0.0021 & -0.0081 & 0.0053 & 0.2466 & 0.0003 \\
0.0001 & 0.3476 & -0.0017 & 0.0200 & -0.3302 & -101.8922 \\
0.0184 & -0.0173 & 1.2318 & 101.7641 & -4.9796 & 0.0360 \\
-0.5659 & -0.0272 & 0.0159 & -4.9801 & -101.7695 & 0.3288 \\
-0.0020 & -101.6925 & -0.0406 & -0.0153 & 0.0281 & 0.0522 \\
0.0001 & 0.0404 & -101.6891 & 0.8283 & -0.0564 & 0.0021 \\
101.6924 & -0.0020 & -0.0001 & -0.0264 & -0.1645 & 0.0006
\end{matrix}\right.
$$

$$
\left.\begin{matrix}
-101.6924 & -0.0768 & 0.2242 \\
-0.0347 & 100.0002 & 18.4763 \\
-0.2345 & 18.4761 & -99.9997 \\
0.0016 & -0.0033 & -0.0018 \\
-0.0442 & -0.0274 & -0.0120 \\
0.0031 & 0.0034 & -0.0466 \\
0.1006 & 0.0180 & 0.0053 \\
0.0037 & -0.1069 & -0.0116 \\
-0.0004 & 0.0065 & 0.0245
\end{matrix}\right].
$$

Appendix 2

The control parameters of the designed centralized controller for the first node:

$$
A_{c1} = \begin{bmatrix}
-102.0399 & 0.0000 & 0.0799 \\
0.0000 & -102.0000 & -0.0000 \\
0.0799 & -0.0000 & -102.1601
\end{bmatrix},
$$

$$
B_{c1} = \begin{bmatrix}
0.1162 & -100.5412 & -0.0000 \\
-0.0000 & -0.0000 & 100.5012 \\
-100.6613 & 0.0438 & -0.0000
\end{bmatrix},
$$

$$C_{c1} = \begin{bmatrix} 0.1157 & -0.0000 & -101.6613 \\ -101.5412 & -0.0000 & 0.0442 \\ -0.0000 & 101.5012 & -0.0000 \end{bmatrix}.$$

The control parameters of the designed centralized controller for the second node:

$$A_{c2} = \begin{bmatrix} -102.0399 & 0.0000 & 0.0799 \\ 0.0000 & -102.0000 & -0.0000 \\ 0.0799 & -0.0000 & -102.1601 \end{bmatrix},$$

$$B_{c2} = \begin{bmatrix} 0.1162 & -100.5412 & -0.0000 \\ -0.0000 & -0.0000 & 100.5012 \\ -100.6613 & 0.0438 & -0.0000 \end{bmatrix},$$

$$C_{c2} = \begin{bmatrix} 0.1157 & -0.0000 & -101.6613 \\ -101.5412 & -0.0000 & 0.0442 \\ -0.0000 & 101.5012 & -0.0000 \end{bmatrix}.$$

The control parameters of the designed centralized controller for the third node:

$$A_{c3} = \begin{bmatrix} -102.0399 & 0.0000 & 0.0799 \\ 0.0000 & -102.0000 & -0.0000 \\ 0.0799 & -0.0000 & -102.1601 \end{bmatrix},$$

$$B_{c3} = \begin{bmatrix} 0.1162 & -100.5412 & -0.0000 \\ -0.0000 & -0.0000 & 100.5012 \\ -100.6613 & 0.0438 & -0.0000 \end{bmatrix},$$

$$C_{c3} = \begin{bmatrix} 0.1157 & -0.0000 & -101.6613 \\ -101.5412 & -0.0000 & 0.0442 \\ -0.0000 & 101.5012 & -0.0000 \end{bmatrix}.$$

References

1. Barabasi AL, Albert R (1999) Emergence of scaling in random networks. Science 286:509–512
2. Strogatz SH (2001) Exploring complex networks. Nature 410:268–276
3. Dorogovtesev SN, Mendes JFF (2002) Evolution of networks. Adv Phys 51:1079–1187
4. Newman MEJ (2003) The structure and function of complex networks. SIAM Rev 45:167–256
5. Erdös P, Rényi A (1959) On random graphs I. Publ Math 6:290–297
6. Erdös P, Rényi A (1960) On the evolution of random graphs. Publ Math Inst Hung Acad Sci 5:17–61
7. Watts DJ, Strogatz SH (1998) Collective dynamics of small-world networks. Nature 393:440–442
8. Newman MEJ, Watts DJ (1999) Renormalization group analysis of the small-world network model. Phys Lett A 263:341–346
9. Tang Z, Park JH, Feng J (2018) Novel approaches to pin cluster synchronization of complex dynamical networks in Lur'e forms. Commun Nonlinear Sci Numer Simul 57:422–438

10. Zhang R, Zeng D, Park JH, Liu Y, Zhong S (2018) Non-fragile sampled-data synchronization for delayed complex dynamical networks with randomly occurring controller gain fluctuations. IEEE Trans Syst Man Cybern: Syst 48:2271–2281
11. Liu Y, Guo BZ, Park JH, Lee SM (2018) Non-fragile exponential synchronization of delayed complex dynamical networks with memory sampled-data control. IEEE Trans Neural Netw Learn Syst 29:118–128
12. Shen H, Park JH, Wu ZG, Zhang Z (2015) Finite-time $\mathscr{H}_\infty$ synchronization for complex networks with semi-Markov jump topology. Commun Nonlinear Sci Numer Simul 24:40–51
13. Lee TH, Wu ZG, Park JH (2012) Synchronization of a complex dynamical network with coupling time-varying delays via sampled-data control. Appl Math Comput 219:1354–1366
14. Lee TH, Park JH, Jung HY, Lee SM, Kwon OM (2012) Synchronization of a delayed complex dynamical network with free coupling matrix. Nonlinear Dyn 69:1081–1090
15. Sheng S, Zhang X, Lu G (2018) Finite-time outer-synchronization for complex networks with Markov jump topology via hybrid control and its application to image encryption. J Frankl Inst 355:6493–6519
16. Lei X, Cai S, Jiang S, Liu Z (2017) Adaptive outer synchronization between two complex delayed dynamical networks via aperiodically intermittent pinning control. Neurocomputing 222:26–35
17. Sun Y, Li W, Ruan J (2013) Generalized outer synchronization between complex dynamical networks with time delay and noise perturbation. Commun Nonlinear Sci Numer Simul 18:989–998
18. Lee TH, Park JH, Wu ZG, Lee SC, Lee DH (2012) Robust $\mathscr{H}_\infty$ decentralized dynamic control for synchronization of a complex dynamical network with randomly occurring uncertainties. Nonlinear Dyn 70:559–570
19. Lee TH, Ji DH, Park JH, Jung HY (2012) Decentralized guaranteed cost dynamic control for synchronization of a complex dynamical network with randomly switching topology. Appl Math Comput 219:996–1010
20. Lee TH, Park JH, Ji DH, Kwon OM, Lee SM (2012) Guaranteed cost synchronization of a complex dynamical network via dynamic feedback control. Appl Math Comput 218:6469–6481
21. Park JH (2004) Design of dynamic output feedback controller for a class of neutral systems with discrete and distributed delays. IEE Proc Control Theory Appl 151:610–614
22. Briat C (2011) Convergence and equivalence results for the Jensen's inequality - application to time-delay and sampled-data systems. IEEE Trans Autom Control 56:1660–1665
23. Lee TH, Park JH, Jung HY, Kwon OM, Lee SM (2014) Improved results on stability of time-delay systems using Wirtinger-based inequality. In: Proceedings of the 19th IFAC world congress, pp 6826–6830
24. Seuret A, Gouaisbaut F (2013) Wirtinger-based integral inequality: application to time-delay systems. Automatica 49:2860–2866
25. Lorenz EN (1963) Deterministic nonperiodic flow. J Atmos Sci 20:130–141

Chapter 10
Reliable Sampled-Data Control for Synchronization of Chaotic Lur'e Systems with Actuator Failures

10.1 Introduction

Chaotic Lur'e systems (CLSs) have received wide attention since many nonlinear systems can be represented by CLSs [1–7]. As well known that the Lur'e system has a feedback connection with a linear dynamical system and a sector bounded nonlinear element. It has been extensively studied since as early as the 1940s when Lur'e and Postnikov first proposed the concept of absolute stability.

Synchronization, as is ubiquitous in nature, has many important applications [8]. Since the pioneering work of Pecora and Carroll [9] was presented for synchronization of two identical chaotic systems, chaotic synchronization has received much attention due to its theoretical and practical importance [10–13]. As a consequence, the investigation of synchronization for CLSs is also hot and far-reaching because various chaotic systems such as Chua's circuit, n-scroll attractors, and hyperchaotic attractors can be modeled as Lur'e systems [14–16].

With digital technology growing, several control strategies are in the spotlight because of easy maintenance, high efficiency, and low cost consumption, such as impulsive control [17], intermittent control [18], and sampled-data control [19–24]. These control strategies make systems a kind of hybrid systems owing to the coexistence of both continuous and discontinuous signals in the system [29]. Among them, the sampled-data control has more advantages than some other existing control mechanisms, since it only uses system sampled information at its sampled instants. Then, sampled-data control is useful to reduce the communication burden. Therefore, it is profound to study the sampled-data synchronization for CLSs. Recently, such the topic is very hot, and many good results appear in the literature [26–29]. Here, it is noted that the most popular method for sampled-data controls is the input delay approach [30] which treats sampled-data control systems as delayed control systems.

In reality, it is well-known that actuator failures are unavoidable because of actuators aging, the damage of actuator's components, the occurrence of input disturbance, and so on. It is also known that the existence of such failures is one of the main causes of poor performance or even instability of control systems [25].

J. H. Park et al., *Dynamic Systems with Time Delays: Stability and Control*,
https://doi.org/10.1007/978-981-13-9254-2_10

However, all the sampled-data control mechanisms in [26–29] are designed under a full reliability assumption that all actuators of CLSs work properly. When failures occur in actuators, the existing synchronization criteria in [26–29] do not work. Therefore, it is significant to design a reliable sampled-data control mechanism for CLSs. However, to the best of our knowledge, few works have been done on this topic.

Motivated by the above discussions, a reliable sampled-data control mechanism is designed for asymptotic synchronization of CLSs with actuator failures. The main contributions are below:

(1) A reliable sampled-data control mechanism is firstly given for CLSs with actuator failures.

(2) A new LKF, which fully deploys the sampled pattern information, is presented for CLSs.

The chapter is organized as follows: a problem statement is described in Sect. 10.2. In Sect. 10.3, a stabilizing condition for chaotic synchronization of CLSs with actuator failures is derived. Section 10.4 contains a numerical example in which the effectiveness of the proposed method will be demonstrated. Conclusions are finally given in Sect. 10.5.

10.2 Problem Description and Preliminaries

Consider the following CLSs of a Master-Slave configuration:

$$\begin{aligned}
&\mathscr{M}:\begin{cases}\dot{y}(t)=\mathscr{A}y(t)+\mathscr{B}f(\mathscr{E}y(t)),\\ p(t)=\mathscr{D}y(t),\end{cases}\\
&\mathscr{S}:\begin{cases}\dot{x}(t)=\mathscr{A}x(t)+\mathscr{B}f(\mathscr{E}x(t))+u(t),\\ q(t)=\mathscr{D}x(t),\end{cases}\\
&\mathscr{C}:u(t)=\mathscr{K}(p(t_k)-q(t_k)),\quad t_k\le t<t_{k+1},
\end{aligned}\tag{10.1}$$

where $\mathscr{M}$ and $\mathscr{S}$ are CLSs called, respectively, the master and controlled slave systems.

In the CLSs (10.1), $\mathscr{C}$ is the controller. $y(t)$ and $x(t)\in\mathbb{R}^n$ are state vectors in master and slave systems, and $p(t)$, $q(t)\in\mathbb{R}^l$ are output vectors. $\mathscr{A}\in\mathbb{R}^{n\times n}$, $\mathscr{D}\in\mathbb{R}^{l\times n}$, $\mathscr{E}\in\mathbb{R}^{m\times n}$, and $\mathscr{B}\in\mathbb{R}^{n\times m}$ are constant matrices. $u(t)\in\mathbb{R}^n$ is the control input with $\mathscr{K}\in\mathbb{R}^{n\times l}$ being the control gain matrix. t_k is the sampled instant produced by a zero-order-hold (ZOH) function.

When the actuators occur failures, the following model is used to express the control signals:

$$u^F(t)=\Lambda u(t),\tag{10.2}$$

where $\Lambda = \mathrm{diag}\{\alpha_1, \alpha_2, \ldots, \alpha_n\}$ is the actuator fault matrix satisfying $0 \le \underline{\alpha}_i \le \alpha_i \le \bar{\alpha}_i \le 1$ $(i = 1, 2, \ldots, n)$. $\underline{\alpha}_i$ and $\bar{\alpha}_i$ stand for the admissible failures of the ith actuator.

Then, Eq. (10.2) can be rewritten as

$$u^F(t) = (\check{\Xi} + \hat{\Xi})u(t), \tag{10.3}$$

where

$$\begin{aligned} \check{\Xi} &= \mathrm{diag}\{\rho_1, \rho_2, \ldots, \rho_n\}, \\ \rho_i &\in \left[-\frac{\bar{\alpha}_i - \underline{\alpha}_i}{2}, \frac{\bar{\alpha}_i - \underline{\alpha}_i}{2}\right], i = 1, 2, \ldots, n, \end{aligned}$$

and

$$\hat{\Xi} = \mathrm{diag}\left\{\frac{\underline{\alpha}_1 + \bar{\alpha}_1}{2}, \frac{\underline{\alpha}_2 + \bar{\alpha}_2}{2}, \ldots, \frac{\underline{\alpha}_n + \bar{\alpha}_n}{2}\right\}.$$

Here, let us define the synchronization error vector as $\delta(t) = y(t) - x(t)$.

Substituting $u^F(t)$ for $u(t)$, one has from (10.1) that

$$\begin{aligned} \dot{\delta}(t) = \mathscr{A}\delta(t) + \mathscr{B}\bar{f}(\mathscr{E}\delta(t)) - (\check{\Xi} + \hat{\Xi})\mathscr{K}\mathscr{D}\delta(t - d(t)), \\ t_k \le t < t_{k+1}, \end{aligned} \tag{10.4}$$

where $d(t) = t - t_k$ is the input delay and $\bar{f}(\mathscr{E}\delta(t)) = f(\mathscr{E}\delta(t) + \mathscr{E}x(t)) - f(\mathscr{E}x(t))$.

The sampling interval d satisfies

$$d = t_{k+1} - t_k > 0, \ k = 1, 2, \ldots. \tag{10.5}$$

Assume the nonlinear function $f(\cdot) : \mathbb{R}^m \to \mathbb{R}^m$ belonging to the sector $[l_i^-, l_i^+]$ with $f_i(\cdot)$ and

$$l_i^- \le \frac{f_i(\alpha_1) - f_i(\alpha_2)}{\alpha_1 - \alpha_2} \le l_i^+, \quad \forall \alpha_1 \ne \alpha_2, i = 1, 2, \ldots, m, \tag{10.6}$$

from which, one has

$$l_i^- \le \frac{\bar{f}_i(\mathscr{E}_i\delta(t))}{\mathscr{E}_i\delta(t)} \le l_i^+, \quad \mathscr{E}_i\delta(t) \ne 0, \ i = 1, 2, \ldots, m, \tag{10.7}$$

where $\mathscr{E}_i$ is the ith row vector of $\mathscr{E}$.

Remark 10.1 As shown in Eq. (10.1), the system is not a time-delay system. However, by the input delay approach, the closed-loop system as expressed in Eq. (10.4) can be considered as a time-delay system. This is why the control problem covered in this chapter can be a topic dealing with this book.

10.3 Main Results

Before presenting main results of this chapter, some variables are first defined for simplicity of long expression in equations.

Define $\mathscr{J}_i = [0_{n,(i-1)n} \quad I_n \quad 0_{n,(7-i)n+2m}]$ $(i = 1, 2, \ldots, 6)$, $\mathscr{J}_7 = [0_{m,6n} \quad I_m \quad 0_{m,n+m}]$, $\mathscr{J}_8 = [0_{n,6n+m} \quad I_n \quad 0_{n,m}]$, $\mathscr{J}_9 = [0_{m,7n+m} \quad I_m]$ and

$$
\begin{aligned}
&\bar{d}(t_k, t) = t_k + \theta(t - t_k),\ \theta \in (0, 1),\\
&l^{(1)} = \mathrm{diag}\{l_1^- l_1^+, l_2^- l_2^+, \ldots, l_m^- l_m^+\},\\
&l^{(2)} = \mathrm{diag}\{\frac{l_1^- + l_1^+}{2}, \frac{l_2^- + l_2^+}{2}, \ldots, \frac{l_m^- + l_m^+}{2}\},\\
&\tilde{\Xi} = \mathrm{diag}\left\{\tilde{\Xi}_1, \tilde{\Xi}_2, \ldots, \tilde{\Xi}_n\right\},\quad \tilde{\Xi}_i = \frac{(\bar{\alpha}_i - \underline{\alpha}_i)^2}{4},\\
&l^- = \mathrm{diag}\{l_1^-, l_2^-, \ldots, l_m^-\},\quad l^+ = \mathrm{diag}\{l_1^+, l_2^+, \ldots, l_m^+\},\\
&T^{(1)}(d(t)) = \frac{d(t)}{d} T_1 + \frac{d - d(t)}{d} T_2,\\
&T^{(2)}(d(t)) = \frac{d(t)}{d} T_3 + \frac{d - d(t)}{d} T_4,\\
&N(d(t)) = \frac{d(t)}{d} N_1 + \frac{d - d(t)}{d} N_2,\\
&\psi(t) = \mathrm{col}\left\{\delta(t),\ \delta(\bar{d}(t_k, t)),\ \delta(t_k),\ \dot{\delta}(t),\ \int_{\bar{d}(t_k,t)}^{t} \delta(s)ds,\ \int_{t_k}^{\bar{d}(t_k,t)} \delta(s)ds\right\},\\
&\beta(t) = \mathrm{col}\Bigg\{\delta(t),\ \delta(\bar{d}(t_k, t)),\ \delta(t_k),\ \dot{\delta}(t),\ \int_{\bar{d}(t_k,t)}^{t} \delta(s)ds,\ \int_{t_k}^{\bar{d}(t_k,t)} \delta(s)ds,\\
&\qquad f(\mathscr{E}\delta(t)),\ \dot{\delta}^T(\bar{d}(t_k, t)),\ f(\mathscr{E}\delta(\bar{d}(t_k, t)))\Bigg\}.
\end{aligned}
$$

Now, the following theorem presents the main result of a control problem covered in this chapter.

Theorem 10.1 *For scalars $d > 0$, $\theta \in (0, 1)$, and τ_i $(i = 1, 2)$, systems $\mathscr{M}$ and $\mathscr{S}$ are globally asymptotically synchronous, if there exist matrices $P \in \mathbb{S}_+^n$, Q, $\hat{R}_{11} \in \mathbb{S}_+^{2n}$, $\Delta = \mathrm{diag}\{\Delta_1, \Delta_2, \ldots, \Delta_m\}$, $\bar{\Delta} = \mathrm{diag}\{\bar{\Delta}_1, \bar{\Delta}_2, \ldots, \bar{\Delta}_m\}$, $\Xi = \mathrm{diag}\{\Xi_1, \Xi_2, \ldots, \Xi_m\}$, $\bar{\Xi} = \mathrm{diag}\{\bar{\Xi}_1, \bar{\Xi}_2, \ldots, \bar{\Xi}_m\} \in \mathbb{D}_+^n$, $T_i \in \mathbb{R}^{m\times m}$ $(i = 1, 2, 3, 4)$, $R = \begin{bmatrix} \hat{R}_{11} & \hat{R}_{12} \\ * & \hat{R}_{22} \end{bmatrix}$, $S \in \mathbb{S}^{3n}$, $\mathscr{K} \in \mathbb{R}^{n\times l}$, $X_i \in \mathbb{R}^{2n\times 6n}$ $(i = 1, 2)$, $N_i \in \mathbb{R}^{6n\times n}$ $(i = 1, 2)$,*

$F \in \mathbb{D}^n$, *and scalar* $\varepsilon > 0$, *such that*

$$\begin{bmatrix} \Psi(d,0) & \mathscr{J}_3^T \mathscr{D}^T \bar{\mathscr{K}}^T \\ * & -\varepsilon I_l \end{bmatrix} < 0, \tag{10.8}$$

$$\begin{bmatrix} \Psi(d,d) & \mathscr{J}_3^T \mathscr{D}^T \bar{\mathscr{K}}^T & \bar{\Pi} \\ * & -\varepsilon I_l & 0 \\ * & * & \Pi \end{bmatrix} < 0, \tag{10.9}$$

where

$$\begin{aligned}
\bar{\Pi} &= [\sqrt{d(1-\theta)}\bar{\Gamma}_2^T X_1^T, \sqrt{d\theta}\bar{\Gamma}_2^T X_2^T], \\
\Pi &= \mathrm{diag}\{-\hat{R}_{11}, -\hat{R}_{11}\}, \\
\Psi(d,d(t)) &= \sum_{i=1}^{4} \Psi_i(d,d(t)),
\end{aligned}$$

with

$$\begin{aligned}
\Psi_1(d,d(t)) = &\begin{bmatrix} \mathscr{E}\mathscr{J}_1 \\ \mathscr{J}_7 \end{bmatrix}^T \begin{bmatrix} -T^{(1)}(d(t))l^{(1)} & T^{(1)}(d(t))l^{(2)} \\ * & -T^{(1)}(d(t)) \end{bmatrix} \begin{bmatrix} \mathscr{E}\mathscr{J}_1 \\ \mathscr{J}_7 \end{bmatrix} \\
&+ \begin{bmatrix} \mathscr{E}\mathscr{J}_2 \\ \mathscr{J}_9 \end{bmatrix}^T \begin{bmatrix} -T^{(2)}(d(t))l^{(1)} & T^{(2)}(d(t))l^{(2)} \\ * & -T^{(2)}(d(t)) \end{bmatrix} \begin{bmatrix} \mathscr{E}\mathscr{J}_2 \\ \mathscr{J}_9 \end{bmatrix},
\end{aligned}$$

$$\begin{aligned}
\Psi_2(d,d(t)) = &\,\mathrm{Sym}\{\mathscr{J}_1^T P \mathscr{J}_4\} + \Gamma_1^T Q \Gamma_1 - \theta \bar{\Gamma}_1^T Q \bar{\Gamma}_1 \\
&+ \mathrm{Sym}\{(\mathscr{J}_7 - l^- \mathscr{E}\mathscr{J}_1)^T \Delta \mathscr{E}\mathscr{J}_4\} \\
&+ \mathrm{Sym}\{(l^+ \mathscr{E}\mathscr{J}_1 - \mathscr{J}_7)^T \Xi \mathscr{E}\mathscr{J}_4\} \\
&+ \theta \mathrm{Sym}\{(\mathscr{J}_9 - l^- \mathscr{E}\mathscr{J}_2)^T \bar{\Delta} \mathscr{E}\mathscr{J}_8\} \\
&+ \theta \mathrm{Sym}\{(l^+ \mathscr{E}\mathscr{J}_2 - \mathscr{J}_9)^T \bar{\Xi} \mathscr{E}\mathscr{J}_8\} + (d - d(t))\Gamma_2^T R \Gamma_2 \\
&- \mathrm{Sym}\{[\mathscr{J}_1^T - \mathscr{J}_3^T, \mathscr{J}_5^T + \mathscr{J}_6^T]\hat{R}_{12}\mathscr{J}_3\} - d(t)\mathscr{J}_3^T \hat{R}_{22} \mathscr{J}_3 \\
&+ \mathrm{Sym}\{\bar{\Gamma}_2^T X_1^T \Gamma_3\} + \mathrm{Sym}\{\bar{\Gamma}_2^T X_2^T \Gamma_4\} \\
&+ \mathrm{Sym}\{(d - d(t))\Gamma_5^T S \dot{\Gamma}_5\} - \Gamma_5^T S \Gamma_5,
\end{aligned}$$

$$\begin{aligned}
\Psi_3(d,d(t)) = &\,\mathrm{Sym}\{\Gamma_6^T F(-\mathscr{J}_4 + \mathscr{A}\mathscr{J}_1 + \mathscr{B}\mathscr{J}_7)\} - \mathrm{Sym}\{\Gamma_6^T \hat{\Xi} \bar{\mathscr{K}} \mathscr{D} \mathscr{J}_3\} \\
&+ \varepsilon \Gamma_6^T \tilde{\Xi} \Gamma_6,
\end{aligned}$$

$$\begin{aligned}
\Psi_4(d,d(t)) = &\,\mathrm{Sym}\big\{\bar{\Gamma}_2^T N(d(t))(-\mathscr{J}_4 + \mathscr{J}_8 + \mathscr{A}\mathscr{J}_1 - \mathscr{A}\mathscr{J}_2 + \mathscr{B}\mathscr{J}_7 \\
&-\mathscr{B}\mathscr{J}_9)\big\},
\end{aligned}$$

and

$$\Gamma_1 = \text{col}\{\mathscr{J}_4, \mathscr{J}_1\}, \ \bar{\Gamma}_1 = \text{col}\{\mathscr{J}_8, \mathscr{J}_2\},$$
$$\Gamma_2 = \text{col}\{\mathscr{J}_4, \mathscr{J}_1, \mathscr{J}_3\}, \ \bar{\Gamma}_2 = \text{col}\{\mathscr{J}_1, \mathscr{J}_2, \mathscr{J}_3, \mathscr{J}_4, \mathscr{J}_5, \mathscr{J}_6\},$$
$$\Gamma_3 = \text{col}\{\mathscr{J}_1 - \mathscr{J}_2, \ \mathscr{J}_5\}, \ \Gamma_4 = \text{col}\{\mathscr{J}_2 - \mathscr{J}_3, \mathscr{J}_6\},$$
$$\Gamma_5 = \text{col}\{\mathscr{J}_1 - \mathscr{J}_3, \mathscr{J}_2 - \mathscr{J}_3, \mathscr{J}_5 + \mathscr{J}_6\},$$
$$\dot{\Gamma}_5 = \text{col}\{\mathscr{J}_4, \theta \mathscr{J}_8, \mathscr{J}_1\}, \ \Gamma_6 = \mathscr{J}_4 + \tau_1 \mathscr{J}_1 + \tau_2 \mathscr{J}_3.$$

In addition, the control gain for chaotic synchronization of the controller given in (10.1) is

$$\mathscr{K} = F^{-1}\bar{\mathscr{K}}. \tag{10.10}$$

Proof First, let us denote $\eta(t) = \text{col}\{\mathscr{E}\delta(t), \bar{f}(\mathscr{E}\delta(t))\}$.

For matrices $T_i \in \mathbb{D}_+^m$ $(i = 1, 2, 3, 4)$, we have $T^{(1)}(d(t)) > 0$ and $T^{(2)}(d(t)) > 0$. Then, from (10.7), one has

$$\eta^T(t)\begin{bmatrix} -T^{(1)}(d(t))l^{(1)} & T^{(1)}(d(t))l^{(2)} \\ * & -T^{(1)}(d(t)) \end{bmatrix}\eta(t) \geq 0, \tag{10.11}$$

and

$$\eta^T(\bar{d}(t_k, t))\begin{bmatrix} -T^{(2)}(d(t))l^{(1)} & T^{(2)}(d(t))l^{(2)} \\ * & -T^{(2)}(d(t)) \end{bmatrix}\eta(\bar{d}(t_k, t)) \geq 0. \tag{10.12}$$

Take the following LKF:

$$V(t) = \sum_{i=1}^{3} V_i(t), \quad t_k \leq t < t_{k+1}, \tag{10.13}$$

where

$$V_1(t) = \delta^T(t)P\delta(t) + \int_{\bar{d}(t_k,t)}^{t} \Gamma_1^T(s)Q\Gamma_1(s)ds,$$
$$V_2(t) = 2\sum_{i=1}^{m}\int_0^{\mathscr{E}_i\delta(t)} \big[\Delta_i(\bar{f}_i(s) - l_i^- s) + \Xi_i(l_i^+ s - \bar{f}_i(s))\big]ds,$$
$$+2\sum_{i=1}^{m}\int_0^{\mathscr{E}_i\delta(\bar{d}(t_k,t))} \big[\bar{\Delta}_i(\bar{f}_i(s) - l_i^- s) + \bar{\Xi}_i(l_i^+ s - \bar{f}_i(s))\big]ds,$$
$$V_3(t) = (d - d(t))\int_{t_k}^{t} \Gamma_2^T(s)R\Gamma_2(s)ds + (d - d(t))\Gamma_3^T(t)S\Gamma_3(t),$$

with $\Gamma_1(t) = \mathrm{col}\{\dot{\delta}(t), \delta(t)\}$, $\Gamma_2(t) = \mathrm{col}\{\dot{\delta}(t), \delta(t), \delta(t_k)\}$, and $\Gamma_3(t) = \mathrm{col}\big\{\delta(t) - \delta(t_k), \delta(\bar{d}(t_k, t)) - \delta(t_k), \int_{t_k}^{t} \delta(s)ds\big\}$.

The time derivative of $V(t)$ is computed as

$$\dot{V}(t) = \sum_{i=1}^{3} \dot{V}_i(t), \tag{10.14}$$

where

$$\dot{V}_1(t) = 2\delta^T(t)P\dot{\delta}(t) + \Gamma_1^T(t)Q\Gamma_1(t) - \theta\Gamma_1^T(\bar{d}(t_k, t))Q\Gamma_1(\bar{d}(t_k, t)), \tag{10.15}$$

$$\begin{aligned} \dot{V}_2(t) = {} & 2(f(\mathscr{E}\delta(t)) - l^-\mathscr{E}\delta(t))^T\Delta\mathscr{E}\dot{\delta}(t) \\ & + 2(l^+\mathscr{E}\delta(t) - f(\mathscr{E}\delta(t)))^T\Xi\mathscr{E}\dot{\delta}(t) \\ & + 2\theta\big(f(\mathscr{E}\delta(\bar{d}(t_k, t))) - l^-\mathscr{E}\delta(\bar{d}(t_k, t))\big)^T\bar{\Delta}\mathscr{E}\dot{\delta}(\bar{d}(t_k, t)) \\ & + 2\theta\big(l^+\mathscr{E}\delta(\bar{d}(t_k, t)) - f(\mathscr{E}\delta(\bar{d}(t_k, t)))\big)^T\bar{\Xi}\mathscr{E}\dot{\delta}(\bar{d}(t_k, t)), \end{aligned} \tag{10.16}$$

$$\begin{aligned} \dot{V}_3(t) = {} & (d - d(t))\Gamma_2^T(t)R\Gamma_2(t) - \int_{t_k}^{t} \Gamma_2^T(s)R\Gamma_2(s)ds \\ = {} & (d - d(t))\Gamma_2^T(t)R\Gamma_2(t) - d(t)e^T(t_k)\hat{R}_{22}\delta(t_k) \\ & - 2\big[(\delta(t) - \delta(t_k))^T, \int_{t_k}^{t} e^T(s)ds\big]\hat{R}_{12}\delta(t_k) \\ & - \int_{\bar{d}(t_k,t)}^{t} \Gamma_1^T(s)\hat{R}_{11}\Gamma_1(s)ds - \int_{t_k}^{\bar{d}(t_k,t)} \Gamma_1^T(s)\hat{R}_{11}\Gamma_1(s)ds \\ & + 2(d - d(t))\Gamma_3^T(t)S\dot{\xi}_3(t) - \Gamma_3^T(t)S\Gamma_3(t). \end{aligned} \tag{10.17}$$

From $\hat{R}_{11} > 0$, it is clear for any matrices $X_i \in \mathbb{R}^{2n \times 6n}$ $(i = 1, 2)$ that

$$\begin{aligned} - \int_{\bar{d}(t_k,t)}^{t} \Gamma_1^T(s)\hat{R}_{11}\Gamma_1(s)ds \le {} & (1 - \theta)d(t)\varphi^T(t)X_1^T\hat{R}_{11}^{-1}X_1\psi(t) \\ & + 2\varphi^T(t)X_1^T\begin{bmatrix} \delta(t) - \delta(\bar{d}(t_k, t)) \\ \int_{\bar{d}(t_k,t)}^{t} \delta(s)ds \end{bmatrix}, \end{aligned} \tag{10.18}$$

and

$$\begin{aligned} - \int_{t_k}^{\bar{d}(t_k,t)} \Gamma_1^T(s)\hat{R}_{11}\Gamma_1(s)ds \le {} & \theta d(t)\varphi^T(t)X_2^T\hat{R}_{11}^{-1}X_2\psi(t) \\ & + 2\varphi^T(t)X_2^T\begin{bmatrix} \delta(\bar{d}(t_k, t)) - e(t_k) \\ \int_{t_k}^{\bar{d}(t_k,t)} \delta(s)ds \end{bmatrix}. \end{aligned} \tag{10.19}$$

From system (10.4), the following equation holds:

$$0 = 2\mho^T(t)F\big[-\dot{\delta}(t) + \mathscr{A}\delta(t) + \mathscr{B}\bar{f}(\mathscr{E}\delta(t)) \\ -(\check{\Xi} + \hat{\Xi})\mathscr{K}\mathscr{D}\delta(t - d(t))\big], \tag{10.20}$$

which could be converted to

$$0 = 2\mho^T(t)F[-\dot{\delta}(t) + \mathscr{A}\delta(t) + \mathscr{B}\bar{f}(\mathscr{E}\delta(t))] \\ -2\mho^T(t)(\check{\Xi} + \hat{\Xi})\bar{\mathscr{K}}\mathscr{D}\delta(t_k), \tag{10.21}$$

where $\bar{\mathscr{K}} = F\mathscr{K}$ and $\mho(t) = \dot{\delta}(t) + \tau_1\delta(t) + \tau_2\delta(t_k)$.

For $\varepsilon > 0$, one obtains

$$-2\mho^T(t)\check{\Xi}\bar{\mathscr{K}}\mathscr{D}\delta(t_k) \le \varepsilon\mho^T(t)\tilde{\Xi}\mho(t) + \varepsilon^{-1}\delta(t_k)^T\mathscr{D}^T\bar{\mathscr{K}}^T\bar{\mathscr{K}}\mathscr{D}\delta(t_k). \tag{10.22}$$

Moreover, from (10.4), one gets

$$\dot{\delta}(\bar{d}(t_k, t)) = \mathscr{A}\delta(\bar{d}(t_k, t)) + \mathscr{B}f(\mathscr{E}\delta(\bar{d}(t_k, t))) - (\check{\Xi} + \hat{\Xi})\mathscr{K}\mathscr{D}\delta(t_k). \tag{10.23}$$

From (10.4) and (10.23), we derive

$$0 = 2\varphi^T(t)N(d(t))\Big[-\dot{\delta}(t) + \dot{\delta}(\bar{d}(t_k, t)) + \mathscr{A}\delta(t) \\ -\mathscr{A}\delta(\bar{d}(t_k, t)) + \mathscr{B}\bar{f}(\mathscr{E}\delta(t)) - \mathscr{B}f(\mathscr{E}\delta(\bar{d}(t_k, t)))\Big]. \tag{10.24}$$

Combining (10.14)–(10.24), we get

$$\dot{V}(t) \le \zeta^T(t)\Psi(d, d(t))\beta(t) \\ = \zeta^T(t)\left(\frac{d - d(t)}{d}\Psi(d, 0) + \frac{d(t)}{d}\Psi(d, d)\right)\beta(t), \tag{10.25}$$

where $\Psi(d, d(t)) = \Psi(d, d(t)) + \Psi^{(1)} + \Psi^{(2)}$ and

$$\Psi^{(1)} = (1 - \theta)d(t)\bar{\Gamma}_2^T X_1^T \hat{R}_{11}^{-1} X_1 \bar{\Gamma}_2 + \theta d(t)\bar{\Gamma}_2^T X_2^T \hat{R}_{11}^{-1} X_2 \bar{\Gamma}_2, \\ \Psi^{(2)} = \varepsilon^{-1}\mathscr{J}_3^T\mathscr{D}^T\bar{\mathscr{K}}^T\bar{\mathscr{K}}\mathscr{D}\mathscr{J}_3.$$

Thus, for $\beta(t) \neq 0$, we can derive from (10.8), (10.9), and (10.25) that

$$\dot{V}(t) < 0, \quad t_k \le t < t_{k+1}. \tag{10.26}$$

For $\beta(t) \neq 0$, from (10.13) and (10.26), one has $\lim_{t\to t_k^-} V(t) \geq V(t_k) \geq 0$ and

$$V(t) > V(t_{k+1}) > 0, \; t_k \leq t < t_{k+1}, \; k = 0, 1, 2, \ldots \tag{10.27}$$

Therefore, by the reliable sampled-data controller (10.2), systems $\mathscr{M}$ and $\mathscr{S}$ are globally asymptotically synchronous. This completes the proof. ■

10.4 Simulating Example

A numerical example is provided to illustrate the validity of the proposed synchronization scheme in this section.

Consider systems $\mathscr{M}$ and $\mathscr{S}$ in (10.1) with the following parameters

$$\mathscr{A} = \begin{bmatrix} -1 & 0 & 0 \\ 0 & -1 & 0 \\ 0 & 0 & -1 \end{bmatrix}, \quad \mathscr{B} = \begin{bmatrix} 1.2 & -1.6 & 0 \\ 1.24 & 1 & 0.9 \\ 0 & 2.2 & 1.5 \end{bmatrix},$$

$$\mathscr{D} = \mathscr{E} = \begin{bmatrix} 1 & 0 & 0 \\ 0 & 1 & 0 \\ 0 & 0 & 1 \end{bmatrix}.$$

It is noted that the model above can be one of neural networks by choosing the nonlinear functions

$$f_i(x_i(t)) = \frac{1}{2}(|x_i(t) + 1| - |x_i(t) - 1|), \; i = 1, 2, 3$$

with satisfying the sector condition: $l_i^- = 0$ and $l_i^+ = 1$.

Choose $x(0) = [0.4\ 0.3\ 0.8]^T$ and $z(0) = [0.2\ 0.4\ 0.9]^T$ as the initial condition for this example.

In order to see the situation of actuator failures, let us choose related parameters. Take $\underline{\alpha}_1 = 0.4$, $\bar{\alpha}_1 = 0.6$, $\underline{\alpha}_2 = 0.6$, $\bar{\alpha}_2 = 0.7$, $\underline{\alpha}_3 = 0.5$, and $\bar{\alpha}_3 = 0.6$.

When $\tau_1 = 0.8$, $\tau_2 = 2.0$, and $\theta = 0.20$, using MATLAB LMI Toolbox to solve the conditions of Theorem 10.1, the maximum allowable upper bound d is 0.9066, and the corresponding gain matrix is

$$\mathscr{K} = \begin{bmatrix} 3.0450 & -1.1922 & -0.2693 \\ 0.7530 & 1.5446 & 0.9829 \\ -0.3732 & 2.5498 & 2.3219 \end{bmatrix}.$$

Then, the state paths of system (10.4) without and with control input are displayed, respectively, in Figs. 10.1 and 10.2. The control input $u(t)$ is depicted in Fig. 10.3. From Fig. 10.2, it is shown that the controlled system (10.4) is stable.

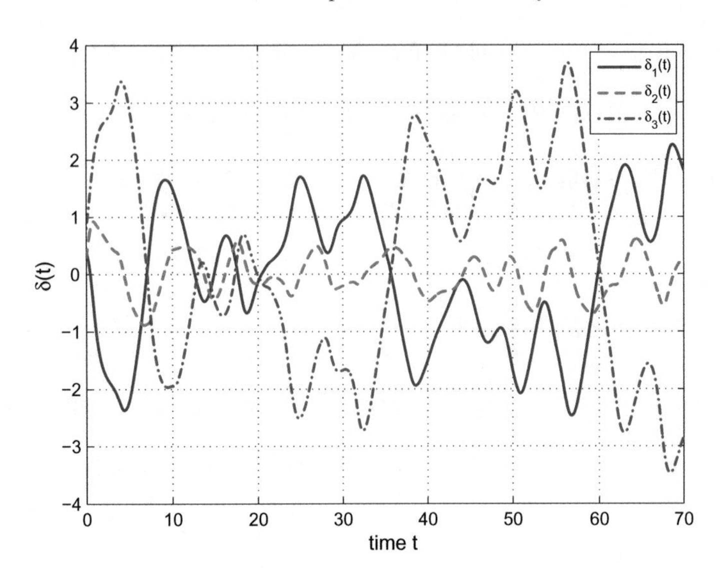

Fig. 10.1 Uncontrolled state paths of system (10.4)

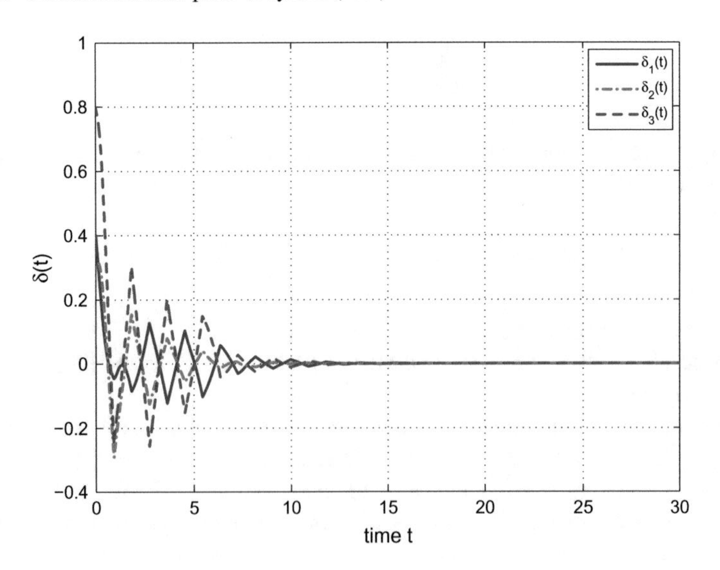

Fig. 10.2 Controlled state paths of system (10.4)

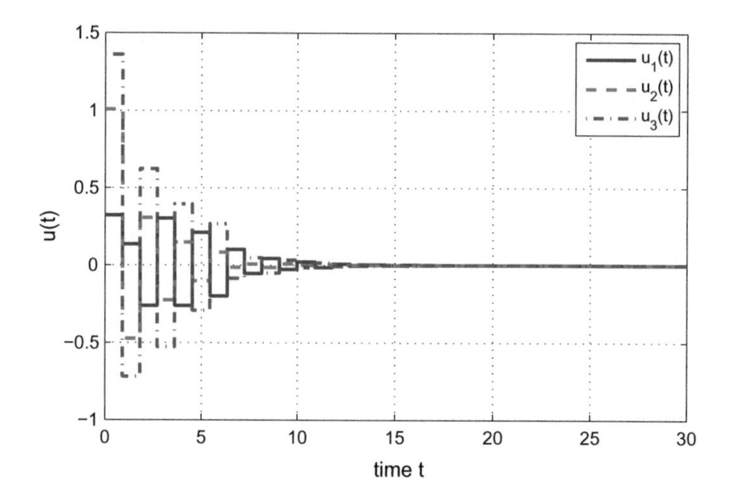

Fig. 10.3 Control inputs $u(t)$

10.5 Conclusion

A reliable sampled-data control mechanism has been put forward for synchronization of CLSs with actuator failures. It is more valid in application than existing control methods. Sufficient conditions have been established to guarantee the synchronization of the master-slave CLSs. Moreover, the desired controller has been designed. A numerical example has been given to verify the theoretical analysis.

Special Acknowledgement Dr. J. H. Park would like to thank Dr. Ruimei Zhang of University of Electronic Science and Technology of China, PR China who contributed greatly to the completion of this chapter, including simulation.

References

1. Liao X, Chen G (2003) Chaos synchronization of general Lur'e systems via time-delay feedback control. Int J Bifurc Chaos 13:207–213
2. Souza FO, Palhares RM, Mendes E, Torres L (2008) Further results on master-slave synchronization of general Lur'e systems with time-varying delay. Int J Bifurc Chaos 18:187–202
3. Huang H, Cao J (2007) Master-slave synchronization of Lur'e systems based on time-varying delay feedback control. Int J Bifurc Chaos 17:4159–4166
4. Lu J, Cao J, Ho DWC (2008) Adaptive stabilization and synchronization for chaotic Lur'e systems with time-varying delays. IEEE Trans Circuits Syst I 55:1347–1356
5. Tang Z, Park JH, Shen H (2018) Finite-time cluster synchronization of Lur'e networks: a nonsmooth approach. IEEE Trans Syst Man Cybern: Syst 48:1213–1224
6. Lee SM, Choi SJ, Ji DH, Park JH, Won SC (2010) Synchronization for chaotic Lur'e systems with sector restricted nonlinearities via delayed feedback control. Nonlinear Dyn 59:277–288

7. Zeng HB, Park JH, Xiao SP, Liu Y (2015) Further results on sampled-data control for master-slave synchronization of chaotic Lur'e systems with time-delay. Nonlinear Dyn 82:851–863
8. Carroll TL, Pecora LM (1991) Synchronizing chaotic systems. IEEE Trans Circuits Syst I: Fundam Theory Appl 38:453–456
9. Pecora LM, Carroll TL (1990) Synchronization in chaotic systems. Phys Rev Lett 64:821–824
10. Park JH (2009) Further results on functional projective synchronization of Genesio-Tesi chaotic system. Mod Phys Lett B 24:1889–1895
11. Park JH, Ji DH, Won SC, Lee SM (2008) $\mathscr{H}_\infty$ synchronization of time-delayed chaotic systems. Appl Math Comput 204:170–177
12. Park JH, Lee SM, Kwon OM (2007) Adaptive synchronization of Genesio-Tesi chaotic system via a novel feedback control. Phys Lett A 371:263–270
13. Park JH (2008) Adaptive control for modified projective synchronization of a four-dimensional chaotic system with uncertain parameters. J Comput Appl Math 213:288–293
14. Lu JG, Hill DJ (2007) Impulsive synchronization of chaotic Lur'e systems by linear static measurement feedback: an LMI approach. IEEE Trans Circuits Syst II: Express Briefs 54:710–714
15. Huijberts H, Nijmeijer H, Oguchi T (2007) Anticipating synchronization of chaotic Lur'e systems. Chaos: Interdiscip 17:013117
16. Zhang CK, He Y, Wu M (2009) Improved global asymptotical synchronization of chaotic Lur'e systems with sampled-data control. IEEE Trans Circuits Syst II: Express Briefs 56:320–324
17. Chen WH, Jiang Z, Lu X, Luo S (2015) $\mathscr{H}_\infty$ synchronization for complex dynamical networks with coupling delays using distributed impulsive control. Nonlinear Anal: Hybrid Syst 17:111–127
18. Cai S, Zhou P, Liu Z (2015) Intermittent pinning control for cluster synchronization of delayed heterogeneous dynamical networks. Nonlinear Anal: Hybrid Syst 18:134–155
19. Shao H, Han QL, Zhang Z, Zhu X (2014) Sampling-interval-dependent stability for sampled-data systems with state quantization. Int J Robust Nonlinear Control 24:2995–3008
20. Deaecto GS, Bolzern P, Galbusera L, Geromel JC (2016) $\mathscr{H}_2$ and $\mathscr{H}_\infty$ control of time-varying delay switched linear systems with application to sampled-data control. Nonlinear Anal: Hybrid Syst 22:43–54
21. Rakkiyappan R, Chandrasekar A, Park JH, Kwon OM (2014) Exponential synchronization criteria for Markovian jumping neural networks with time-varying delays and sampled-data control. Nonlinear Anal: Hybrid Syst 14:16–37
22. Wu ZG, Shi P, Su H, Chu J (2013) Sampled-data synchronization of chaotic Lure systems with time delays. IEEE Trans Neural Netw Learn Syst 24:410–421
23. Lee TH, Wu ZG, Park JH (2012) Synchronization of a complex dynamical network with coupling time-varying delays via sampled-data control. Appl Math Comput 219:1354–1366
24. Lee TH, Park JH, Lee SM, Kwon OM (2013) Robust synchronisation of chaotic systems with randomly occurring uncertainties via stochastic sampled-data control. Int J Control 86:107–119
25. Shen H, Wu ZG, Park JH, Zhang Z (2015) Extended dissipativity-based synchronization of uncertain chaotic neural networks with actuator failures. J Frankl Inst 352:1722–1738
26. Wu ZG, Shi P, Su H, Chu J (2012) Exponential synchronization of neural networks with discrete and distributed delays under time-varying sampling. IEEE Trans Neural Netw Learn Syst 23:1368–1375
27. Hua C, Ge C, Guan X (2015) Synchronization of chaotic Lur'e systems with time delays using sampled-data control. IEEE Trans Neural Netw Learn Syst 26:1214–1221
28. Zhang R, Zeng D, Zhong S (2017) Novel master-slave synchronization criteria of chaotic Lur'e systems with time delays using sampled-data control. J Frankl Inst 354:4930–4954
29. Lee TH, Park JH (2017) Improved criteria for sampled-data synchronization of chaotic Lur'e systems using two new approaches. Nonlinear Anal: Hybrid Syst 24:132–145
30. Fridman E, Seuret A, Richard JP (2004) Robust sampled-data stabilization of linear systems: an input delay approach. Automatica 40:1441–1446

Part IV
Applications of Dynamics Systems with Time-Delays

Chapter 11
$\mathscr{H}_\infty$ Filtering for Discrete-Time Nonlinear Systems

11.1 Introduction

In the past few decades, the issue of $\mathscr{H}_\infty$ problem has been received considerable attention due to the wide application of various fields such as astronautics, signal processing, and military [1, 2]. There has two main major methods to be employed to deal with the filtering problem: the Kalman filtering method and the $\mathscr{H}_\infty$ filtering method. The Kalman filtering requires the external disturbance to be white Gaussian noise which is not always satisfied in many engineering applications. As an effective alternative, the $\mathscr{H}_\infty$ filtering does not require the external noise to be white Gaussian. The advantage of using an $\mathscr{H}_\infty$ filter rather than the conventional Kalman filter is that no statistical assumptions are required regarding the exogenous signals. In addition, $\mathscr{H}_\infty$ filtering is known to be robust against unmodeled dynamics [3]. It is clear that the aim of the problem is to design a suitable and stable filter to minimize the $\mathscr{H}_\infty$ norm such that the mapping from exogenous input to estimation error is less than a prescribed level. Some methods to this topic, such as interpolation, algebraic Riccati equations, and linear matrix inequalities (LMIs), have been established in recent years.

A large number of practical systems are unavoidable subject to time delays, which usually leads to instability and poor performance of the system [4, 5]. Therefore, the $\mathscr{H}_\infty$ problems for nonlinear systems with time delays have been investigated in the last decades.

For delayed discrete-time systems, $\mathscr{H}_\infty$ filtering problems are investigated in [6, 7] and some sufficient conditions for the filter design are derived. However, these conditions are delay-independent, which are generally conservative, especially, for the case of small size delay. Thus, most of the research in this area has evolved to obtain delay-dependent results. For instance, some improved and delay-dependent results are derived by employing a model transformation approach [8, 9]. These results can be improved because the estimation of difference upper bound is tighter.

On the other hand, the $\mathscr{H}_\infty$ filters developed in [6–9] are only focused on linear systems. Similar to time delay, the nonlinearity is inevitable in many real systems and

J. H. Park et al., *Dynamic Systems with Time Delays: Stability and Control*,
https://doi.org/10.1007/978-981-13-9254-2_11

makes system analysis more difficult [10, 11]. Hence, it is important and significant to consider the $\mathscr{H}_\infty$ filtering problem for nonlinear systems [12–17]. Moreover, it should be noted that the results need to satisfy the condition that the filters should be implemented precisely [6–9, 12–17]. This condition is difficult to satisfy in practical situations. Actually, inaccuracy or uncertainty always occur in filter implementation in real world. As a result, non-fragile filtering design has been investigated in [18–24]. Moreover, the filter gain variations may subject to random changes because of environmental circumstance in the process of filter implementation among networks. Thus, it is important to consider the situation of randomly occurring gain variations in filter design [25]. To the best of our knowledge, the problem of non-fragile $\mathscr{H}_\infty$ filtering for nonlinear discrete-time systems with randomly occurring gain variations has not been well addressed.

In view of the above issue, the $\mathscr{H}_\infty$ filtering problem of a class of discrete nonlinear systems with interval time-varying delays and randomly occurring gain variations. Based on the Lyapunov functional method, a sufficient condition for non-fragile $\mathscr{H}_\infty$ filtering of the resulting closed-loop nonlinear system is established. By the condition, an improved filtering design for linear systems is achieved. Finally, two numerical examples are illustrated to demonstrate the validity and less conservativeness of the proposed method.

11.2 Problem Statement

Consider the following nonlinear discrete-time delay system with time-varying delay:

$$
\begin{aligned}
x(k+1) &= Ax(k) + A_d x(k-h(k)) + Ff(x(k)) + B_w w(k), \\
y(k) &= Cx(k) + C_d x(k-h(k)) + H_1 h(x(k)) + H_2 h(x(k-h(k)) \\
&\quad + D_w w(k), \\
z(k) &= Lx(k) + L_d x(k-h(k)) + G_w w(k),
\end{aligned}
\tag{11.1}
$$

where $x(k) \in \mathbb{R}^n$ is state vector and $y(k) \in \mathbb{R}^m$ is measured output, $w(k) \in \mathbb{R}^q$ denotes a disturbance input belonging to $\mathscr{L}_2[0, \infty)$, $z(k) \in \mathbb{R}^p$ is the signal to be estimated, and $A, A_d, F, B_w, C, C_d, H_1, H_2, D_w, L, L_d, G_w$ are known constant matrices with appropriate dimensions.

Assumption 11.1 The time delay $h(k)$ is assumed to be time-varying and satisfies

$$h_1 \le h(k) \le h_2,$$

where $h_1 > 0$ and $h_2 > 0$ are constants representing the lower and upper bounds of the delay, respectively.

Assumption 11.2 $f(\cdot)$ and $h(\cdot)$ are sector-bounded functions and satisfy $f(0) = h(0) = 0$.

Assumption 11.3 The nonlinear vector-valued functions are assumed to satisfy

$$[f(x) - f(y) - U_1(x-y)][f(x) - f(y) - U_2(x-y)] \leq 0, \forall x, y \in \mathbb{R}^n, \quad (11.2)$$
$$[h(x) - h(y) - V_1(x-y)][h(x) - h(y) - V_2(x-y)] \leq 0, \forall x, y \in \mathbb{R}^m, \quad (11.3)$$

where U_1, U_2, V_1, and V_2 are known real constant matrices, and $U_2 - U_1$ and $V_2 - V_1$ are positive definite matrices, respectively.

In this study, our main goal is to estimate the signals $z(k)$.

To accomplish this purpose, the following non-fragile full-order filter is designed:

$$\begin{aligned} x_f(k+1) &= (A_f + \alpha(k)\Delta A_f(k))x_f(k) + (B_f + \beta(k)\Delta B_f(k))y(k) \\ z_f(k) &= C_f x_f(k) + D_f y(k), \quad x_f(0) = 0, \end{aligned} \quad (11.4)$$

where $x_f(k) \in \mathbb{R}^n$ is the state of filter, $z_f(k) \in \mathbb{R}^m$ is the estimated output, the matrices A_f, B_f, C_f, D_f are appropriately dimensioned filter gains to be determined, and $\alpha(k)$ and $\beta(k)$ are mutually independent Bernoulli-distributed white sequences.

Assumption 11.4 A natural assumption on $\alpha(k)$ and $\beta(k)$ is made as follows:

$$\begin{aligned} &\Pr\{\alpha(\mathrm{k}) = 1\} = \mathbb{E}\{\alpha(k)\} = \bar{\alpha}, \\ &\Pr\{\alpha(\mathrm{k}) = 0\} = 1 - \bar{\alpha}, \\ &\Pr\{\beta(\mathrm{k}) = 1\} = \mathbb{E}\{\beta(k)\} = \bar{\beta}, \\ &\Pr\{\beta(\mathrm{k}) = 0\} = \bar{\beta}. \end{aligned}$$

Assumption 11.5 The uncertain perturbation matrices $\Delta A_f(k)$ and $\Delta B_f(k)$ are defined as follows:

$$\begin{aligned} \Delta A_f(k) &= M_a \Delta_1(k) N_a, \\ \Delta B_f(k) &= M_b \Delta_2(k) N_b, \end{aligned}$$

where M_a, M_b, N_a and N_b are known constant matrices with appropriate dimensions, and $\Delta_1(k)$ and $\Delta_2(k)$ are unknown matrix functions satisfying

$$\Delta_s^T(k)\Delta_s(k) \leq I, s = 1, 2.$$

Remark 11.1 It is inevitable that the additive gain variations have resulted from the unexpected errors occurred by the filter implementation, e.g., round off errors in numerical computation and programming errors [18–20]. In addition, the filter parameters may change in the Bernoulli distribution pattern [25]. Thus, the following description is proposed to compensate the parameter variations of the filter in a random fashion, which has not been implemented in nonlinear discrete-time systems.

Denote the variables

$$\zeta(k) = \begin{bmatrix} x(k) \\ x_f(k) \end{bmatrix},$$
$$e(k) = z(k) - z_f(k).$$

Then, the following error system can be obtained

$$\begin{aligned}
\zeta(k+1) &= (\bar{A} + \bar{\alpha}\hat{A}(k) + \hat{\alpha}(k)\hat{A}(k) + \bar{\beta}\hat{B}(k) + \hat{\beta}(k)\hat{B}(k))\zeta(k) \\
&\quad + (\bar{A}_d + \bar{\beta}\hat{A}_h(k) + \hat{\beta}(k)\hat{A}_h(k))\zeta(k-h(k)) + \bar{F}f(x(k)) \\
&\quad + (\bar{H}_1 + \bar{\beta}\hat{H}_1(k) + \hat{\beta}(k)\hat{H}_1(k))h(x(k)) \\
&\quad + (\bar{H}_2 + \bar{\beta}\hat{H}_2(k) + \hat{\beta}(k)\hat{H}_2(k))h(x(k-h(k))) \\
&\quad + (\bar{B}_w + \bar{\beta}\hat{B}_w(k) + \hat{\beta}(k)\hat{B}_w(k))w(k), \\
e(k) &= \bar{C}\zeta(k) + \bar{C}_d\zeta(k-h(k)) \\
&\quad + \bar{D}_1 h(x(k)) + \bar{D}_2 h(x(k-h(k))) + \bar{D}_w w(k),
\end{aligned} \tag{11.5}$$

where

$$\begin{aligned}
\bar{A} &= \begin{bmatrix} A & 0 \\ B_f C & A_f \end{bmatrix}, \ \hat{A}(k) = \begin{bmatrix} 0 & 0 \\ 0 & \Delta A_f(k) \end{bmatrix}, \ \hat{B}(k) = \begin{bmatrix} 0 & 0 \\ \Delta B_f(k) C & 0 \end{bmatrix}, \\
\bar{A}_d &= \begin{bmatrix} A_d & 0 \\ B_f C_d & 0 \end{bmatrix}, \ \hat{A}_h(k) = \begin{bmatrix} 0 & 0 \\ \Delta B_f(k) C_d & 0 \end{bmatrix}, \ \bar{F} = \begin{bmatrix} F \\ 0 \end{bmatrix}, \\
\bar{H}_1 &= \begin{bmatrix} 0 \\ B_f H_1 \end{bmatrix}, \ \hat{H}_1(k) = \begin{bmatrix} 0 \\ \Delta B_f(k) H_1 \end{bmatrix}, \ \bar{H}_2 = \begin{bmatrix} 0 \\ B_f H_2 \end{bmatrix}, \\
\hat{H}_2(k) &= \begin{bmatrix} 0 \\ \Delta B_f(k) H_2 \end{bmatrix}, \ \bar{B}_w = \begin{bmatrix} B_w \\ B_f D_w \end{bmatrix}, \ \hat{B}_w(k) = \begin{bmatrix} 0 \\ \Delta B_f(k) D_w \end{bmatrix}, \\
\bar{C} &= \begin{bmatrix} L - D_f C & -C_f \end{bmatrix}, \ \bar{C}_d = \begin{bmatrix} L_d - D_f C_d & 0 \end{bmatrix}, \\
\bar{D}_1 &= -D_f H_1, \ \bar{D}_2 = -D_f H_2, \ \bar{D}_w = G_w - D_f D_w.
\end{aligned}$$

Definition 11.1 The filtering error system (11.5) with $w(t) = 0$ is said to be asymptotically stable in the mean-square sense, if for any initial condition, there holds

$$\lim_{k \to \infty} \mathbb{E}\left\{\|\zeta(k)\|^2\right\} = 0. \tag{11.6}$$

Definition 11.2 For a given positive scalar γ, the filtering error system (11.5) is said to be asymptotically stable in mean square with a guaranteed $\mathscr{H}_\infty$ performance γ if it is asymptotically stable and the filtering error $e(t)$ satisfies

$$\sum_{k=0}^{\infty} \mathbb{E}\left\{\|e(k)\|^2\right\} \le \gamma^2 \sum_{k=0}^{\infty} \mathbb{E}\left\{\|w(k)\|^2\right\} \tag{11.7}$$

for all nonzero $w(\cdot) \in \mathscr{L}[0, \infty)$ subject to the zero initial condition.

Now, the filtering problem can be stated as follows:
Filtering problem: Design an $\mathscr{H}_\infty$ filter of the form (11.4) such that the filtering error system (11.5) is asymptotically stable in the mean-square sense and achieves a prescribed $\mathscr{H}_\infty$ performance level.

Lemma 11.1 ([26]) *For a given matrix $Z \in \mathbb{S}_n^+$ and any sequence of discrete-time variable x in $[-h, 0] \bigcap \mathbb{Z} \to \mathbb{R}^n$, where $h \geq 1$, the following inequality holds:*

$$-\sum_{i=-h+1}^{0} y^T(i)Zy(i) \leq -\frac{1}{h}\begin{bmatrix}\Theta_0\\ \Theta_1\end{bmatrix}^T \begin{bmatrix} Z & 0\\ 0 & 3\left(\frac{h+1}{h-1}\right)Z\end{bmatrix}\begin{bmatrix}\Theta_0\\ \Theta_1\end{bmatrix},$$

where $y(i) = x(i) - x(i-1)$ and

$$\Theta_0 = x(0) - x(-h),$$
$$\Theta_1 = x(0) + x(-h) - \frac{2}{h+1}\sum_{i=-h}^{0} x(i).$$

Remark 11.2 When Lemma 11.1 is employed to time-varying delay, the term $\frac{h+1}{h-1}$ might to difficult to be dealt with. As pointed out in [26], the following Lemma 11.2 is provided to make the term $\frac{h+1}{h-1}$ disappear from the inequality.

Lemma 11.2 ([26]) *For a given matrix $Z \in \mathbb{S}_n^+$ and any sequence of discrete-time variable x in $[-h, 0] \bigcap \mathbb{Z} \to \mathbb{R}^n$, where $h \geq 1$, the following inequality holds:*

$$-\sum_{i=-h+1}^{0} y^T(i)Zy(i) \leq -\frac{1}{h}\begin{bmatrix}\Theta_0\\ \Theta_1\end{bmatrix}^T \begin{bmatrix} Z & 0\\ 0 & 3\left(\frac{h+1}{h-1}\right)Z\end{bmatrix}\begin{bmatrix}\Theta_0\\ \Theta_1\end{bmatrix},$$

where $y(i)$, Θ_0 and Θ_1 are defined in Lemma 11.1.

Lemma 11.3 ([27]) *For any matrix* $\begin{bmatrix} M & S\\ * & M\end{bmatrix} \geq 0$, *scalars* $\eta_1(k) > 0, \eta_2(k) > 0$ *satisfying* $\eta_1(k) + \eta_2(k) = 1$, *vector functions* $m_1(k)$ *and* $m_2(k) : \mathscr{N} \to \mathbb{R}^n$, *the following inequality holds*

$$-\frac{1}{\eta_1(k)}m_1^T(k)Mm_1(k) - \frac{1}{\eta_2(k)}m_2^T(k)Mm_2(k)$$
$$\leq \begin{bmatrix} m_1(k)\\ m_2(k)\end{bmatrix}^T \begin{bmatrix} M & S\\ * & M\end{bmatrix}\begin{bmatrix} m_1(k)\\ m_2(k)\end{bmatrix}^T.$$

11.3 Main Results

11.3.1 $\mathscr{H}_\infty$ Filtering for Nonlinear Discrete-Time Systems

In view of augmented system (11.5), the problem of $\mathscr{H}_\infty$ filtering for nonlinear discrete time systems with interval time-varying delay and randomly occurring gain variations will be addressed in this section.

For the sake of simplicity for matrix representation, we define $\hat{e}_i \in \mathbb{R}^{(19n+q)\times n}$ to be block entry matrix. For example, $\hat{e}_2 = [0\ I\ \underbrace{0\ \ldots\ 0}_{18}]^T$. Some other notations are given as

$$
\begin{aligned}
\eta^T(k) = \Bigg[&\zeta^T(k)\ \zeta^T(k-h_1)\ \zeta^T(k-h(k))\ \zeta^T(k-h_2)\ \Delta\zeta^T(k)\ \frac{1}{h_1+1}\sum_{i=k-h_1}^{k}\zeta^T(i)\\
&\frac{1}{h(k)-h_1+1}\sum_{i=k-h_2}^{k-h(k)}\zeta^T(i)\ \frac{1}{h_2-h(k)+1}\sum_{i=k-h(k)}^{k-h_1}\zeta^T(i)\\
&f^T(x(k))\ h^T(x(k))\ h^T(x(k-h(k)))\ w^T(k)\Bigg],
\end{aligned}
$$

$$
\begin{aligned}
e_1 &= [\hat{e}_1\ \hat{e}_2], e_2 = [\hat{e}_3\ \hat{e}_4], e_3 = [\hat{e}_5\ \hat{e}_6], e_4 = [\hat{e}_7\ \hat{e}_8],\\
e_5 &= [\hat{e}_9\ \hat{e}_{10}], e_6 = [\hat{e}_{11}\ \hat{e}_{12}], e_7 = [\hat{e}_{13}\ \hat{e}_{14}], e_8 = [\hat{e}_{15}\ \hat{e}_{16}],\\
e_9 &= \hat{e}_{17}, e_{10} = \hat{e}_{18}, e_{11} = \hat{e}_{19}, e_{12} = \hat{e}_{20},\\
K &= [I\ 0], \bar{M}_a = \begin{bmatrix} 0 \\ M_a \end{bmatrix}, \bar{N}_a = \begin{bmatrix} 0 & N_a \end{bmatrix}, \bar{M}_b = \begin{bmatrix} 0 \\ M_b \end{bmatrix},\\
\bar{N}_{bc} &= \begin{bmatrix} N_b C & 0 \end{bmatrix}, \bar{N}_{bc_d} = \begin{bmatrix} N_b C_d & 0 \end{bmatrix},\\
\bar{U}_1 &= \frac{U_1^T U_2 + U_2^T U_1}{2}, \bar{U}_2 = -\frac{U_1^T + U_2^T}{2},
\end{aligned}
$$

$$
\begin{aligned}
\bar{V}_1 &= \frac{V_1^T V_2 + V_2^T V_1}{2}, \bar{U}_2 = -\frac{V_1^T + V_2^T}{2},\\
\bar{W}_1 &= \frac{W_1^T W_2 + W_2^T W_1}{2}, \bar{W}_2 = -\frac{W_1^T + W_2^T}{2},\\
\Sigma_1 &= [e_1 + e_5\ \ (h_1+1)e_6 - e_2\ \ (h(k)-h_1+1)e_7 + (h_2-h(k)+1)e_8 - e_3 - e_4],\\
\Sigma_2 &= [e_1\ \ (h_1+1)e_6 - e_1\ \ (h(k)-h_1+1)e_7 + (h_2-h(k)+1)e_8 - e_3 - e_2],\\
\Sigma_3 &= [e_1 - e_2\ \ e_1 + e_2 - 2e_6],\\
\Sigma_4 &= [e_3 - e_4\ \ e_3 + e_4 - 2e_8\ \ e_2 - e_3\ \ e_2 + e_3 - 2e_7],\\
H\bar{A} &= \begin{bmatrix} H_1 A + \hat{B}_f C & \hat{A}_f \\ H_2^T A + \hat{B}_f C & \hat{A}_f \end{bmatrix}, \ H\bar{A}_d = \begin{bmatrix} H_1 A_d + \hat{B}_f C_d & 0 \\ H_2^T A_d + \hat{B}_f C_d & 0 \end{bmatrix},\\
H\bar{F} &= \begin{bmatrix} H_1 F \\ H_2^T F \end{bmatrix}, \ H\bar{H}_1 = \begin{bmatrix} \hat{B}_f H_1 \\ \hat{B}_f H_1 \end{bmatrix}, \ H\bar{H}_2 = \begin{bmatrix} \hat{B}_f H_2 \\ \hat{B}_f H_2 \end{bmatrix},\\
H\bar{B}_w &= \begin{bmatrix} H_1 B_w + \hat{B}_f D_w \\ H_2^T B_w + \hat{B}_f D_w \end{bmatrix}, \bar{R}_2 = \begin{bmatrix} R_2 & 0 \\ 0 & 3R_2 \end{bmatrix}
\end{aligned}
$$

$$
\begin{aligned}
\Phi &= [H\bar{A}-H \;\; 0 \;\; H\bar{A}_d \;\; 0 \;\; -H \;\; 0 \;\; 0 \;\; 0 \;\; 0 \;\; H\bar{H}_1 \;\; H\bar{H}_2 \;\; H\bar{B}_w],\\
\Psi &= [\bar{C} \;\; 0 \;\; \bar{C}_d \;\; 0 \;\; 0 \;\; 0 \;\; 0 \;\; 0 \;\; 0 \;\; \bar{D}_1 \;\; \bar{D}_2 \;\; \bar{D}_w],\\
\varepsilon_a &= \varepsilon_2+\varepsilon_3+\varepsilon_4+\varepsilon_5+\varepsilon_6, \quad \varepsilon_b = \varepsilon(\varepsilon_8+\varepsilon_8+\varepsilon_{10}+\varepsilon_{11}+\varepsilon_{12}),\\
\Xi &= \frac{1}{2}(\Sigma_1-\Sigma_2)\mathscr{P}(\Sigma_1+\Sigma_2)^T+\frac{1}{2}(\Sigma_1+\Sigma_2)\mathscr{P}(\Sigma_1-\Sigma_2)^T+e_1Q_1e_1^T-e_2Q_1e_2^T\\
&\quad +e_1Q_2e_1-e_4Q_2e_4^T+h_1^2e_5R_1e_5^T-\Sigma_3\begin{bmatrix} R_1 & 0\\ 0 & 3\lambda(h_1)R_1\end{bmatrix}\Sigma_3^T\\
&\quad +(h_2-h_1)^2e_5R_2e_5^T-\Sigma_4\begin{bmatrix} \bar{R}_2 & \mathscr{X}\\ * & 3\bar{R}_2\end{bmatrix}\Sigma_4^T\\
&\quad +[e_1 \;\; e_9]\begin{bmatrix} -K^T\bar{U}_1K & -K^T\bar{U}_2\\ * & -I\end{bmatrix}[e_1 \;\; e_9]^T\\
&\quad +[e_1 \;\; e_{10}]\begin{bmatrix} -K^T\bar{V}_1K & -K^T\bar{V}_2\\ * & -I\end{bmatrix}[e_1 \;\; e_{10}]^T\\
&\quad +[e_3 \;\; e_{11}]\begin{bmatrix} -K^T\bar{W}_1K & -K^T\bar{W}_2\\ * & -I\end{bmatrix}[e_3 \;\; e_{11}]^T\\
&\quad +[e_1+\varepsilon e_5]\Phi+\Phi^T[e_1+\varepsilon e_5]^T+e_1[\bar{\alpha}(\varepsilon_1^{-1}\varepsilon\varepsilon_7^{-1})\bar{N}_a^T\bar{N}_a\\
&\quad ++\;\bar{\beta}(\varepsilon_2^{-1}+\varepsilon\varepsilon_8^{-1})\bar{N}_{bc}^T\bar{N}_{bc}]e_1^T+e_3\bar{\beta}(\varepsilon_3^{-1}+\varepsilon\varepsilon_9^{-1})\bar{N}_{bcd}^T\bar{N}_{bcd}e_3^T\\
&\quad +e_{10}\bar{\beta}(\varepsilon_4^{-1}+\varepsilon\varepsilon_{10}^{-1})H_1^TN_b^TN_bH_1e_{10}^T+e_{11}\bar{\beta}(\varepsilon_5^{-1}+\varepsilon\varepsilon_{11}^{-1})H_2^TN_b^TN_bH_2e_{11}^T\\
&\quad +e_{12}\bar{\beta}(\varepsilon_6^{-1}+\varepsilon\varepsilon_{12}^{-1})D_w^TN_b^TN_bD_we_{12}^T-\gamma^2e_{12}e_{12}^T.
\end{aligned}
$$

Here is the first main result for the filter design problem.

Theorem 11.1 *For given scalars $h_2 > h_1 > 0, \gamma > 0, \varepsilon > 0, \varepsilon_i > 0\,(i = 1, 2\ldots, 12)$ and matrices $U_1, U_1, V_1, V_2, W_1, W_2$, there exists an admissible non-fragile $\mathscr{H}_\infty$ filter (11.4) such that the filtering error system (11.5) is asymptotically stable in mean square for $w(t) = 0$ and the filtering error $e(k)$ satisfies (11.7) under the zero initial condition for any nonzero $w(t) \in \mathscr{L}[0,\infty)$, if there exist matrices $\mathscr{P} > 0$, $Q_1 > 0$, $Q_2 > 0$, $R_1 > 0$, $R_2 > 0$, $\mathscr{X}$, H, $\hat{A}_f$, $\hat{B}_f$, $\hat{C}_f$, $\hat{D}_f$, $H = \begin{bmatrix} H_1 & H_2\\ H_2^T & H_2\end{bmatrix}$, such that the following LMIs hold:*

$$
\begin{bmatrix}
\Xi & \Psi^T & e_1\bar{\alpha}\varepsilon_1H\bar{M}_a & e_1\varepsilon_a\bar{\beta}H\bar{M}_b & e_5\bar{\alpha}\varepsilon\varepsilon_7H\bar{M}_a & e_5\varepsilon_b\bar{\beta}T\bar{M}_b\\
* & -I & 0 & 0 & 0 & 0\\
* & * & -\varepsilon_1\bar{\alpha}I & 0 & 0 & 0\\
* & * & * & -\varepsilon_a\bar{\beta}I & 0 & 0\\
* & * & * & * & -\varepsilon\varepsilon_7\bar{\alpha}I & 0\\
* & * & * & * & * & -\varepsilon_b\bar{\beta}I
\end{bmatrix} < 0, \quad (11.8)
$$

$$
\begin{bmatrix} \bar{R}_2 & \mathscr{X}\\ * & \bar{R}_2\end{bmatrix} \geq 0. \quad (11.9)
$$

Moreover, if the LMIs admit feasible solutions, the matrices of the admissible $\mathscr{H}_\infty$ filter in the form (11.4) can be obtained by

$$A_f = H_2^{-1}\hat{A}_f,\ B_f = H_2^{-1}\hat{B}_f,\ C_f = \hat{C}_f,\ D_f = \hat{D}_f.$$

Proof Construct the following Lyapunov functional:

$$V(k) = V_1(k) + V_2(k) + V_3(k) + V_4(k) + V_5(k) \tag{11.10}$$

with

$$\begin{aligned}
V_1(k) &= \begin{bmatrix} \zeta(k) \\ \sum\limits_{i=k-h_1}^{k-1} \zeta(i) \\ \sum\limits_{i=k-h_2}^{k-h_1-1} \zeta(i) \end{bmatrix}^T \mathscr{P} \begin{bmatrix} \zeta(k) \\ \sum\limits_{i=k-h_1}^{k-1} \zeta(i) \\ \sum\limits_{i=k-h_2}^{k-h_1-1} \zeta(i) \end{bmatrix}, \\
V_2(k) &= \sum_{i=k-h_1}^{k-1} \zeta^T(k) Q_1 \zeta(k), \\
V_3(k) &= \sum_{i=k-h_2}^{k-1} \zeta^T(k) Q_2 \zeta(k), \\
V_4(k) &= h_1 \sum_{i=-h_1+1}^{0} \sum_{j=k+i}^{k} \Delta\zeta^T(j) R_1 \Delta\zeta(j), \\
V_5(k) &= (h_2 - h_1) \sum_{i=-h_2+1}^{-h_1} \sum_{j=k+i}^{k} \Delta\zeta^T(j) R_2 \Delta\zeta(j).
\end{aligned}$$

The difference of $V(k)$ along the error system (11.5) with $w(t) = 0$ will be computed. Notice that

$$\begin{bmatrix} \zeta(k+1) \\ \sum\limits_{i=k-h_1+1}^{k} \zeta(i) \\ \sum\limits_{i=k-h_2+1}^{k-h_1} \zeta(i) \end{bmatrix} = \begin{bmatrix} \zeta(k) + \Delta\zeta(k) \\ \sum\limits_{i=k-h_1}^{k} \zeta(i) - \zeta(k-h_1) \\ \sum\limits_{i=k-h_2}^{k-h(k)} \zeta(i) + \sum\limits_{i=k-h(k)}^{k-h_1} \zeta(i) - \zeta(k-h(k)) - \zeta(k-h_2) \end{bmatrix},$$

$$\begin{bmatrix} \zeta(k) \\ \sum\limits_{i=k-h_1}^{k-1} \zeta(i) \\ \sum\limits_{i=k-h_2}^{k-h_1-1} \zeta(i) \end{bmatrix} = \begin{bmatrix} \zeta(k) \\ \sum\limits_{i=k-h_1}^{k} \zeta(i) - \zeta(k) \\ \sum\limits_{i=k-h_2}^{k-h(k)} \zeta(i) + \sum\limits_{i=k-h(k)}^{k-h_1} \zeta(i) - \zeta(k-h(k)) - \zeta(k-h_1) \end{bmatrix}.$$

The differences of $V_i(k) (i = 1, 2, \ldots, 5)$ are calculated as

$$\begin{aligned}\mathbb{E}\{\Delta V_1(k)\} &= \mathbb{E}\left\{\xi^T(k)(\Sigma_1\mathscr{P}\Sigma_1^T - \Sigma_2\mathscr{P}\Sigma_2^T)\xi(k)\right\}\\ &= \mathbb{E}\left\{\xi^T(k)(\Sigma_1+\Sigma_2)\mathscr{P}(\Sigma_1-\Sigma_2)^T\xi(k)\right\}, \quad (11.11)\\ \mathbb{E}\{\Delta V_2(k)\} &= \mathbb{E}\left\{\zeta^T(k)Q_1\zeta(k) - \zeta^T(k-h_1)Q_1\zeta(k-h_1)\right\}, \quad (11.12)\\ \mathbb{E}\{\Delta V_3(k)\} &= \mathbb{E}\left\{\zeta^T(k)Q_2\zeta(k) - \zeta^T(k-h_2)Q_2\zeta(k-h_2)\right\}, \quad (11.13)\\ \mathbb{E}\{\Delta V_4(k)\} &= \mathbb{E}\Bigg\{h_1^2\Delta\zeta^T(k)R_1\Delta\zeta(k)\\ &\quad -h_1\sum_{i=k-h_1+1}^{k}\Delta\zeta^T(i)R_1\Delta\zeta(i)\Bigg\}, \quad (11.14)\\ \mathbb{E}\{\Delta V_5(k)\} &= \mathbb{E}\Bigg\{(h_2-h_1)^2\Delta\zeta^T(k)R_2\Delta\zeta(k)\\ &\quad -(h_2-h_1)\sum_{i=k-h_2+1}^{k-h_1}\Delta\zeta^T(i)R_2\Delta\zeta(i)\Bigg\}. \quad (11.15)\end{aligned}$$

Based on Lemma 11.1, one has

$$\begin{aligned}&-h_1\sum_{i=k-h_1+1}^{k}\Delta\zeta^T(i)R_1\Delta\zeta(i)\\ &\le -\begin{bmatrix}\eta_1(k)\\ \eta_2(k)\end{bmatrix}^T\begin{bmatrix}R_1 & 0\\ 0 & 3\lambda(h_1)R_1\end{bmatrix}\begin{bmatrix}\eta_1(k)\\ \eta_2(k)\end{bmatrix}, \quad (11.16)\end{aligned}$$

where $\eta_1(k) = \zeta(k) - \zeta(k-h_1)$, $\eta_2(k) = \zeta(k) + \zeta(k-h_1) - \frac{2}{h_1+1}\sum_{i=k-h_1}^{k}\zeta(i)$, $\lambda(h_1) = 1(h_1 = 1)$, and $\lambda(h_1) = \frac{h_1+1}{h_1-1}(h_1 > 1)$.

By using Lemmas 11.2 and 11.3, we can get

$$\begin{aligned}&-(h_2-h_1)\sum_{i=k-h_2+1}^{k-h_1}\Delta\zeta^T(i)R_2\Delta\zeta(i)\\ &= -(h_2-h_1)\sum_{i=k-h_2+1}^{k-h(k)}\Delta\zeta^T(i)R_2\Delta\zeta(i) - (h_2-h_1)\sum_{i=k-h(k)+1}^{k-h_1}\Delta\zeta^T(i)R_2\Delta\zeta(i)\\ &\le -\frac{h_2-h_1}{h_2-h(k)}\begin{bmatrix}c\gamma_1(k)\\ \gamma_2(k)\end{bmatrix}^T\begin{bmatrix}R_2 & 0\\ 0 & 3R_2\end{bmatrix}\begin{bmatrix}\gamma_1(k)\\ \gamma_2(k)\end{bmatrix}\\ &\quad -\frac{h_2-h_1}{h(k)-h_1}\begin{bmatrix}\gamma_3(k)\\ \gamma_4(k)\end{bmatrix}^T\begin{bmatrix}R_2 & 0\\ 0 & 3R_2\end{bmatrix}\begin{bmatrix}\gamma_3(k)\\ \gamma_4(k)\end{bmatrix}\\ &\le -\gamma^T(k)\begin{bmatrix}\bar{R}_2 & \mathscr{X}\\ * & \bar{R}_2\end{bmatrix}\gamma(k), \quad (11.17)\end{aligned}$$

where

$$\gamma(k) = [\gamma_1^T(k)\ \gamma_2^T(k)\ \gamma_3^T(k)\ \gamma_4^T(k)]^T,$$
$$\gamma_1(k) = \zeta(k-h(k)) - \zeta(k-h_2),$$
$$\gamma_2(k) = \zeta(k-h(k)) + \zeta(k-h_2) - \frac{2}{h_2 - h(k) + 1}\sum_{i=k-h_2}^{k-h(k)} \zeta(i),$$
$$\gamma_3(k) = \zeta(k-h_1) - \zeta(k-h(k)),$$
$$\gamma_4 = \zeta(k-h_1) + \zeta(k-h(k)) - \frac{2}{h(k) - h_1 + 1}\sum_{k-h(k)}^{i=k-h_1} \zeta(i),$$
$$\bar{R}_2 = \begin{bmatrix} R_2 & 0 \\ 0 & 3R_2 \end{bmatrix}.$$

From Assumptions 11.2 and 11.3, it follows readily that

$$\begin{bmatrix} \zeta(k) \\ f(x(k)) \end{bmatrix}^T \begin{bmatrix} -K^T\bar{U}_1K & -K^T\bar{U}_2 \\ * & -I \end{bmatrix}\begin{bmatrix} \zeta(k) \\ f(x(k)) \end{bmatrix} \geq 0, \tag{11.18}$$
$$\begin{bmatrix} \zeta(k) \\ h(x(k)) \end{bmatrix}^T \begin{bmatrix} -K^T\bar{V}_1K & -K^T\bar{V}_2 \\ * & -I \end{bmatrix}\begin{bmatrix} \zeta(k) \\ h(x(k)) \end{bmatrix} \geq 0, \tag{11.19}$$
$$\begin{bmatrix} \zeta(k-h(k)) \\ h(x(k-h(k))) \end{bmatrix}^T \begin{bmatrix} -K^T\bar{W}_1K & -K^T\bar{W}_2 \\ * & -I \end{bmatrix}\begin{bmatrix} \zeta(k-h(k)) \\ h(x(k-h(k))) \end{bmatrix} \geq 0. \tag{11.20}$$

In view of (11.5), for any matrix H, we can obtain

$$\begin{aligned} &2\mathbb{E}\Big\{[\zeta(k) + \varepsilon\Delta\zeta(k)]H[(-\Delta\zeta(k) + (\bar{A} - I + \bar{\alpha}\hat{A}(k) + \hat{\alpha}(k)\hat{A}(k) \\ &+ \bar{\beta}\hat{B}(k) + \hat{\beta}(k)\hat{B}(k)))\zeta(k) \\ &+ (\bar{A}_d + \bar{\beta}\hat{A}_h(k) + \hat{\beta}(k)\hat{A}_h(k))\zeta(k-h(k)) + \bar{F}f(x(k)) \\ &+ (\bar{H}_1 + \bar{\beta}\hat{H}_1(k) + \hat{\beta}(k)\hat{H}_1(k))h(x(k)) \\ &+ (\bar{H}_2 + \bar{\beta}\hat{H}_2(k) + \hat{\beta}(k)\hat{H}_2(k))h(x(k-h(k))) \\ &+ (\bar{B}_w + \bar{\beta}\hat{B}_w(k) + \hat{\beta}(k)\hat{B}_w(k))w(k)]\Big\} = 0. \end{aligned} \tag{11.21}$$

For positive constants $\varepsilon_i (i = 1, 2, \ldots, 12)$, the following inequalities hold:

$$\begin{aligned} &\mathbb{E}\{2\zeta^T(k)H\hat{A}(k)\zeta(k)\} \leq \mathbb{E}\{\zeta^T(k)(\varepsilon_1 H\bar{M}_a\bar{M}_a^T H + \varepsilon_1^{-1}\bar{N}_a^T\bar{N}_a)\zeta(k)\}, \\ &\mathbb{E}\{2\zeta^T(k)H\hat{B}(k)\zeta(k)\} \leq \mathbb{E}\{\zeta^T(k)(\varepsilon_2 H\bar{M}_b\bar{M}_b^T H + \varepsilon_2^{-1}\bar{N}_{bc}^T\bar{N}_{bc})\zeta(k)\}, \\ &\mathbb{E}\{2\zeta^T(k)H\hat{A}_h(k)\zeta(k-h(k))\} \leq \mathbb{E}\{\zeta^T(k)\varepsilon_3 H\bar{M}_b\bar{M}_b^T H\zeta(k) \\ &\quad + \varepsilon_3^{-1}\zeta^T(k-h(k))\bar{N}_{bc_d}^T\bar{N}_{bc_d}\zeta(k-h(k))\}, \\ &\mathbb{E}\{2\zeta^T(k)H\hat{H}_1(k)h(x(k))\} \leq \mathbb{E}\{\zeta^T(k)\varepsilon_4 H\bar{M}_b\bar{M}_b^T H\zeta(k) \\ &\quad + \varepsilon_4^{-1}h^T(x(k))H_1^T N_b^T N_b H_1 h(x(k))\}, \end{aligned}$$

$$\begin{aligned}
&\mathbb{E}\{2\zeta^T(k)H\hat{H}_2(k)h(x(k-h(k)))\} \leq \mathbb{E}\{\zeta^T(k)\varepsilon_5 H\bar{M}_b\bar{M}_b^T H\zeta(k)\\
&\quad+\varepsilon_5^{-1}h^T(x(k-h(k)))H_2^T N_b^T N_b H_2 h(x(k-h(k)))\},\\
&\mathbb{E}\{2\zeta^T(k)H\hat{B}_w(k)w(k)\} \leq \mathbb{E}\{\zeta^T(k)\varepsilon_6 H\bar{M}_b\bar{M}_b^T H\zeta(k)\\
&\quad+\varepsilon_6^{-1}w^T(k)D_w^T N_b^T N_b D_w w(k)\},\\
&\mathbb{E}\{2\Delta\zeta^T(k)H\hat{A}(k)\zeta(k)\} \leq \mathbb{E}\{\Delta\zeta^T(k)\varepsilon_7 H\bar{M}_a\bar{M}_a^T H\Delta\zeta(k)\\
&\quad+\varepsilon_7^{-1}\zeta^T(k)\bar{N}_a^T\bar{N}_a\zeta(k)\},\\
&\mathbb{E}\{2\Delta\zeta^T(k)H\hat{B}(k)\zeta(k)\} \leq \mathbb{E}\{\Delta\zeta^T(k)\varepsilon_8 H\bar{M}_b\bar{M}_b^T H\Delta\zeta(k)\\
&\quad+\varepsilon_8^{-1}\zeta^T(k)\bar{N}_{bc}^T\bar{N}_{bc}\zeta(k)\},\\
&\mathbb{E}\{2\Delta\zeta^T(k)H\hat{A}_h(k)\zeta(k-h(k))\} \leq \mathbb{E}\{\Delta\zeta^T(k)\varepsilon_9 H\bar{M}_b\bar{M}_b^T H\Delta\zeta(k)\\
&\quad+\varepsilon_9^{-1}\zeta^T(k-h(k))\bar{N}_{bc_d}^T\bar{N}_{bc_d}\zeta(k-h(k))\},\\
&\mathbb{E}\{2\Delta\zeta^T(k)H\hat{H}_1(k)h(x(k))\} \leq \mathbb{E}\{\Delta\zeta^T(k)\varepsilon_{10} H\bar{M}_b\bar{M}_b^T H\Delta\zeta(k)\\
&\quad+\varepsilon_{10}^{-1}h^T(x(k))H_1^T N_b^T N_b H_1 h(x(k))\},\\
&\mathbb{E}\{2\Delta\zeta^T(k)H\hat{H}_2(k)h(x(k-h(k)))\} \leq \mathbb{E}\{\Delta\zeta^T(k)\varepsilon_{11} H\bar{M}_b\bar{M}_b^T H\Delta\zeta(k)\\
&\quad+\varepsilon_{11}^{-1}h^T(x(k-h(k)))H_2^T N_b^T N_b H_2 h(x(k-h(k)))\},\\
&\mathbb{E}\{2\Delta\zeta^T(k)H\hat{B}_w(k)w(k)\} \leq \mathbb{E}\{\Delta\zeta^T(k)\varepsilon_{12} H\bar{M}_b\bar{M}_b^T H\Delta\zeta(k)\\
&\quad+\varepsilon_{12}^{-1}w^T(k)D_w^T N_b^T N_b D_w w(k)\}.
\end{aligned} \tag{11.22}$$

Now, the $\mathscr{H}_\infty$ performance for filtering error system (11.5) will be established.

If the difference of $V(k)$ is negative, then $z(k) \to 0$ as $k \to \infty$. Next, by assuming zero initial conditions for the filtering error system, the performance index is

$$\begin{aligned}
J &= \sum_{k=0}^{\infty}\mathbb{E}\left\{e(k)^T e(k) - \gamma^2 w(k)^T w(k)\right\}\\
&= \sum_{k=0}^{\infty}\mathbb{E}\left\{\{e(k)^T e(k) - \gamma^2 w(k)^T w(k) + \Delta V(k)\} + V(0) - V(\infty)\right\}
\end{aligned}$$

If $\mathbb{E}\left\{e(k)^T e(k) - \gamma^2 w(k)^T w(k) + \Delta V(k)\right\} < 0$, then $V(k) \to 0$ as $k \to \infty$. Taking H as

$$H = \begin{bmatrix} H_1 & H_2 \\ H_2^T & H_2 \end{bmatrix},$$

and by a simple matrix calculation, it is straightforward to verify that

$$H\bar{A} = \begin{bmatrix} H_1 A + H_2 B_f C & H_2 A_f \\ H_2^T A + H_2 B_f C & H_2 A_f \end{bmatrix}, \; H\bar{A}_d = \begin{bmatrix} H_1 A_d + H_2 B_f C_d & 0 \\ H_2^T A_d + H_2 B_f C_d & 0 \end{bmatrix},$$

$$
H\bar{F} = \begin{bmatrix} H_1 F \\ H_2^T F \end{bmatrix}, \quad H\bar{H}_1 = \begin{bmatrix} H_2 B_f H_1 \\ H_2 B_f H_1 \end{bmatrix},
$$
$$
H\bar{H}_2 = \begin{bmatrix} H_2 B_f H_2 \\ H_2 B_f H_2 \end{bmatrix}, \quad H\bar{B}_w = \begin{bmatrix} H_1 B_w + H_2 B_f D_w \\ H_2^T B_w + H_2 B_f D_w \end{bmatrix}.
$$

Define a set of variables as follows:

$$
\hat{A}_f = H_2 A_f, \ \hat{B}_f = H_2 B_f, \ \hat{C}_f = C_f, \ \hat{D}_f = D_f.
$$

Then, by combining Eqs. (11.11)–(11.22), one can obtain

$$
\begin{aligned}
&\mathbb{E}\Big\{\Delta V(k) + e(k)^T e(k) - \gamma^2 w(k)^T w(k)\Big\} \le \mathbb{E}\Big\{\xi^T(k)(\Xi + \Psi^T\Psi \\
&+\bar{\alpha}\varepsilon_1 e_1 H\bar{M}_a\bar{M}_a^T H e_1^T + \bar{\beta}\varepsilon_a e_1 H\bar{M}_b\bar{M}_b^T H e_1^T \\
&+\bar{\alpha}\varepsilon_7 e_5 H\bar{M}_a\bar{M}_a^T H e_5^T + \bar{\beta}\varepsilon_b e_5 H\bar{M}_b\bar{M}_b^T H e_5^T)\xi(k)\Big\}.
\end{aligned} \tag{11.23}
$$

By Schur complement, $\Xi + \Psi^T\Psi + \bar{\alpha}\varepsilon_1 e_1 H\bar{M}_a\bar{M}_a^T H e_1^T + \bar{\beta}\varepsilon_a e_1 H\bar{M}_b\bar{M}_b^T H e_1^T + \bar{\alpha}\varepsilon\varepsilon_7 e_5 H\bar{M}_a\bar{M}_a^T H e_5^T + \bar{\beta}\varepsilon\varepsilon_b e_5 H\bar{M}_b\bar{M}_b^T H e_5^T < 0$ is equivalent to (11.8), which means $J < 0$ for any nonzero $w(k) \in \mathscr{L}_2$, i.e., the filtering error system has a guaranteed γ level of disturbance attenuation. This completes the proof of the theorem. ■

11.3.2 $\mathscr{H}_\infty$ Filtering for Linear Systems

In this subsection, the $\mathscr{H}_\infty$ filtering for linear systems will be established.

If $F = H_1 = H_2 = 0$, the system (11.1) is reduced to the following linear system:

$$
\begin{aligned}
x(k+1) &= Ax(k) + A_d x(k-h(k)) + B_w w(k), \\
y(k) &= Cx(k) + C_d x(k-h(k)) + D_w w(k), \\
z(k) &= Lx(k) + L_d x(k-h(k)) + G_w w(k).
\end{aligned} \tag{11.24}
$$

And, by taking $\Delta_1(k) = \Delta_2(k) = 0$, the filter of the form (11.4) is reduced to

$$
\begin{aligned}
x_f(k+1) &= A_f x_f(k) + B_f y(k), \\
z_f(k) &= C_f x_f(k) + D_f y(k), \quad \hat{x}(0) = 0.
\end{aligned} \tag{11.25}
$$

In this case, the corresponding error system is written as

$$
\begin{aligned}
\zeta(k+1) &= \bar{A}\zeta(k) + \bar{A}_d\zeta(k-h(k)) + \bar{B}_w w(k), \\
e(k) &= \bar{C}\zeta(k) + \bar{C}_d\zeta(k-h(k)) + \bar{D}_w w(k),
\end{aligned} \tag{11.26}
$$

where $\bar{A}$, $\bar{A}_d$, $\bar{B}_w$, $\bar{C}$, $\bar{C}_d$ and $\bar{D}_w$ are defined in (11.5).

Now, we focus on the $\mathscr{H}_\infty$ performance analysis for system (11.24). By Theorem 11.1, we have the following Corollary 11.1 which says that the error system (11.26) is asymptotically mean-square stable with $\mathscr{H}_\infty$ performance γ. To do this, we define $\bar{e}_i \in \mathbb{R}^{(16n+q)\times n}$ to be the block entry matrix. For example, $\bar{e}_2 = [0\ I\ \underbrace{0\ \ldots\ 0}_{15}]^T$.

And, some of scalar, vectors, and matrices are defined as

$$
\begin{aligned}
\tilde{\eta}^T(k) &= [\zeta^T(k)\ \zeta^T(k-h_1)\ \zeta^T(k-h(k))\ \zeta^T(k-h_2)\ \Delta\zeta^T(k)\ \frac{1}{h_1+1}\sum_{i=k-h_1}^{k}\zeta^T(i) \\
&\quad \frac{1}{h(k)-h_1+1}\sum_{i=k-h_2}^{k-h(k)}\zeta^T(i)\ \frac{1}{h_2-h(k)+1}\sum_{i=k-h(k)}^{k-h_1}\zeta^T(i)\ w^T(k)], \\
\tilde{e}_1 &= [\bar{e}_1\ \bar{e}_2], \tilde{e}_2 = [\bar{e}_3\ \bar{e}_4], \tilde{e}_3 = [\bar{e}_5\ \bar{e}_6], \tilde{e}_4 = [\bar{e}_7\ \bar{e}_8], \\
\tilde{e}_5 &= [\bar{e}_9\ \bar{e}_{10}], \tilde{e}_6 = [\bar{e}_{11}\ \bar{e}_{12}], \tilde{e}_7 = [\bar{e}_{13}\ \bar{e}_{14}], \tilde{e}_8 = [\bar{e}_{15}\ \bar{e}_{16}], \tilde{e}_9 = \bar{e}_{17} \\
\tilde{\Phi} &= [H\bar{A} - H\ 0\ H\bar{A}_d\ 0\ -H\ 0\ 0\ 0\ H\bar{B}_w], \\
\tilde{\Psi} &= [\bar{C}\ 0\ \bar{C}_d\ 0\ 0\ 0\ 0\ 0\ \bar{D}_w], \\
\tilde{\Sigma}_1 &= [\tilde{e}_1 + \tilde{e}_5\ (h_1+1)\tilde{e}_6 - \tilde{e}_2\ (h(k)-h_1+1)\tilde{e}_7 + (h_2-h(k)+1)\tilde{e}_8 - \tilde{e}_3 - \tilde{e}_4], \\
\tilde{\Sigma}_2 &= [\tilde{e}_1\ (h_1+1)\tilde{e}_6 - \tilde{e}_1\ (h(k)-h_1+1)\tilde{e}_7 + (h_2-h(k)+1)\tilde{e}_8 - \tilde{e}_3 - \tilde{e}_2], \\
\tilde{\Sigma}_3 &= [\tilde{e}_1 - \tilde{e}_2\ \tilde{e}_1 + \tilde{e}_2 - 2\tilde{e}_6],
\end{aligned}
$$

$$
\begin{aligned}
\tilde{\Sigma}_4 &= [\tilde{e}_3 - \tilde{e}_4\ \tilde{e}_3 + \tilde{e}_4 - 2\tilde{e}_8\ \tilde{e}_2 - \tilde{e}_3\ \tilde{e}_2 + \tilde{e}_3 - 2\tilde{e}_7], \\
\tilde{\Xi} &= \frac{1}{2}(\tilde{\Sigma}_1 - \tilde{\Sigma}_2)\mathscr{P}(\tilde{\Sigma}_1 + \tilde{\Sigma}_2)^T + \frac{1}{2}(\tilde{\Sigma}_1 + \tilde{\Sigma}_2)\mathscr{P}(\tilde{\Sigma}_1 - \tilde{\Sigma}_2)^T + \tilde{e}_1 Q_1 \tilde{e}_1^T \\
&\quad -\tilde{e}_2 Q_1 \tilde{e}_2^T + \tilde{e}_1 Q_2 \tilde{e}_1 - \tilde{e}_4 Q_2 \tilde{e}_4^T + h_1^2 \tilde{e}_5 R_1 \tilde{e}_5^T - \tilde{\Sigma}_3 \begin{bmatrix} R_1 & 0 \\ 0 & 3\lambda(h_1)R_1 \end{bmatrix} \tilde{\Sigma}_3^T \\
&\quad +(h_2-h_1)^2 \tilde{e}_5 R_2 \tilde{e}_5^T - \tilde{\Sigma}_4 \begin{bmatrix} \bar{R}_2 & \mathscr{X} \\ * & 3\bar{R}_2 \end{bmatrix} \tilde{\Sigma}_4^T \\
&\quad +[\tilde{e}_1 + \varepsilon\tilde{e}_5]\tilde{\Phi} + \tilde{\Phi}^T[\tilde{e}_1 + \varepsilon\tilde{e}_5]^T - \gamma^2 \tilde{e}_9 \tilde{e}_9^T
\end{aligned} \tag{11.27}
$$

Corollary 11.1 *For given constants $h_2 > h_1 > 0, \gamma > 0$ and $\varepsilon > 0$, there exists an admissible non-fragile $\mathscr{H}_\infty$ filter (11.25) such that the filtering error system (11.26) is asymptotically stable in mean square for $w(t) = 0$ and the filtering error $e(k)$ satisfies (11.7) under the zero initial condition for any nonzero $w(t) \in \mathscr{L}[0,\infty)$, if there exist matrices $\mathscr{P} > 0, Q_1 > 0, Q_2 > 0, R_1 > 0, R_2 > 0, \mathscr{X}, \hat{A}_f, \hat{B}_f, \hat{C}_f, \hat{D}_f, H = \begin{bmatrix} H_1 & H_2 \\ H_2^T & H_2 \end{bmatrix}$, satisfying the following LMIs*

$$
\begin{bmatrix} \tilde{\Xi} & \tilde{\Psi}^T \\ * & -I \end{bmatrix} < 0, \tag{11.28}
$$

$$
\begin{bmatrix} \bar{R}_2 & \mathscr{X} \\ * & \bar{R}_2 \end{bmatrix} \geq 0. \tag{11.29}
$$

Moreover, if the above LMIs admit feasible solutions, the matrices of the admissible $\mathscr{H}_\infty$ filter in the form (11.25) can be obtained by

$$A_f = H_2^{-1}\hat{A}_f,\ B_f = H_2^{-1}\hat{B}_f,\ C_f = \hat{C}_f,\ D_f = \hat{D}_f.$$

Remark 11.3 Compared with the result in [8, 9], the less conservative results of this chapter come from two aspects. On one hand, the augmented vector, which includes the terms such as $\sum_{i=k-h_1}^{k-1} \zeta(i),\ \sum_{i=k-h_2}^{k-h_1-1} \zeta(i)]$, is included in (11.10).

On the other hand, Lemma 11.1 is used to estimate the upper bound of $-h_1 \sum_{i=k-h_1+1}^{k} \Delta\zeta^T(i)R_1\Delta\zeta(i)$ instead of using Jensen's inequality. Also, Lemmas 11.2 and 11.3 are employed to tackle with the term $-(h_2-h_1)\sum_{i=k-h_2+1}^{k-h_1} \Delta\zeta^T(i)R_2\Delta\zeta(i)$.

Remark 11.4 In Theorem 11.1 (Corollary 11.1), the filter parameters can be derived by solving the following optimization problem such that the $\mathscr{H}_\infty$ disturbance attenuation level is minimized:

$$\min \gamma^2, \text{ subject to the LMIs (11.8) and (11.9) ((11.28). and (11.29)).}$$

The designed filters can ensure that filtering error system is exponentially stable and achieves a prescribed $\mathscr{H}_\infty$ performance level γ if the optimal γ is derived from the above optimization problem.

11.4 Numerical Examples

In this section, two examples are given to show the effectiveness of the proposed method on the design of the $\mathscr{H}_\infty$ filter.

Example 1 Consider the following simplified longitudinal flight system [12]:

$$x_1(k+1) = 0.9944x_1(k) - 0.1203x_2(k) - 0.4302x_3(k), \tag{11.30}$$
$$x_2(k+2) = 0.0017x_1(k) + 0.9902x_2(k) - (0.0747 + 0.01r(k))x_3(k), \tag{11.31}$$
$$x_3(k+1) = 0.8187x_2(k) + 0.1w(k), \tag{11.32}$$

where $r(k)$ is unknown but satisfies $|r(k)| \le 1$.

The measurement and estimated signals are assumed to be

$$\begin{aligned} y(k) &= 0.2x_1(k) + 0.1x_2(k) + (0.1 + 0.01r(k))x_3(k) \\ &\quad +0.1x_1(k-d(k)) + (0.1 + 0.01r(k))x_2(k-d(k)) \\ &\quad +0.02r(k)x_2(k-d(k)) + 0.04r(k)x_3(k) + 0.1w(k), \end{aligned} \tag{11.33}$$

$$z(k) = 0.1x_2(k) + 0.2x_3(k) + 0.1x_1(k-d(k)) + 0.2x_2(k-d(k)). \tag{11.34}$$

It can be see that the system expressed by Eqs. (11.30)–(11.34) satisfies Eqs. (11.2) and (11.3) and has the following form (11.1)

$$A = \begin{bmatrix} 0.9944 & -0.1203 & -0.4302 \\ 0.0017 & 0.9902 & -0.0747 \\ 0 & 0.8187 & 0 \end{bmatrix}, \ A_d = 0, \ F = 0,$$

$$B_w = \begin{bmatrix} 0 \\ 0 \\ 0.1 \end{bmatrix}, C = \begin{bmatrix} 0.2 & 0.1 & 0 \end{bmatrix}, \ C_d = \begin{bmatrix} 0.1 & 0.1 & 0 \end{bmatrix},$$

$$H_1 = H_2 = \begin{bmatrix} 0.2 & 0.2 & 0.2 \end{bmatrix}, \ D_w = 0.1,$$

$$L = \begin{bmatrix} 0 & 0.1 & 0.2 \end{bmatrix}, \ L_d = \begin{bmatrix} 0 & 0.1 & 0 \end{bmatrix}, G_w = 0.$$

The parameters of nonlinear functions are given as

$$U_1 = \begin{bmatrix} 0 & 0 & 0 \\ 0 & 0 & -0.01 \\ 0 & 0 & 0 \end{bmatrix}, \ U_2 = \begin{bmatrix} 0 & 0 & 0 \\ 0 & 0 & 0.01 \\ 0 & 0 & 0 \end{bmatrix},$$

$$V_1 = \begin{bmatrix} 0 & 0 & -0.01 \\ 0 & 0 & 0 \\ 0 & 0 & -0.04 \end{bmatrix}, \ V_2 = \begin{bmatrix} 0 & 0 & 0.01 \\ 0 & 0 & 0 \\ 0 & 0 & 0.04 \end{bmatrix},$$

$$W_1 = \begin{bmatrix} 0 & 0 & 0 \\ 0 & -0.01 & 0 \\ 0 & -0.02 & 0 \end{bmatrix}, \ W_2 = \begin{bmatrix} 0 & 0 & 0 \\ 0 & 0.01 & 0 \\ 0 & 0.02 & 0 \end{bmatrix}.$$

The uncertain parameters of the filter (11.4) can be described as:

$$M_a = M_b = \begin{bmatrix} 0.1 \\ 0.2 \\ 0.3 \end{bmatrix}, N_a = \begin{bmatrix} 0.1 & 0.1 & 0.1 \end{bmatrix},$$

$$N_b = 0.2, \ \Delta_1(k) = \Delta_2(k) = \sin(k).$$

Let $d_1 = 1, d_2 = 5, \alpha = 0.2, \beta = 0.5, \varepsilon = 1, \varepsilon_i = 0.5 \ (i = 1, 2, \ldots, 12)$.

Then, by simple computation, we can get that the minimum $\mathscr{H}_\infty$ performance is $\gamma = 0.1754$ by Theorem 11.1 and the corresponding filter parameters are

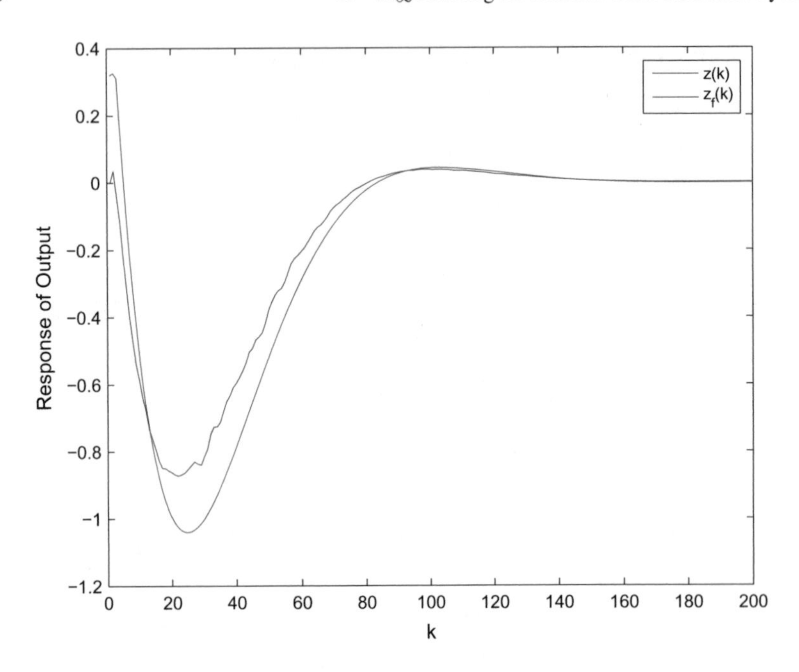

Fig. 11.1 Responses of $z(t)$ and $z_f(k)$ in Example 1

$$A_f = \begin{bmatrix} 0.8056 & -0.1376 & 0.1675 \\ 0.0134 & 0.8593 & -0.1038 \\ -0.0004 & 0.0576 & 0.4599 \end{bmatrix}, \ B_f = \begin{bmatrix} -0.5180 \\ 0.0365 \\ -0.0008 \end{bmatrix},$$
$$C_f = \begin{bmatrix} 0.0159 & -0.0970 & -0.0393 \end{bmatrix}, \ D_f = 0.3243.$$

For initial condition $x(0) = [-0.5 \ \ 0 \ \ 0.5]^T$ and $w(k) = \sin(k)e^{-10k}$, the response of $z(k)$ and $z_f(k)$, and the error $e(k)$ are shown in Figs. 11.1 and 11.2, respectively. From the simulation results, we can see that the desired system performance can be achieved by using the filter designed method.

Example 2 Consider the system (11.24) with the following parameters:

$$A = \begin{bmatrix} 0.85 & 0.1 \\ -0.1 & 0.7 \end{bmatrix}, \ A_d = \begin{bmatrix} 0.2 & 0 \\ -0.2 & 0.1 \end{bmatrix}, B_w = \begin{bmatrix} 0.1 \\ 0.4 \end{bmatrix},$$
$$C = \begin{bmatrix} 0.2 & 2.5 \end{bmatrix}, \ C_d = \begin{bmatrix} -0.5 & 0.5 \end{bmatrix}, \ D_w = -1,$$
$$L = \begin{bmatrix} 0 & 2.2 \end{bmatrix}, \ L_d = \begin{bmatrix} 1.5 & -0.4 \end{bmatrix}, G_w = -0.1.$$

The $\mathscr{H}_\infty$ performance γ got by different methods are expressed in Table 11.1.

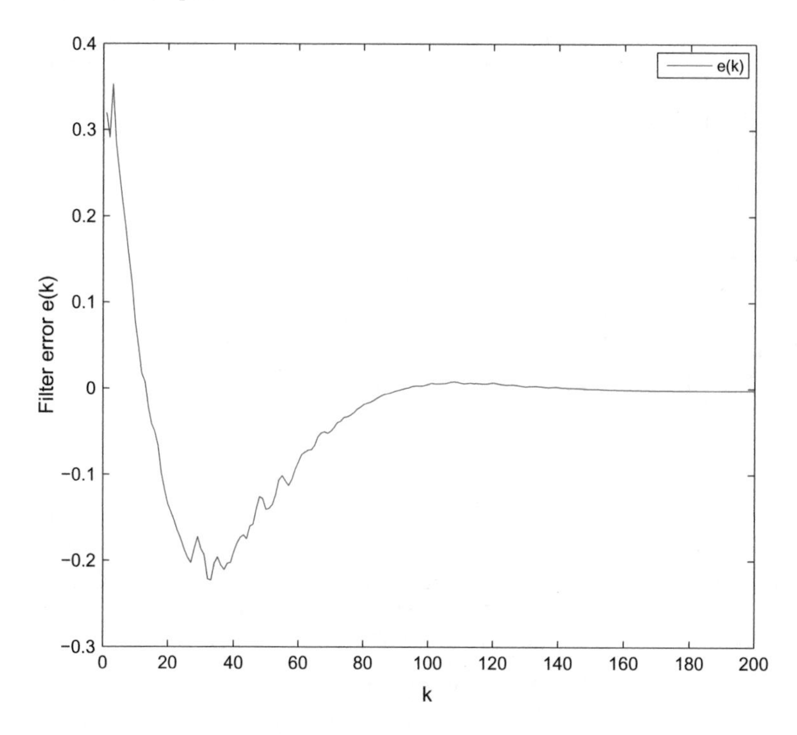

Fig. 11.2 Estimated error of $e(k)$ in Example 1

Table 11.1 Comparison of $\mathscr{H}_\infty$ performance γ

$d(k)$	$d_1 = 1$	$d_1 = 1$	$d_1 = 2$	$d_1 = 2$
	$d_2 = 4$	$d_2 = 5$	$d_2 = 5$	$d_2 = 6$
[8]	5.0782	6.4910	5.5151	7.0533
[9]	4.9431	6.1604	5.3551	6.7581
Corollary 1	4.2786	4.8368	4.4473	4.9473

When $d_1 = 1$, $d_2 = 4$ and $\varepsilon = 2$, we can get the minimal $\mathscr{H}_\infty$ performance level $\gamma = 4.2786$ by Corollary 11.1, which is much smaller than 5.0782 and 4.9431.

Thus, the our result in this chapter is less conservative than that proposed in [8, 9]. By solving the LMIs in (11.28) and (11.29), the corresponding parameters of the filter gains are as follows

$$A_f = \begin{bmatrix} 1.3242 & 2.9175 \\ -0.2274 & -0.3790 \end{bmatrix}, \; B_f = \begin{bmatrix} 0.7747 \\ -0.3693 \end{bmatrix},$$
$$C_f = \begin{bmatrix} -1.0532 & -4.1632 \end{bmatrix}, \; D_f = -0.8552.$$

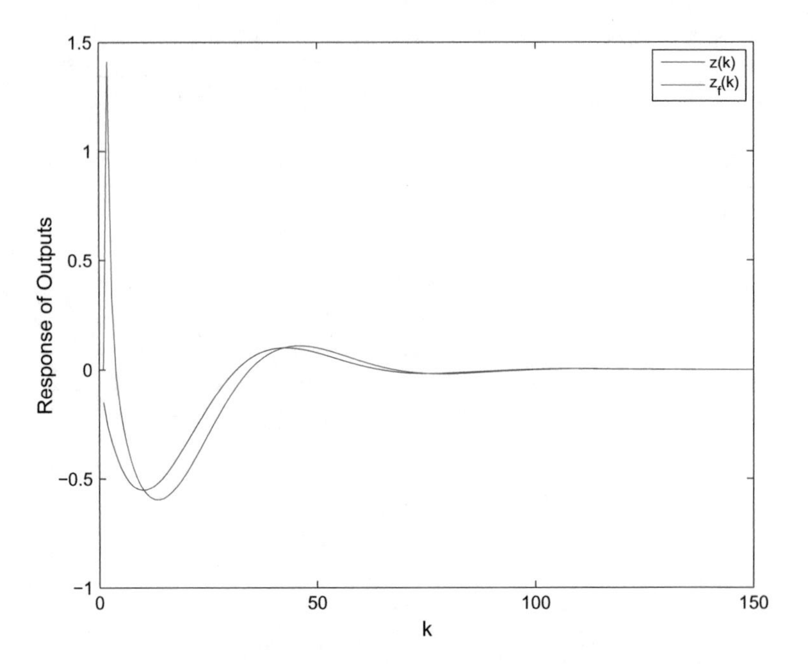

Fig. 11.3 Responses of $z(t)$ and $z_f(k)$ in Example 2

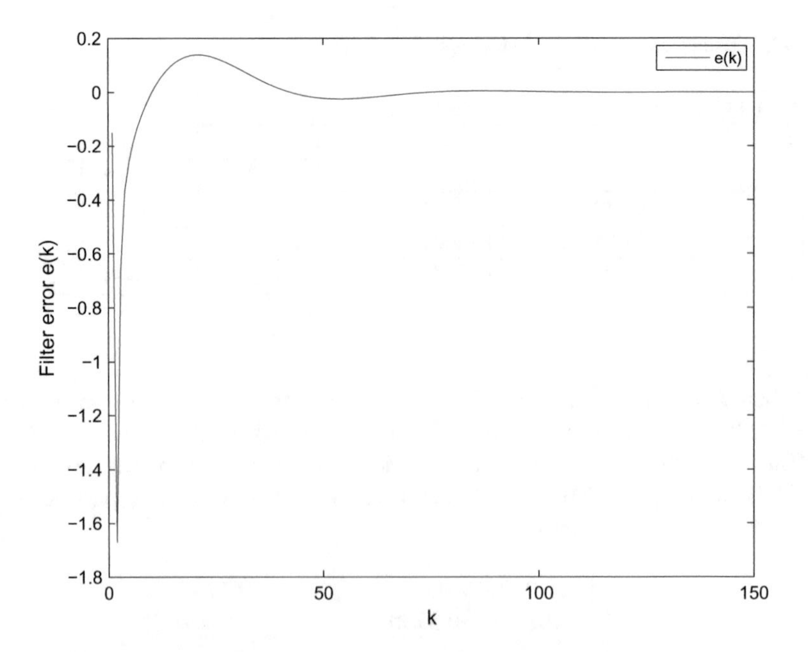

Fig. 11.4 Estimated error of $e(k)$ in Example 2

For initial condition $x(0) = [0.5 \quad -0.5]^T$ and $w(k) = \sin(k)e^{-10k}$, the response of $z(k)$ and $z_f(k)$, and the error $e(k)$ are shown in Figs. 11.3 and 11.4, respectively. From these simulation results, it can be seen that the disturbance is effectively attenuated by designed $\mathscr{H}_\infty$ filter.

11.5 Conclusion

In this chapter, we have studied the problem of the delay-dependent non-fragile $\mathscr{H}_\infty$ filtering with nonlinear discrete-time delay. Combining with the Lyapunov functional method, we have got a sufficient condition to ensure the filtering error system mean-square stable asymptotically with a prescribed $\mathscr{H}_\infty$ performance under the consideration of a zero equality. Thus, a novel $\mathscr{H}_\infty$ filter is presented to improve the linear system. Furthermore, the model transformation method is not employed. Finally, the numerical simulations are proposed to illustrate the effectiveness of the improved approach.

References

1. Geromel JC, de Oliveria MC (2001) $\mathscr{H}_2$ and $\mathscr{H}_\infty$ robust filtering for convex bounded uncertain systems. IEEE Trans Autom Control 46:100–107
2. Shi P, Bonkas EK, Agarwal RK (1999) Kalman filtering for continuous time uncertain systems with Markovian jumping parameters. IEEE Trans Autom Control 44:1592–1597
3. Zhang J, Xia Y, Shi P (2009) Parameter-dependent robust $\mathscr{H}_\infty$ filtering for uncertain discrete-time systems. Automatica 45:560–565
4. Li J, Zhang Y, Pan Y (2015) Mean-square exponential stability and stabilization of stochastic singular systems with multiple time-varying delays. Circuits Syst Signal Process 34:1187–1210
5. Li J, Li L (2015) Mean-square exponential stability for stochastic discrete-time recurrent neural networks with mixed time delays. Neuroocmputing 151:790–797
6. Kim JH, Ahn SJ, Ahn S (2005) Guaranteed cost and $\mathscr{H}_\infty$ filtering for discrete-time polytopic uncertain systems with time-delay. J Frankl Inst 342:365–378
7. Liu H, Sun F, He K, Sun Z (2001) Robust $\mathscr{H}_\infty$ filtering for uncertain discrete-time state-delayed systems. IEEE Trans Signal Process 49:1696–1703
8. He Y, Liu GP, Rees D, Mu W (2009) $\mathscr{H}_\infty$ filtering for discrete-time systems with time-varying delay. Signal Process 89:275–282
9. You J, Gao H, Basin M (2013) Further improved results on $\mathscr{H}_\infty$ filtering for discrete time-delay systems. Signal Process 93:1845–1852
10. Wang Z, Ho DWC, Liu Y, Liu X (2009) Robust $\mathscr{H}_\infty$ control for a class of nonlinear discrete-time delay stochastic systems with missing measurements. Automatica 45:684–691
11. Li H, Gao Y, Shi P, Lam H (2015) Observer-based fault detection for nonlinear system with sensor fault and limited communication capacity. IEEE Trans Autom Control 61:2745–2751
12. Xu S (2002) Robust $\mathscr{H}_\infty$ filtering for a class of discrete-time uncertain nonlinear systems with state delay. IEEE Trans Circuits Syst I: Fundam Theory and Appl 49:1853–1859
13. Gao H, Lam J, Wang C (2005) Induced $\mathscr{L}_2$ and generalized $\mathscr{H}_2$ filtering for systems with repeated scalar nonlinearities. IEEE Trans Signal Process 53:4215–4226
14. Yang R, Shi P, Liu G (2011) Filtering for discrete-time networked nonlinear systems with mixed random delays and packet dropouts. IEEE Trans Autom Control 56:2655–2660

15. Zhang J, Xia Y, Tao R (2009) New Results on $\mathscr{H}_\infty$ filtering for fuzzy time-delay systems. IEEE Trans Fuzzy Syst 17:128–137
16. Huang S, He X, Zhang N (2011) New results on $\mathscr{H}_\infty$ filter for nonlinear systems with time delay via T-S fuzzy model. IEEE Trans Fuzzy Syst 19:193–199
17. Su Y, Chen B, Lin C, Zhang H (2009) A new fuzzy $\mathscr{H}_\infty$ filter design for nonlinear continuous-time dynamic systems with time-varying delays. Fuzzy Sets Syst 160:3539–3549
18. Wu YQ, Su H, Lu R, Wu ZG, Shu Z (2015) Passivity-based non-fragile control for Markovian-jump systems with aperiodic sampling. Syst Control Lett 84:35–43
19. Zhang D, Cai WJ, Xie LH, Wang QG (2015) Non-fragile distributed filtering for T-S fuzzy systems in sensor networks. IEEE Trans Fuzzy Syst 23:1883–1980
20. Zhang D, Shi P, Zhang WA (2016) Non-fragile distributed filtering for fuzzy systems with multiplicative gain variation. Signal Process 121:102–110
21. Liu Y, Park JH, Guo B (2016) Non-fragile $\mathscr{H}_\infty$ filtering for nonlinear discrete-time delay systems with randomly occurring gain variations. ISA Trans 63:196–203
22. Liu G, Park JH, Xu S, Zhuang G (2019) Robust non-fragile $\mathscr{H}_\infty$ fault detection filter design for delayed singular Markovian jump systems with linear fractional parametric uncertainties. Nonlinear Anal: Hybrid Syst 32:65–78
23. Shen M, Park JH, Fei S (2018) Event-triggered non-fragile $\mathscr{H}_\infty$ filtering of Markov jump systems with imperfect transmissions. Signal Process 149:204–213
24. Liu Y, Guo BZ, Park JH (2017) Non-fragile $\mathscr{H}_\infty$ filtering for delayed Takagi-Sugeno fuzzy systems with randomly occurring gain variations. Fuzzy Sets Syst 316:99–116
25. Shen H, Wu ZG, Park JH (2015) Reliable mixed passive and $\mathscr{H}_\infty$ for semi-Markov jump systems with randomly occurring uncertainties and sensor failures. Int J Robust Nonlinear Control 25:3231–3235
26. Seuet A, Gouaisbaut F, Fridman E (2015) Stability of discrete-time systems with time-varying delay via a novel summation inequality. IEEE Trans Autom Control 60:2740–2745
27. Park PG, Ko JW, Jeong C (2011) Reciprocally convex approach to stability of systems with time-varying delays. Automatica 47:235–238

Chapter 12
Design of Dissipative Filter for Delayed Nonlinear Interconnected Systems via Takagi-Sugeno Fuzzy Modelling

12.1 Introduction

As we know, modeling of many real systems, such as power systems, aerospace systems, mobile robots, and process control systems, depends on large-scale interconnected nonlinear systems [1, 2]. In the control of interconnected systems, the stabilizing controller in the decentralized method is designed independently for each subsystem, which only use the local signals for feedback, at the same time. Hence, it can be an efficient and effective technique to be applied for achieving a control objective of the overall system. Decentralized control technology can not only reduce the amount of calculation, but also improve the robustness and reliability of interactive operation fault. Thus, decentralized control of large-scale interconnected nonlinear systems has been received considerable attention in the past few decades [3–9].

What's more, a great amount of effort has been focussing on filtering problem due to it widely exists in the field of control theory and signal processing. $\mathscr{H}_\infty$ filtering is the most attractive one among many kinds of filtering techniques, because its advantages that no statistical assumptions on the exogenous noises are needed. Hence, the $\mathscr{H}_\infty$ filter has stronger robustness than many other methods, and a lot of significant and important results on $\mathscr{H}_\infty$ filter have been published in the literature. In addition, the $\mathscr{H}_\infty$ filtering problem has been studied in various system, for example, linear system [10], descriptor systems [11], Markov system [12–15], and so on. Due to the nonlinearity of large-scale interconnected systems, the problem of decentralized $\mathscr{H}_\infty$ filtering cannot be solved at present.

In order to solve this problem, Takagi-Sugeno (T-S) fuzzy model can be an efficient way to approximate nonlinear systems [16–24]. Many methods have been proposed in stability analysis and controller design for large-scale interconnected systems, which are based on T-S fuzzy models. For example, the stability problem and $\mathscr{H}_\infty$ controller and filter design were investigated in [25–29]. The decentralized fuzzy $\mathscr{H}_\infty$ filtering for nonlinear interconnected with multiple time delays was investigated in [30]. The problem of decentralized $\mathscr{H}_\infty$ for nonlinear interconnected systems with time-varying delay was studied in [31]. Besides, For the design of decentralized $\mathscr{H}_\infty$

J. H. Park et al., *Dynamic Systems with Time Delays: Stability and Control*,
https://doi.org/10.1007/978-981-13-9254-2_12

filters for time-varying nonlinear interconnected systems, the performance index can be improved by applying the Lyapunov function of a fuzzy line integral [32]. It is worth noting that common Lyapunov-Krasovskii function [30, 31], Jensen's inequality [32], and free-weighting matrix [30] in existing works are used to design the $\mathscr{H}_\infty$ filter for large-scale interconnected systems, which may lead to conservatism to some degree. Hence, it is significant to study further the $\mathscr{H}_\infty$ filter design of large-scale interconnected systems.

In addition, the dissipative theory has been studied by many researchers after the concept of dissipative systems was first proposed in [33, 34]. For example, the dissipative reliable filtering problem of T-S fuzzy systems was presented in [35]. The problem of extended dissipative state estimation of the Markov jump neural networks was investigated in [36]. The extended dissipative analysis for generalized Markovian switching neural networks with two delay components was considered in [37]. The event-triggered reliable dissipative filtering for a T-S fuzzy system is developed in [38]. So far, there are still many problems to be solved in the design of decentralized dissipative $\mathscr{H}_\infty$ filters for interconnected systems with interval time-varying delays.

In this chapter, the problem of decentralized dissipative filtering for nonlinear interconnected systems with time-varying delays is studied. Our main purpose is to establish a delay-dependent criterion such that the obtained closed-loop system is asymptotically stable with a strict $(Q, S, R)-\alpha-$ dissipativity. The gain of the filter can be obtained by solving a set of linear matrix inequalities (LMIs). A simulation example is given to show the effectiveness of the proposed method.

12.2 Problem Formulation

Consider a nonlinear large-scale interconnected systems with interval time-varying delay composed of K subsystems where the ith subsystem is expressed by

Plant rule j:
IF $\theta_{i1}(t)$ is M_{i1j}, $\theta_{i2}(t)$ is M_{i2j}, ..., $\theta_{ip}(t)$ is M_{ipj},
THEN

$$\begin{aligned}
\dot{x}_i(t) &= A_{ij}x_i(t) + A_{dij}x_i(t-h_i(t)) + \sum_{n=1,n\neq i}^{K} B_{ni}x_n(t) + C_{ij}v_i(t),\\
y_i(t) &= L_{ij}x_i(t) + L_{dij}x_i(t-h_i(t)) + E_{ij}v_i(t),\\
z_i(t) &= F_{ij}x_i(t),\\
x_i(t) &= \phi_i(t), \forall t \in [-d_2, 0],
\end{aligned} \tag{12.1}$$

where $\theta_i(t) = (\theta_{i1}, \theta_{i2}, \ldots, \theta_{ip})$ denote some measurable premise variables, and $M_{igj}(g = 1, 2, \ldots, p)$ are fuzzy sets; $x_i(t) \in \mathbb{R}^{n_{ix}}$ denotes state vector, $y_i(t) \in \mathbb{R}^{n_{iy}}$ is output vector, and $z_i(t) \in \mathbb{R}^{n_{iz}}$ is the signal to be estimated; $v_i(t) \in \mathbb{R}^{n_{iv}}$ is the disturbance which belongs to $L_2[0, \infty)$; A_{ij}, A_{dij}, C_{ij}, L_{ij}, L_{dij}, E_{ij}, and F_{ij} are constant matrices with appropriate dimensions; B_{ni} is the interconnection between the ith and nth subsystem, and $B_{ni} = 0$ for $i = n$; $\phi_i(t)$ denotes the given initial condition sequence.

The time-varying delay $h_i(t)$ satisfies:

$$h_{i1} \leq h_i(t) \leq h_{i2}, \quad \mu_{i1} \leq \dot{h}_i(t) \leq \mu_{i2}. \tag{12.2}$$

Throughout the use of "fuzzy blending", the final output of the interconnected systems with interval time-varying delay can be inferred as follows

$$\begin{aligned}
\dot{x}_i(t) &= \sum_{j=1}^{r_i} h_{ij}(\theta_i(t)[A_{ij}x_i(t) + A_{dij}x_i(t - h_i(t)) + C_{ij}v_i(t)] + \sum_{n=1,n\neq i}^{K} B_{ni}x_n(t), \\
y_i(t) &= \sum_{j=1}^{r_i} h_{ij}(\theta_i(t))[L_{ij}x_i(t) + L_{dij}x_i(t - h_i(t)) + E_{ij}v_i(t)], \\
z_i(t) &= \sum_{j=1}^{r_i} h_{ij}(\theta_i(t))F_{ij}x_i(t), \\
x_i(t) &= \phi_i(t), \forall t \in [-d_2, 0],
\end{aligned} \tag{12.3}$$

with

$$h_{ij}(\theta_i(t)) = \frac{w_{ij}(\theta_i(t))}{\sum_{i=1}^{r_i} w_{ij}(\theta_i(t))}, \quad w_{ij}(\theta_i(t)) = \prod_{g=1}^{p} M_{igj}(\theta_{ig}(t))$$

In (12.3), $h_{ij}(\theta_i(t))$ is the fuzzy basic function, and $M_{ipj}(\theta_{ip}(t))$ is the grade of membership of $\theta_{ip}(t)$ in M_{ipj}.

Then, it can be seen that

$$w_{ij}(\theta_i(t)) \geq 0, \quad \text{and} \quad \sum_{j=1}^{r_i} w_{ij}(\theta_i(t)) \geq 0$$

for all t.

Therefore, we have

$$h_{ij}(\theta_i(t)) \geq 0, \quad \text{and} \quad \sum_{j=1}^{r_i} h_{ij}(\theta_i(t)) = 1.$$

In this chapter, the filter for the fuzzy systems S_i is give as:

Filter rule j:
IF $\theta_{i1}(t)$ is M_{i1j}, $\theta_{i2}(t)$ is M_{i2j}, ..., $\theta_{ip}(t)$ is M_{ipj},
THEN

$$
\begin{aligned}
x_{fi}(t) &= A_{fij}x_{fi}(t) + B_{fij}y_i(t) \\
z_{fi}(t) &= F_{fij}x_{fi}(t), \\
x_{fi}(t) &= \phi_{fi}(t), \forall t \in [-d_2, 0],
\end{aligned}
\tag{12.4}
$$

where $x_{fi}(t) \in \mathbb{R}^{n_{fix}}$ is the state vector of the filter; $z_{fi}(t) \in \mathbb{R}^{n_{fiz}}$ denotes an estimation of $z_i(t)$; A_{fij}, B_{fij} and F_{fij} are filter parameters to be designed.

The defuzzified output of the subsystem (12.4) is described by

$$
\begin{aligned}
x_i(t) &= \sum_{j=1}^{r_i} h_{ij}(\theta_i(t))[A_{fij}x_{fi}(t) + B_{fij}y_i(t)], \\
z_i(t) &= \sum_{j=1}^{r_i} h_{ij}(\theta_i(t))F_{fij}x_{fi}(t), \\
x_{fi}(t) &= \phi_{fi}(t), \forall t \in [-h_2, 0].
\end{aligned}
\tag{12.5}
$$

Combining (12.3) and (12.5), and selecting a state vector $\zeta^T(t) = [x_i^T(t), x_{fi}^T(t)]^T$ and $e_i(t) = z_i(t) - z_{fi}(t)$, then filter error subsystem can be derived as:

$$
\begin{aligned}
\dot{\zeta}_i(t) &= \bar{A}_i(t)\zeta_i(t) + \bar{A}_{di}(t)\zeta_i(t - h_i(t)) + \sum_{n=1, n\neq i}^{K} \bar{B}_{ni}\zeta_n(t) + \bar{C}_i(t)v_i(t), \\
e_i(t) &= \bar{F}_i(t)\zeta(t), \\
\zeta_i(t) &= [\phi_i^T(t) \ \ \phi_{fi}^T(t)]^T, \forall t \in [-h_2, 0]
\end{aligned}
\tag{12.6}
$$

where

$$
\begin{aligned}
\bar{A}_i(t) &= \sum_{j=1}^{r_i} h_{ij}(\theta_i(t)) \sum_{k=1}^{r_i} h_{ik}(\theta_i(t)) \begin{bmatrix} A_{ij} & 0 \\ B_{fik}L_{ij} & A_{fik} \end{bmatrix} \\
&= \begin{bmatrix} A_i(t) & 0 \\ B_{fi}(t)L_i(t) & A_{fi}(t) \end{bmatrix}, \\
\bar{A}_{di}(t) &= \sum_{j=1}^{r_i} h_{ij}(\theta_i(t)) \sum_{k=1}^{r_i} h_{ik}(\theta_i(t)) \begin{bmatrix} A_{dij} & 0 \\ B_{fik}L_{dij} & 0 \end{bmatrix} \\
&= \begin{bmatrix} A_{di}(t) & 0 \\ B_{fi}(t)L_{di}(t) & 0 \end{bmatrix},
\end{aligned}
$$

$$
\begin{aligned}
\bar{C}_i(t) &= \sum_{j=1}^{r_i} h_{ij}(\theta_i(t)) \sum_{k=1}^{r_i} h_{ik}(\theta_i(t)) \begin{bmatrix} C_{ij} \\ B_{fik}E_{ij} \end{bmatrix} \\
&= \begin{bmatrix} C_i(t) \\ B_{fi}(t)E_i(t) \end{bmatrix},
\end{aligned}
$$

$$\bar{F}_i(t) = \sum_{j=1}^{r_i} h_{ij}(\theta_i(t)) \sum_{k=1}^{r_i} h_{ik}(\theta_i(t)) \Big[F_{ij} \ -F_{fik} \Big]$$
$$= \Big[F_i(t) \ -F_{fi}(t) \Big],$$
$$\bar{B}_{ni} = \begin{bmatrix} B_{ni} & 0 \\ 0 & 0 \end{bmatrix}.$$

Let $\zeta(t) = [\zeta_1^T(t), \zeta_2^T(t), \ldots, \zeta_N^T(t)]^T$ and $v(t) = [v_1^T(t), v_2^T(t), \ldots, v_N^T(t)]^T$. Before formulating the problem, the following definition, which is called generalized dissipativity, is first given.

Definition 12.1 For a given scalar $\beta > 0$ and constant matrices $Q \leq 0$, S and a symmetric matrix R, the filtering error system (12.6) is said to be strictly $(Q, S, R) - \beta$ dissipative and β is called the dissipativity performance bound, if the following inequality is satisfied under the zero initial condition and $v(t) \in \mathscr{L}_2[0, \infty)$:

$$\int_0^t J(s)ds \geq \beta \int_0^t v^T(s)v(s)ds, \tag{12.7}$$

where $J(t) = e^T(t)Qe(t) + 2e^T(t)Sv(t) + v^T(t)R\omega(t)$.

In general, it is assumed that $Q \leq 0$ and $-Q = \bar{Q}^T\bar{Q}$ for some $\bar{Q} \geq 0$.

Remark 12.1 The dissipativity in Definition 12.1 includes some special cases as follows:

(1) When $Q = 0$, $S = I$, and $R = 2\alpha I$, the strict $(Q, S, R) - \alpha$ dissipative becomes passivity performance.

(2) When $Q = -I$, $S = 0$, and $R = (\alpha^2 + \alpha)I$, the strict $(Q, S, R) - \alpha$ dissipative reduces to $\mathscr{H}_\infty$ performance.

(3) When $Q = -vI$, $S = (1 - v)I$, and $R = ((\alpha^2 - \alpha)v + 2\alpha)I$, the strict $(Q, S, R) - \alpha$ dissipative means mixed passivity/$\mathscr{H}_\infty$ performance.

In order to propose the main result of this chapter, the following lemmas will be introduced.

Lemma 12.1 *Given $R \in \mathbb{R}^{n \times n} > 0$, a vector $\omega : [a, b] \in \mathbb{R}^n$, and an auxiliary scalar function $\{f_i(u), i \in [0, n] | u \in [a, b], f_0(u) = 1\}$ satisfying $\int_a^b f_i(s)f_j(s)ds = 0 (0 \leq i, j \leq n, i \neq j)$ with $f_i(s)(i = 1, \ldots, n)$ not identically zero. Then for any matrices $N_i \in \mathbb{R}^{k \times n} (i = 0, \ldots, n)$ and a vector $\xi \in \mathbb{R}^k$ the following inequality hold*

$$-\int_a^b \omega^T(s)R\omega(s)ds$$
$$\leqslant \xi^T \left\{ \sum_{i=0}^{n} \int_a^b f_i^2(s)ds N_i R^{-1} N_i^T + \text{Sym}(\sum_{i=0}^{n} N_i \eta_i) \right\} \xi,$$

where $\eta_i \in \mathbb{R}^{n \times k} (i = 0, \ldots, n)$, and $\int_a^b f_i(s)\omega(s)ds = \eta_i \zeta$.

Proof Lemma 12.1 can be easily obtained by integrating the following inequality from a to b

$$-2\varpi^T(s)N\varpi(s) \leq \varpi^T(s)NR^{-1}N^T\varpi(s) + \omega^T(s)R\omega(s),$$

where $N = [N_0^T, N_1^T, \ldots, N_n^T]^T$ and $\varpi(s) = [f_0(s)\zeta^T, f_1(s)\zeta^T, \ldots, f_n(s)\zeta^T]^T$.

The rest of the proof is omitted. ■

Lemma 12.2 ([39]) *For a given matrix $R \in \mathbb{R}^{n\times n}$, any differentiable function x in $[a, b] \to \mathbb{R}^n$, the equality*

$$\int_a^b \dot{x}^T(s)R\dot{x}(s)ds \geqslant \frac{1}{b-a}\Omega^T \mathrm{diag}(R, 3R, 5R)\Omega,$$

holds, where

$$\Omega = \begin{bmatrix} x(b) - x(a) \\ x(b) + x(a) - \frac{2}{b-a}\int_a^b x(s)ds \\ x(b) - x(a) + \frac{6}{b-a}\int_a^b x(s)ds - \frac{12}{(b-a)^2}\int_a^b\int_s^b x^T(u)duds \end{bmatrix}.$$

Lemma 12.3 ([40]) *For any matrices $R_1 \in \mathbb{R}^n, R_2 \in \mathbb{R}^n$, $Y_1 \in \mathbb{R}^{2n\times n}$ and $Y_2 \in \mathbb{R}^{2n\times n}$, the following inequality holds*

$$\begin{bmatrix} \frac{1}{\beta}R_1 & 0 \\ 0 & \frac{1}{1-\beta}R_2 \end{bmatrix} \geqslant \Theta_M(\beta), \quad \forall \beta \in (0, 1)$$

where

$$\Theta_M(\beta) = \mathrm{Sym}\{Y_1[I_n, 0_n] + Y_2[0_n, I_n]\} - \varphi Y_1 R_1^{-1} Y_1^T - (1-\varphi)Y_2 R_2^{-1} Y_2^T.$$

12.3 Main Results

12.3.1 Asymptotic Stability for the Closed Loop Interconnected Systems

In this subsection, a delay-dependent condition for asymptotic stability of closed-loop system for nonlinear interconnected systems with interval time-varying delay will be established.

For the sake of simplicity, let us define $\bar{e}_i \in \mathbb{R}^{(20n_{ix}+n_{iv})\times n_{ix}}$ to be block entry matrix. For example, $\bar{e}_2 = [0, I, \underbrace{0, \ldots, 0}_{18}, 0]^T$. Some other vectors and matrices are given as

$$\xi^T(t) = \Big[\zeta^T(t)\ \ \zeta^T(t-h_{i1})\ \ \zeta^T(t-h_i(t))\ \ \zeta^T(t-h_{i2})\ \ \frac{1}{h_{i1}}\int_{t-h_{i1}}^{t}\zeta^T(s)ds$$
$$\frac{1}{d_i(t)-h_{i1}}\int_{t-d_i(t)}^{t-h_{i1}}\zeta^T(s)ds\ \ \frac{1}{h_{i2}-h_i(t)}\int_{t-h_{i2}}^{t-h_i(t)}\zeta^T(s)ds$$
$$\frac{1}{(d_i(t)-h_{i1})^2}\int_{t-d_i(t)}^{t-h_{i1}}\int_{\beta}^{t-h_{i1}}\zeta^T(s)dsd\beta$$
$$\frac{1}{(h_{i2}-h_i(t))^2}\int_{t-h_{i2}}^{t-h_i(t)}\int_{\beta}^{t-h_i(t)}\zeta^T(s)dsd\beta$$
$$\dot{\zeta}^T(t)\ \ v_i^T(t)\Big]$$

$$\begin{aligned}
e_1 &= [\bar{e}_1, \bar{e}_2], e_2 = [\bar{e}_3, \bar{e}_4],\ e_3 = [\bar{e}_5, \bar{e}_6],\\
e_4 &= [\bar{e}_7, \bar{e}_8],\ e_5 = [\bar{e}_9, \bar{e}_{10}],\ e_6 = [\bar{e}_{11}, \bar{e}_{12}],\\
e_7 &= [\bar{e}_{13}, \bar{e}_{14}], e_8 = [\bar{e}_{15}, \bar{e}_{16}], e_9 = [\bar{e}_{17}, \bar{e}_{18}],\\
e_{10} &= [\bar{e}_{19}, \bar{e}_{20}],\ e_{11} = \bar{e}_{21},\\
e_0 &= [0, \underbrace{0, \ldots, 0}_{19}, 0]^T,\\
\Pi_{i1} &= [e_1\ \ h_{i1}e_5\ \ (h_i(t)-h_{i1})e_6\ \ (h_{i2}-h_i(t))e_7],\\
\Pi_{i2} &= [e_{10}\ \ e_1-e_2\ \ e_2-(1-\dot{h}_i(t))e_3\ \ (1-\dot{h}_i(t))e_3-e_4],\\
\eta_{i1} &= [(h_i(t)-h_{i1})e_6\ \ e_2-e_3],\\
\eta_{i2} &= [(h_i(t)-h_{i1})(-e_6+2e_8)\ \ e_2+e_3-2e_6],\\
\sigma_{i1} &= [(h_{i2}-h_i(t))e_7\ \ e_3-e_4],\\
\sigma_{i2} &= [(h_{i2}-h_i(t))(-e_7+2e_9)\ \ e_3+e_4-2e_7],\\
\Pi_{i3} &= [e_2-e_3\ \ e_2+e_3-2e_6\ \ e_2-e_3+6e_6-12e_8],\\
\Pi_{i4} &= [e_3-e_4\ \ e_3+e_4-2e_7\ \ e_3-e_4+6e_7-12e_9],\\
\Phi_i(t) &= [G_i\bar{A}_i(t)\ \ 0\ \ G_i\bar{A}_{di}(t)\ \ \underbrace{0, \ldots, 0}_{6}\ \ -G_i\ \ G_i\bar{C}_i(t)],\\
\Psi_i(t) &= [\bar{F}_i(t)\bar{Q}\ \ \underbrace{0, \ldots, 0}_{10}],\\
\bar{e} &= [e_2\ \ e_3\ \ e_4\ \ e_6\ \ e_7\ \ e_8\ \ e_9],\\
\bar{M}_{im} &= \bar{e}M_{im}, \bar{N}_{im} = \bar{e}N_{im}, (m = 1, 2),\\
\tilde{S}_i &= \mathrm{diag}\{S_i, 3S_i, 5S_i\},\\
\mathscr{R}_{iaug1} &= \mathscr{R}_i + \begin{bmatrix} 0 & X_{i1}\\ * & 0\end{bmatrix},\\
\mathscr{R}_{iaug2} &= \mathscr{R}_i + \begin{bmatrix} 0 & X_{i2}\\ * & 0\end{bmatrix},
\end{aligned}$$

$$\begin{aligned}
\Sigma_{i1[h_i(t),\dot{h}_i(t)]} &= \mathrm{Sym}\{\Pi_{i1}\mathscr{P}_i\Pi_{i2}^T\} + [e_1\ \ e_2]Q_{i1}[e_1\ \ e_2]^T\\
&\quad - (1-\dot{h}_i(t))[e_1\ \ e_3]Q_{i1}[e_1\ \ e_3]^T + (h_i(t)-h_{i1})\mathrm{Sym}\{[e_1\ \ e_6]Q_{i1}[e_{10}\ \ e_0]^T\}
\end{aligned}$$

$$
\begin{aligned}
&+ (1 - \dot{h}_i(t))[e_1 \ \ e_3]Q_{i2}[e_1 \ \ e_3]^T - [e_1 \ \ e_4]Q_{i2}[e_1 \ \ e_4]^T + h_{i1}^2 e_{10} W_i e_{10}^T \\
&+ (h_{i2} - h_i(t))\mathrm{Sym}\{[e_1 \ \ e_7]Q_{i2}[e_{10} \ \ e_0]^T\} \\
&- [e_1 + e_2 - 2e_5]W_i[e_1 + e_2 - 2e_5]^T + (h_{i2} - h_{i1})[e_1 \ \ e_{10}]R_i[e_1 \ \ e_{10}]^T \\
&+ \mathrm{Sym}\{\bar{N}_{i1}\eta_{i1}^T + \bar{N}_{i2}\eta_{i2}^T + \bar{M}_{i1}\sigma_{i1}^T + \bar{M}_{i2}\sigma_{i2}^T\} + (h_{i2} - h_{i1})^2 e_{10} S_i e_{10}^T \\
&- \mathrm{Sym}\{Y_{i1}\Pi_{i3}^T + Y_{i2}\Pi_{i4}^T\} + \mathrm{Sym}\{(e_1 + \varepsilon_i e_{10})\Phi_i(t)\} + \varepsilon_{i1}^{-1} e_1 G_i e_1^T \\
&+ \varepsilon_{i2}^{-1} e_{10} G_i e_{10}^T + (K-1)e_1 \sum_{n=1, n\neq i}^{K} (\varepsilon_{n1} + \varepsilon_{n2})\bar{B}_{ni}^T G_i \bar{B}_{ni} e_1^T \\
&- \mathrm{Sym}\{e_1 \bar{F}_i^T(t) S e_{11}^T\} - e_{11}[R - \beta I]e_{11}^T, \\
\Sigma_{i2} &= e_2 X_{i1} e_2^T - e_3 X_{i1} e_3^T + e_3 X_{i2} e_3^T - e_4 X_{i2} e_4^T, \\
\Sigma_{i3[h_i(t)]} &= (h_i(t) - d_{1i})(\bar{N}_{i1}\mathscr{R}_{iaug1}^{-1}\bar{N}_{i1}^T + \frac{1}{3}\bar{N}_{i2}\mathscr{R}_{iaug1}^{-1}\bar{N}_{i2}^T) \\
&+ (d_{2i} - h_i(t))(\bar{M}_{i1}\mathscr{R}_{iaug2}^{-1}\bar{M}_{i1}^T + \frac{1}{3}\bar{M}_{i2}\mathscr{R}_{iaug2}^{-1}\bar{M}_{i2}^T), \\
\Sigma_{i4[h_i(t)]} &= \varphi_i Y_{i1}\tilde{S}_i^{-1} Y_{i1}^T + (1 - \varphi_i) Y_{i2}\tilde{S}_i^{-1} Y_{i2}^T,
\end{aligned}
$$

Here is the first theorem for a main result.

Theorem 12.1 *For given scalars h_{i1}, h_{i2}, μ_{i1}, μ_{i2}, β, $\varepsilon_{i1} > 0$, $\varepsilon_{i2} > 0$, and ε_i, the filtering error system composed of K filtering-error subsystem (12.6) satisfying the condition of time delay (12.2) is asymptotically stable with strict $(Q, S, R) - \beta-$ dissipative, if there exist matrices $\mathscr{P}_i > 0$, $\mathscr{Q}_{i1} > 0$, $\mathscr{Q}_{i2} > 0$, $\mathscr{W}_i > 0$, $\mathscr{R}_i > 0$, $\mathscr{S}_i > 0$, $\mathscr{X}_{i1}$, $\mathscr{X}_{i2}$, M_{i1}, M_{i2}, N_{i1}, N_{i2}, Y_{i1}, Y_{i2}, and G_i with appropriate dimensions satisfying the following LMIs:*

$$
\begin{bmatrix} \Sigma_{i1[h_i(t)=h_{i1}, \dot{h}_i(t)\in\{\mu_{1i},\mu_{2i}\}]} + \Sigma_{i2} & \Psi_i^T(t) & \Upsilon_{i1} \\ * & -I & 0 \\ * & * & \Upsilon_{i2} \end{bmatrix} < 0, \tag{12.8}
$$

$$
\begin{bmatrix} \Sigma_{i1[h_i(t)=h_{i2}, \dot{h}_i(t)\in\{\mu_{1i},\mu_{2i}\}]} + \Sigma_{i2} & \Psi_i^T(t) & \Upsilon_{i3} \\ * & -I & 0 \\ * & * & \Upsilon_{i4} \end{bmatrix} < 0, \tag{12.9}
$$

$$
\mathscr{R}_{iaug1} > 0, \quad \mathscr{R}_{iaug2} > 0. \tag{12.10}
$$

where

$$
\begin{aligned}
\Upsilon_{i1} &= [\sqrt{h_{i2} - h_{i1}}\bar{M}_{i1} \ \ \sqrt{h_{i2} - h_{i1}}\bar{M}_{i2} \ \ Y_{i2}], \\
\Upsilon_{i2} &= = \mathrm{diag}\left\{-\mathscr{R}_{iaug2}, -3\mathscr{R}_{iaug2}, -\tilde{S}_i\right\}. \\
\Upsilon_{i3} &= [\sqrt{h_{i2} - h_{i1}}\bar{N}_{i1} \ \ \sqrt{h_{i2} - h_{i1}}\bar{N}_{i2} \ \ Y_{i1}], \\
\Upsilon_{i4} &= = \mathrm{diag}\left\{-\mathscr{R}_{iaug1}, -3\mathscr{R}_{iaug1}, -\tilde{S}_i\right\}.
\end{aligned}
$$

Proof Let us consider the following Lyapunov-Krasovskii functional candidate

$$V(t) = \sum_{i=1}^{K}(V_{i1}(t) + V_{i2}(t) + V_{i3}(t) + V_{i4}(t) + V_{i5}(t)), \tag{12.11}$$

where

$$V_{i1}(t) = \begin{bmatrix} \zeta_i(t) \\ \int_{t-h_{i1}}^{t} \zeta_i(s)ds \\ \int_{t-h_i(t)}^{t-h_{i1}} \zeta_i(s)ds \\ \int_{t-h_{i2}}^{t-h_i(t)}(s)\zeta_i(s)ds \end{bmatrix}^T \mathscr{P}_i \begin{bmatrix} \zeta_i(t) \\ \int_{t-h_{i1}}^{t} \zeta_i(s)ds \\ \int_{t-h_i(t)}^{t-h_{i1}} \zeta_i(s)ds \\ \int_{t-h_{i2}}^{t-h_i(t)}(s)\zeta_i(s)ds \end{bmatrix},$$

$$V_{i2}(t) = \int_{t-h_i(t)}^{t-h_{i1}} \begin{bmatrix} \zeta_i(t) \\ \zeta_i(s) \end{bmatrix}^T \mathscr{Q}_{i1} \begin{bmatrix} \zeta_i(t) \\ \zeta_i(s) \end{bmatrix} ds,$$

$$V_{i3}(t) = \int_{t-h_{i2}}^{t-h_i(t)} \begin{bmatrix} \zeta_i(t) \\ \zeta_i(s) \end{bmatrix}^T \mathscr{Q}_{i2} \begin{bmatrix} \zeta_i(t) \\ \zeta_i(s) \end{bmatrix} ds,$$

$$V_{i4}(t) = h_{i1} \int_{-h_{i1}}^{0} \int_{t+\beta}^{t} \dot{\zeta}_i(s) W_i \dot{\zeta}_i(s) ds d\beta,$$

$$V_{i5}(t) = \int_{-h_{i2}}^{-h_{i1}} \int_{t+\beta}^{t} \begin{bmatrix} \zeta_i(s) \\ \dot{\zeta}_i(s) \end{bmatrix}^T \mathscr{R}_i \begin{bmatrix} \zeta_i(s) \\ \dot{\zeta}_i(s) \end{bmatrix} ds d\beta,$$

$$V_{i6}(t) = (h_{i2} - h_{i1}) \int_{-h_{i2}}^{-h_{i1}} \int_{t+\beta}^{t} \dot{\zeta}_i(s) S_i \dot{\zeta}_i(s) ds d\beta.$$

The time derivative of $V_{i1}(t)$ is given as

$$\dot{V}_{i1}(t) = 2 \begin{bmatrix} \zeta_i(t) \\ \int_{t-h_{i1}}^{t} \zeta_i(s)ds \\ \int_{t-h_i(t)}^{t-h_{i1}} \zeta_i(s)ds \\ \int_{t-h_{i2}}^{t-h_i(t)}(s)\zeta_i(s)ds \end{bmatrix}^T \mathscr{P}_i \times \begin{bmatrix} \dot{\zeta}_i(t) \\ \zeta_i(t) - \zeta_i(t - h_{i1}) \\ \zeta_i(t - h_{i1}) - (1 - \dot{h}_i(t))\zeta_i(t - h_i(t)) \\ (1 - \dot{h}_i(t))\zeta_i(t - h_i(t)) - \zeta_i(t - h_{i2}) \end{bmatrix}. \tag{12.12}$$

Calculating the time derivative of $V_{is}(t) (s = 2, 3, \ldots, 6)$ yields

$$\dot{V}_{i2}(t) = \begin{bmatrix} \zeta_i(t) \\ \zeta_i(t - h_{i1}) \end{bmatrix}^T \mathscr{Q}_{i1} \begin{bmatrix} \zeta_i(t) \\ \zeta_i(t - h_{i1}) \end{bmatrix}$$

$$-(1-\dot{h}_i(t))\begin{bmatrix}\zeta_i(t)\\ \zeta_i(t-h_i(t))\end{bmatrix}^T \mathscr{Q}_{i1}\begin{bmatrix}\zeta_i(t)\\ \zeta_i(t-h_i(t))\end{bmatrix}$$
$$+2\begin{bmatrix}(h_i(t)-h_{i1})\zeta_i(t)\\ \int_{t-h_i(t)}^{t-h_{i1}}\zeta_i(s)ds\end{bmatrix}^T \mathscr{Q}_{i1}\begin{bmatrix}\dot{\zeta}_i(t)\\ 0\end{bmatrix}^T, \tag{12.13}$$

$$\dot{V}_{i3}(t)=(1-\dot{h}_i(t))\begin{bmatrix}\zeta_i(t)\\ \zeta_i(t-h_i(t))\end{bmatrix}^T \mathscr{Q}_{i2}\begin{bmatrix}\zeta_i(t)\\ \zeta_i(t-h_i(t))\end{bmatrix}$$
$$-\begin{bmatrix}\zeta_i(t)\\ \zeta_i(t-h_{i2})\end{bmatrix}^T \mathscr{Q}_{i2}\begin{bmatrix}\zeta_i(t)\\ \zeta_i(t-h_{i2})\end{bmatrix}$$
$$+2\begin{bmatrix}(h_{i2}-h_i(t))\zeta_i(t)\\ \int_{t-h_{i2}}^{t-h_i(t)}\zeta_i(s)ds\end{bmatrix}^T \mathscr{Q}_{i2}\begin{bmatrix}\dot{\zeta}_i(t)\\ 0\end{bmatrix}^T, \tag{12.14}$$

$$\dot{V}_{i4}(t)=h_{i1}^2\dot{\zeta}_i^T(t)W_i\dot{\zeta}_i(t)-d_{1i}\int_{t-h_{i1}}^{t}\dot{\zeta}_i^T(s)W_i\dot{\zeta}_i(s)ds, \tag{12.15}$$

$$\dot{V}_{i5}(t)=(h_{i2}-h_{i1})\begin{bmatrix}\zeta_i(t)\\ \dot{\zeta}_i(t)\end{bmatrix}^T \mathscr{R}_i\begin{bmatrix}\zeta_i(t)\\ \dot{\zeta}_i(t)\end{bmatrix}$$
$$-\int_{t-h_{i2}}^{t-h_{i1}}\begin{bmatrix}\zeta_i(s)\\ \dot{\zeta}_i(s)\end{bmatrix}^T \mathscr{R}_i\begin{bmatrix}\zeta_i(s)\\ \dot{\zeta}_i(s)\end{bmatrix}ds, \tag{12.16}$$

$$\dot{V}_{i6}(t)=(h_{i2}-h_{i1})^2\dot{\zeta}_i^T(t)S_i\dot{\zeta}_i(t)$$
$$-(h_{i2}-h_{i1})\int_{t-d_{i2}}^{t-h_{i1}}\dot{\zeta}_i^T(s)S_i\dot{\zeta}_i(s)ds. \tag{12.17}$$

In view of Lemma 12.1, one has

$$-h_{i1}\int_{t-h_{i1}}^{t}\dot{\zeta}_i^T(s)W_i\dot{\zeta}_i(s)ds$$
$$\leq\left[\zeta(t)+\zeta(t-h_{i1})-\frac{2}{h_{i1}}\int_{t-h_{i1}}^{t}\zeta(s)ds\right]^T\times$$
$$W_i\left[\zeta(t)+\zeta(t-h_{i1})-\frac{2}{h_{i1}}\int_{t-h_{i1}}^{t}\zeta(s)ds\right]. \tag{12.18}$$

For any symmetric matrices $X_{im}(m=1,2)$, the following two zero equalities hold:

$$\zeta_i^T(t-h_{i1})X_{i1}\zeta_i(t-h_{i1})-\zeta_i^T(t-h_i(t))X_{i1}\zeta_i(t-h_i(t))$$
$$-2\int_{t-h_i(t)}^{t-h_{i1}}\zeta_i^T(s)X_{i1}\dot{\zeta}_i(s)ds=0,$$
$$\zeta_i^T(t-h_i(t))X_{i2}\zeta_i(t-h_i(t))-\zeta_i^T(t-h_{i2})X_{i2}\zeta_i(t-h_{i2})$$
$$-2\int_{t-h_{i2}}^{t-h_i(t)}\zeta_i^T(s)X_{i2}\dot{\zeta}_i(s)ds=0. \tag{12.19}$$

Summing the two zero equalities of (12.19), we can obtain

$$
\begin{aligned}
0 = \xi_i^T(t)\Upsilon_{2i}\xi_i(t) - 2\int_{t-h_i(t)}^{t-h_{i1}} \zeta_i^T(s)X_{i1}\dot{\zeta}_i(s)ds \\
-2\int_{t-h_{i2}}^{t-d_i(t)} \zeta_i^T(s)X_{i2}\dot{\zeta}_i(s)ds.
\end{aligned}
\tag{12.20}
$$

Combing (12.16) and (12.20), one has

$$
\begin{aligned}
& -\int_{t-h_i(t)}^{t-h_{i1}} \begin{bmatrix} \zeta_i(s) \\ \dot{\zeta}_i(s) \end{bmatrix}^T \mathscr{R}_i \begin{bmatrix} \zeta_i(s) \\ \dot{\zeta}_i(s) \end{bmatrix} ds \\
& -\int_{t-h_{i2}}^{t-h_i(t)} \begin{bmatrix} \zeta_i(s) \\ \dot{\zeta}_i(s) \end{bmatrix}^T \mathscr{R}_i \begin{bmatrix} \zeta_i(s) \\ \dot{\zeta}_i(s) \end{bmatrix} ds \\
& -2\int_{t-h_i(t)}^{t-h_{i1}} \zeta_i^T(s)X_{i1}\dot{\zeta}_i(s)ds \\
& -2\int_{t-h_{i2}}^{t-h_i(t)} \zeta_i^T(s)X_{i2}\dot{\zeta}_i(s)ds \\
= & -\int_{t-h_i(t)}^{t-h_{i1}} \begin{bmatrix} \zeta_i(s) \\ \dot{\zeta}_i(s) \end{bmatrix}^T \mathscr{R}_{iaug1} \begin{bmatrix} \zeta_i(s) \\ \dot{\zeta}_i(s) \end{bmatrix} ds \\
& -\int_{t-h_{i2}}^{t-h_i(t)} \begin{bmatrix} \zeta_i(s) \\ \dot{\zeta}_i(s) \end{bmatrix}^T \mathscr{R}_{iaug2} \begin{bmatrix} \zeta_i(s) \\ \dot{\zeta}_i(s) \end{bmatrix} ds.
\end{aligned}
\tag{12.21}
$$

If $\mathscr{R}_{iaug1} > 0, \mathscr{R}_{iaug2} > 0$, based on Lemma 12.1, we have

$$
\begin{aligned}
& -\int_{t-h_i(t)}^{t-h_{i1}} \begin{bmatrix} \zeta_i(s) \\ \dot{\zeta}_i(s) \end{bmatrix}^T \mathscr{R}_{iaug1} \begin{bmatrix} \zeta_i(s) \\ \dot{\zeta}_i(s) \end{bmatrix} ds \\
& -\int_{t-h_{i2}}^{t-h_i(t)} \begin{bmatrix} \zeta_i(s) \\ \dot{\zeta}_i(s) \end{bmatrix}^T \mathscr{R}_{iaug2} \begin{bmatrix} \zeta_i(s) \\ \dot{\zeta}_i(s) \end{bmatrix} ds \\
\le & \xi_i^T(t) \Big\{ (h_i(t) - h_{i1})(\bar{N}_{i1}\mathscr{R}_{iaug1}^{-1}\bar{N}_{i1}^T + \frac{1}{3}\bar{N}_{i2}\mathscr{R}_{iaug1}^{-1}\bar{N}_{i2}^T) \\
& +(h_{i2} - h_i(t))(\bar{M}_{i1}\mathscr{R}_{iaug2}^{-1}\bar{M}_{i1}^T + \frac{1}{3}\bar{M}_{i2}\mathscr{R}_{iaug2}^{-1}\bar{M}_{i2}^T) \Big\} \xi_i(t) \\
& +\xi_i^T(t) \left\{ \mathrm{Sym}(\bar{N}_{i1}\eta_{i1}^T + \bar{N}_{i2}\eta_{i2}^T + \bar{M}_{i1}\sigma_{i1}^T + \bar{M}_{i2}\sigma_{i2}^T) \right\} \xi_i(t).
\end{aligned}
$$

By applying Lemmas 12.2 and 12.3, it can obtain

$$
-(h_{i2} - h_{i1})\int_{t-h_{i2}}^{t-h_{i1}} \dot{\zeta}_i^T(s)S_i\dot{\zeta}_i(s)ds
$$

$$
\begin{aligned}
&= -(h_{i2}-h_{i1})\int_{t-h_i(t)}^{t-h_{i1}} \dot{\zeta}_i^T(s)S_i x\dot{\zeta}_i(s)ds \\
&\quad -(h_{i2}-h_{i1})\int_{t-h_{i2}}^{t-h_i(t)} \dot{\zeta}_i^T(s)S_i\dot{\zeta}_i(s)ds \\
&\le -\xi_i^T(t)\left\{\frac{h_{i2}-h_{i1}}{h_i(t)-h_{i1}}\Omega_{i1}\tilde{S}_i\Omega_{i1}^T + \frac{h_{i2}-h_{i1}}{h_{i2}-h_i(t)}\Omega_{i2}\tilde{S}_i\Omega_{i2}^T\right\}\xi_i(t) \\
&= -\xi_i^T(t)\left\{[\Omega_{i1}\ \ \Omega_{i2}]\begin{bmatrix}\frac{1}{\beta}\tilde{S}_i & 0\\ 0 & \frac{1}{1-\beta}\tilde{S}_i\end{bmatrix}[\Omega_{i1}\ \ \Omega_{i2}]^T\right\}\xi_i(t) \\
&\le -\xi_i^T(t)\big\{\mathrm{Sym}\left\{Y_{i1}\Omega_{i1}^T + Y_{i2}\Omega_{i2}^T\right\} \\
&\quad -\varphi_i Y_{i1}\tilde{S}_i^{-1}Y_{i1}^T - (1-\varphi_i)Y_{i2}\tilde{S}_i^{-1}Y_{i2}^T\big\}\xi_i(t)
\end{aligned} \tag{12.22}
$$

where $\varphi_i = \frac{h_i(t)-h_{i1}}{h_{i2}-h_{i1}}$, $h_{i1} \le h_i(t) \le h_{i2}$.

Also, one has

$$
\begin{aligned}
&2\sum_{i=1}^{K}\left[\zeta_i^T(t)G_i + \varepsilon_i\dot{\zeta}_i^T(t)G_i\right]\Big[-\dot{\zeta}_i(t) + \bar{A}_i(t)\zeta_i(t) \\
&+\bar{A}_{di}(t)\zeta_i(t-h_i(t)) + \sum_{n=1,n\neq i}^{K}\bar{B}_{ni}\zeta_n(t) + \bar{C}_i(t)v_i(t)\Big] = 0.
\end{aligned}
$$

Note that

$$
\begin{aligned}
&2\sum_{i=1}^{K}\zeta_i^T(t)G_i\sum_{n=1,n\neq i}^{N}\bar{B}_{ni}\zeta_n(t) \\
&\le \sum_{i=1}^{K}\Big\{\varepsilon_{i1}\sum_{n=1,n\neq i}^{K}\zeta_n^T(t)\bar{B}_{ni}^T G_i\sum_{n=1,n\neq i}^{K}\bar{B}_{ni}\zeta_n(t) \\
&+\varepsilon_{i1}^{-1}\zeta_i^T(t)G_i\zeta_i(t)\Big\} \\
&\le \sum_{i=1}^{K}\Big\{(K-1)\varepsilon_{i1}\sum_{n=1,n\neq i}^{K}\zeta_n^T(t)\bar{B}_{ni}^T G_i\bar{B}_{ni}\zeta_n(t) \\
&+\varepsilon_{i1}^{-1}\zeta_i^T(t)G_i\zeta_i(t)\Big\}. \\
&\le \sum_{i=1}^{K}\Big\{(K-1)\sum_{n=1,n\neq i}^{K}\varepsilon_{n1}\zeta_i^T(t)\bar{B}_{ni}^T G_i\bar{B}_{ni}\zeta_i(t) \\
&+\varepsilon_{i1}^{-1}\zeta_i^T(t)G_i\zeta_i(t)\Big\},
\end{aligned} \tag{12.23}
$$

and using the similar process of (12.23), one can derive

$$
\begin{aligned}
&2\sum_{i=1}^{K}\dot{\zeta}_i^T(t)G_i\sum_{n=1,n\neq i}^{N}\bar{B}_{ni}\zeta_n(t)\\
&\leq\sum_{i=1}^{K}\left\{(K-1)\sum_{n=1,n\neq i}^{K}\varepsilon_{n2}\zeta_i^T(t)\bar{B}_{ni}^TG_i\bar{B}_{ni}\zeta_i(t)\right.\\
&\left.+\varepsilon_{i2}^{-1}\dot{\zeta}_i^T(t)G_i\dot{\zeta}_i(t)\right\}.
\end{aligned}
\tag{12.24}
$$

By combing (12.12)–(12.24), and adding $-e_i^T(t)Qe_i(t)-2e_i^T(t)Sv_i(t)-v_i^T(t)$ $[R-\beta I]v_i(t)$ on both sides, one has

$$
\begin{aligned}
&\dot{V}_i(t)-e_i^T(t)Qe_i(t)-2e_i^T(t)Sv_i(t)-v_i^T(t)[R-\beta I]v_i(t)\\
&\leq\sum_{i=1}^{K}\xi_i^T(t)\left\{\Sigma_{i1[h_i(t),\dot{h}_i(t)]}+\Sigma_{i2}+\Sigma_{i3[h_i(t)]}+\Sigma_{i4[h_i(t)]}\right.\\
&\left.-\Psi_i^T(t)Q\Psi_i(t)\right\}\xi_i(t).
\end{aligned}
\tag{12.25}
$$

From (12.8)–(12.10) and Schur complement, we have

$$
\dot{V}_i(t)-E_i(t)\leq 0,
\tag{12.26}
$$

where $E_i(t)=e_i^T(t)Qe_i(t)+2e_i^T(t)Sv_i(t)+v_i^T(t)[R-\beta I]v_i(t)$.

Integrating both sides of (12.26) from 0 to t yields

$$
V_i(t)-V_i(0)\leq\int_0^t E_i(s)ds,
\tag{12.27}
$$

which implies that $\int_0^t E_i(s)ds>0$ under the zero initial condition.

Hence, from Definition 12.1, it can be get the conclusion that the closed-loop system (12.6) is strictly $(Q, S, R)-\beta$ dissipative. Moreover, when $v_i(t)=0$, $\dot{V}_i(t)\leq e_i^T(t)Qe_i(t)\leq 0$, which means that the closed loop system (12.6) is asymptotically stable. This completes the proof. ■

12.3.2 Filter Design for the Interconnected Nonlinear Systems

In this subsection, a delay-dependent decentralized dissipative filter design for interconnected nonlinear systems with interval time-varying delay is addressed. Based on Theorem 12.1, the following result can be derived.

Theorem 12.2 *For given scalars h_{i1}, h_{i2}, μ_{i1}, μ_{i2}, β, $\varepsilon_{i1} > 0$, $\varepsilon_{i2} > 0$, and ε_i, the filtering error system composed of K filtering-error subsystem (12.6) satisfying the condition of time delay (12.2) is asymptotically stable with strict $(Q, S, R) - \beta-$dissipative, if there exist matrices $\mathscr{P}_i > 0$, $\mathscr{Q}_{i1} > 0$, $\mathscr{Q}_{i2} > 0$, $\mathscr{W}_i > 0$, $\mathscr{R}_i > 0$, $\mathscr{S}_i > 0$, $\mathscr{X}_{i1}$, $\mathscr{X}_{i2}$, M_{i1}, M_{i2}, N_{i1}, N_{i2}, Y_{i1}, Y_{i2}, and G_i, $\hat{A}_{fij}$, $\hat{B}_{fij}$ and $\hat{F}_{fij}$ with appropriate dimensions satisfying the following LMIs:*

$$\Lambda_{ijk1} + \Lambda_{ikj1} < 0,\ j \leq k, \tag{12.28}$$

$$\Lambda_{ijk2} + \Lambda_{ikj2} < 0,\ j \leq k, \tag{12.29}$$

$$\mathscr{R}_{iaug1} > 0, \quad \mathscr{R}_{iaug2} > 0. \tag{12.30}$$

where

$$\Lambda_{ijk1} = \begin{bmatrix} \bar{\Sigma}_{ijk1[h_i(t)=h_{i1},\dot{h}_i(t)\in\{\mu_{i1},\mu_{i2}\}]} + \Sigma_{i2} & \bar{\Psi}_{ijk}^T & \Upsilon_{i1} \\ * & -I & 0 \\ * & * & \Upsilon_{i2} \end{bmatrix},$$

$$\Lambda_{ijk2} = \begin{bmatrix} \bar{\Sigma}_{ijk1[h_i(t)=h_{i2},\dot{h}_i(t)\in\{\mu_{i1},\mu_{i2}\}]} + \Sigma_{i2} & \bar{\Psi}_{ijk}^T & \Upsilon_{i3} \\ * & -I & 0 \\ * & * & \Upsilon_{i4} \end{bmatrix},$$

$$\begin{aligned}
\bar{\Sigma}_{ijk1[h_i(t),\dot{h}_i(t)]} &= \Sigma_{i1[h_i(t),\dot{h}_i(t)]} - \text{Sym}\{(e_1 + \varepsilon_i e_{10})\Phi_i(t)\} \\
&+ \text{Sym}\{(e_1 + \varepsilon_i e_{10})\bar{\Phi}_{ijk}\} + \text{Sym}\{e_1 \bar{F}_i^T(t) S e_{11}^T\} \\
&- \text{Sym}\{e_1 \hat{F}_{ijk}^T S e_{11}^T\}, \\
\bar{\Phi}_{ijk} &= [G_i\hat{A}_{ijk}\ \ 0\ \ G_i\hat{A}_{dijk}\ \ \underbrace{0, \ldots, 0}_{6}\ \ -G_i\ \ G_i\hat{C}_{ijk}], \\
\bar{\Psi}_{ijk} &= [\hat{F}_{ijk}\bar{Q}\ \ \underbrace{0, \ldots, 0}_{10}],
\end{aligned}$$

$$\begin{aligned}
G_i\hat{A}_{ijk} &= \begin{bmatrix} G_{i1}A_{ik} + \hat{B}_{fij}L_{ik} & \hat{A}_{fij} \\ G_{i3}A_{ik} + \hat{B}_{fij}L_{ik} & \hat{A}_{fij} \end{bmatrix}, \\
G_i\hat{A}_{dijk} &= \begin{bmatrix} G_{i1}A_{dik} + \hat{B}_{fij}L_{dik} & 0 \\ G_{i3}A_{dik} + \hat{B}_{fij}L_{dik} & 0 \end{bmatrix}, \\
G_i\hat{C}_{ijk} &= \begin{bmatrix} G_{i1}C_{ik} + \hat{B}_{fij}E_{ik} \\ G_{i3}C_{ik} + \hat{B}_{fij}E_{ik} \end{bmatrix}, \\
\hat{F}_{ijk} &= \begin{bmatrix} F_{ij} & -F_{fik} \end{bmatrix}.
\end{aligned}$$

Furthermore, the filtering matrices in (12.5) are given by

$$A_{fij} = G_{i2}^{-1}\hat{A}_{fij}, \quad B_{fij} = G_{i2}^{-1}\hat{B}_{fij}, \quad F_{fij} = \hat{F}_{fij}. \tag{12.31}$$

Proof Let the matrix G_i be

$$G_i = \begin{bmatrix} G_{i1} & G_{i2} \\ G_{i3} & G_{i4} \end{bmatrix}.$$

Then, pre- and post- multiply G_i by diag$\{I_i \ \ G_{i2}G_{i4}^{-1}\}$ to derive

$$\begin{bmatrix} I_i & 0 \\ 0 & G_{i2}G_{i4}^{-1} \end{bmatrix} G_i \begin{bmatrix} I_i & 0 \\ 0 & G_{i4}^{-1}G_{i2}^T \end{bmatrix} = \begin{bmatrix} G_{i1} & G_{i2}G_{i4}^{-1}G_{i2}^T \\ G_{i2}G_{i4}^T G_{i3} & G_{i2}G_{i4}^{-1}G_{i2}^T \end{bmatrix}. \quad (12.32)$$

Consequently, the G_I matrix can be defined, without loss of generality, as $G_i = \begin{bmatrix} G_{i1} & G_{i2} \\ G_{i3} & G_{i2} \end{bmatrix}$.
By a simple matrix calculation, it is straightforward to obtain that

$$\begin{aligned} G_i\bar{A}_i(t) &= \sum_{k=1}^{r_i} h_{ik}(\theta_i(t)) \sum_{j=1}^{r_i} h_{ij}(\theta_i(t)) \times \\ &\quad \begin{bmatrix} G_{i1}A_{ik} + G_{i2}B_{fij}L_{ik} & G_{i2}A_{fij} \\ G_{i3}A_{ik} + G_{i2}B_{fij}L_{ik} & G_{i2}A_{fij} \end{bmatrix}, \\ G_i\bar{A}_{di}(t) &= \sum_{k=1}^{r_i} h_{ik}(\theta_i(t)) \sum_{j=1}^{r_i} h_{ij}(\theta_i(t)) \times \\ &\quad \begin{bmatrix} G_{i1}A_{dik} + G_{i2}B_{fij}L_{dik} & 0 \\ G_{i3}A_{dik} + G_{i2}B_{fij}L_{dik} & 0 \end{bmatrix}, \\ G_i\bar{C}_i(t) &= \sum_{k=1}^{r_i} h_{ik}(\theta_i(t)) \sum_{j=1}^{r_i} h_{ij}(\theta_i(t)) \times \\ &\quad \begin{bmatrix} G_{i1}C_{ik} + G_{i2}B_{fij}E_{ik} \\ G_{i3}C_{ik} + G_{i2}B_{fij}E_{ik} \end{bmatrix}. \end{aligned} \quad (12.33)$$

Define a set of variables as

$$\hat{A}_{fij} = G_{i2}A_{fij}, \quad \hat{B}_{fij} = G_{i2}B_{fij}. \quad (12.34)$$

Also, denote the left side of (12.8), (12.9) as $\Lambda_{i1}(t)$, $\Lambda_{i2}(t)$, respectively, then

$$\begin{aligned} \Lambda_{i1}(t) &= \sum_{j=1}^{r_i} h_{ij}^2(\theta_i(t))\Lambda_{ijk1} \\ &\quad + \sum_{j<k}^{r_i} h_{ij}(\theta_i(t))h_{ik}(\theta_i(t))(\Lambda_{ijk1} + \Lambda_{ikj1}) < 0, \end{aligned} \quad (12.35)$$

and

$$\Lambda_{i2}(t) = \sum_{j=1}^{r_i} h_{ij}^2(\theta_i(t))\Lambda_{ijk2} \\ + \sum_{j<k}^{r_i} h_{ij}(\theta_i(t))h_{ik}(\theta_i(t))(\Lambda_{ijk2} + \Lambda_{ikj2}) < 0. \tag{12.36}$$

The filter error system including K subsystems is asymptotically stable and satisfies the $(Q, S, R) - \beta-$ dissipativity, which completes the proof. ■

Remark 12.2 The improvement of the proposed approach may rely on three aspects. First, the augmented Lyapunov-Krasovskii functional is constructed in this chapter, which is different from [31, 32]. Second, the new inequality is employed to estimate the upper bound of the cross term. Last, zero inequalities are used d to the design the filter of the interconnected nonlinear systems.

12.4 Numerical Example

In this section, a numerical example is used to demonstrate effectiveness of the proposed design approach.

We consider a double-inverted pendulums system connected by a spring [29–31]. Two interconnected subsystems are included, and the system equation is the same as the one in [29–31].

Taking $d_{i1} = 0$, $d_{i2} = 0.1$, and $\mu_{i2} = -\mu_{i1} = 0.2$, $i = 1, 2$, as expressed in [31].

In order to show the effectiveness of the proposed approach, the large-scale interconnected systems can be described the following T-S fuzzy model, and the parameters are given as

$$A_{11} = \begin{bmatrix} 0 & 1 \\ -44.75 & -20 \end{bmatrix}, A_{12} = \begin{bmatrix} 0 & 1 \\ -41.19 & -20 \end{bmatrix},$$
$$A_{21} = \begin{bmatrix} 0 & 1 \\ -42.55 & -20 \end{bmatrix}, A_{22} = \begin{bmatrix} 0 & 1 \\ -38.99 & -20 \end{bmatrix},$$
$$A_{d11} = A_{d12} = \begin{bmatrix} 0 & 0 \\ -1 & -0.4 \end{bmatrix},$$
$$A_{d21} = A_{d22} = \begin{bmatrix} 0 & 0 \\ -1.2 & -0.5 \end{bmatrix},$$
$$B_{12} = \begin{bmatrix} 0 & 0 \\ 0.8 & 0 \end{bmatrix}, B_{21} = \begin{bmatrix} 0 & 0 \\ 1 & 0 \end{bmatrix},$$
$$C_{11} = C_{12} = \begin{bmatrix} 0 \\ 0.5 \end{bmatrix}, C_{21} = C_{22} = \begin{bmatrix} 0 \\ 0.4 \end{bmatrix},$$

$$
\begin{aligned}
L_{11} &= L_{12} = L_{21} = L_{22}\begin{bmatrix}1 & 0\end{bmatrix}, \\
L_{d11} &= L_{d12} = \begin{bmatrix}0.1 & 0\end{bmatrix}, \\
L_{d21} &= L_{d22} = \begin{bmatrix}0.8 & 0\end{bmatrix}, \\
F_{11} &= F_{12} = F_{21} = F_{22}\begin{bmatrix}0 & 1\end{bmatrix}, \begin{bmatrix}-0.4 & 0.3\end{bmatrix}, \\
F_{21} &= \begin{bmatrix}1 & -0.5\end{bmatrix}, F_{22} = \begin{bmatrix}-0.2 & 0.3\end{bmatrix}, \\
E_{11} &= E_{12} = E_{21} = E_{22} = 0.
\end{aligned}
$$

In this example, the membership functions are selected as $h_{i1}(x_{i1}) = |0.637x_{i1}|$, then $h_{i2}(x_{i1}) = 1 - |0.637x_{i1}|$.

Consider $d_{i1} = 0$, when $d_{i2} = 0.1$, and $\mu_{i2} = 0.2$, the $\mathscr{H}_\infty$ performance index obtained in [30] and [31] are 0.0287 and 0.0106, respectively, while the $\mathscr{H}_\infty$ performance index is 0.0936 by using our proposed method, which means that the derived result can provide better performance than the existing works.

When $d_i(t) = 0.6 + 0.3\sin(0.5t)$, then the lower bound of time delay is 0.3, which is not considered in [29–31]. The disturbance is given as $v_1(t) = \sin(4\pi t)e^{-0.5t}$ and $v_2(t) = \sin(2\pi t)e^{-0.5t}$. The trajectory of the state response $x_1(t)$ and $x_2(t)$ are shown in Figs. 12.1 and 12.2 under the initial $x_1(0) = [-0.5, 0.5]^T$ and $x_2(0) = [-0.5, 0.5]^T$.

Then, by solving the LMIs in Theorem 12.2, the following two cases will be taken into consideration in next part.

(1) $\mathscr{H}_\infty$ performance case, Let $Q = -I$, $S = 0$, and $R = (\alpha^2 + \alpha)I$, the $\mathscr{H}_\infty$ performance index is 0.0864 by solving LMIs in (12.28)–(12.30), and the corresponding filter parameter matrices can be derived as follows:

$$
A_{f11} = \begin{bmatrix} -1.8535 & -0.3518 \\ 1.8851 & -17.5782 \end{bmatrix}, A_{f12} = \begin{bmatrix} -1.8233 & -0.2952 \\ -1.2725 & -10.1269 \end{bmatrix},
$$

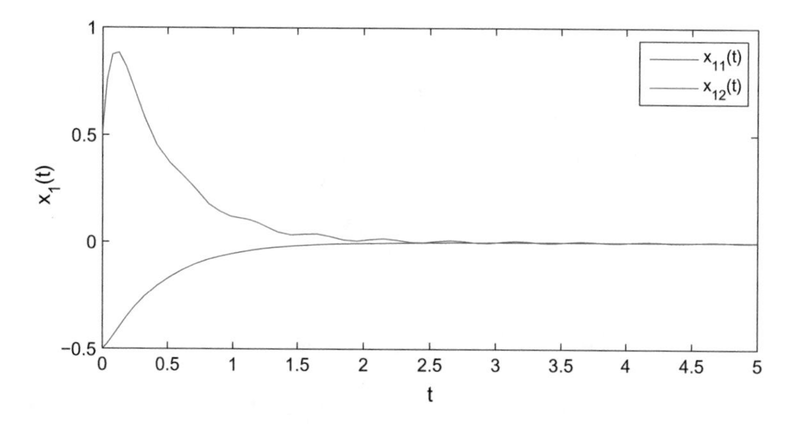

Fig. 12.1 Response of the states of the system $x_1(t)$

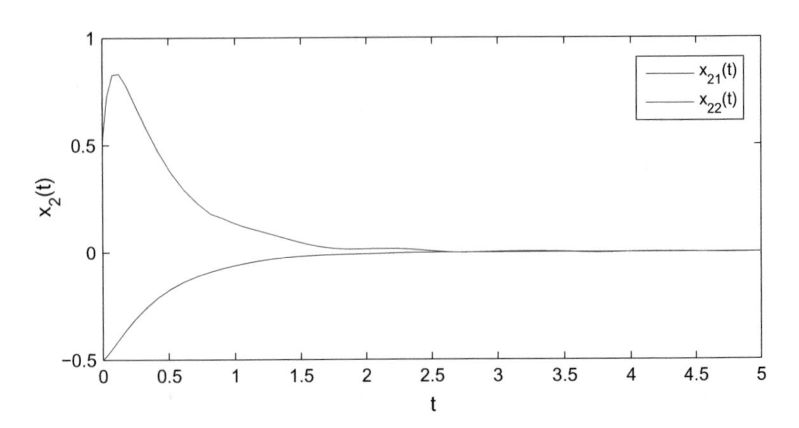

Fig. 12.2 Response of the states of the system $x_2(t)$

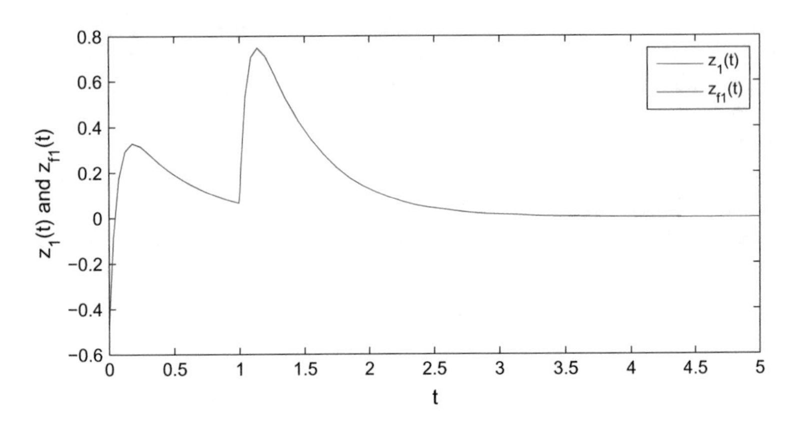

Fig. 12.3 $z_1(t)$ and its estimation $z_{f1}(t)$ for Case 1

$$
\begin{aligned}
A_{f21} &= \begin{bmatrix} -1.8520 & 0.2688 \\ -2.5516 & -12.2916 \end{bmatrix}, A_{f22} = \begin{bmatrix} -1.4555 & -0.0196 \\ -2.5606 & -12.2777 \end{bmatrix}, \\
B_{f11} &= \begin{bmatrix} -1.2764 \\ 21.5233 \end{bmatrix}, B_{f12} = \begin{bmatrix} -1.5826 \\ 17.8615 \end{bmatrix}, \\
B_{f21} &= \begin{bmatrix} -0.2572 \\ 25.2754 \end{bmatrix}, B_{f22} = \begin{bmatrix} -0.4368 \\ 22.3471 \end{bmatrix}, \\
F_{f11} &= \begin{bmatrix} 0.1310 & -0.5175 \end{bmatrix}, F_{f12} = \begin{bmatrix} 0.1563 & -0.5394 \end{bmatrix}, \\
F_{f21} &= \begin{bmatrix} 0.0283 & -0.7732 \end{bmatrix}, F_{f22} = \begin{bmatrix} 0.0459 & -0.7736 \end{bmatrix}.
\end{aligned}
$$

For the above gain matrices, $z_i(t)$ and $z_{fi}(t)$ are given in Figs. 12.3 and 12.4, and the trajectory of the system error are presented in Figs. 12.5 and 12.6. From these figures, it can be seen that the designed $\mathscr{H}_\infty$ fuzzy filter can stabilize this large-scale nonlinear system.

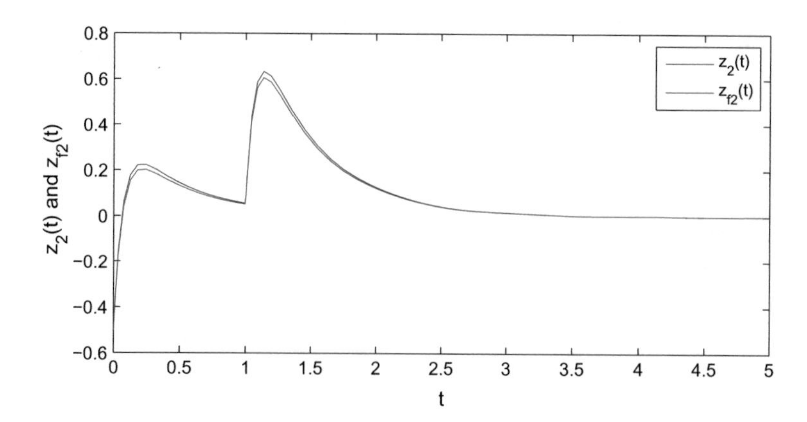

Fig. 12.4 $z_2(t)$ and its estimation $z_{f2}(t)$ for Case 1

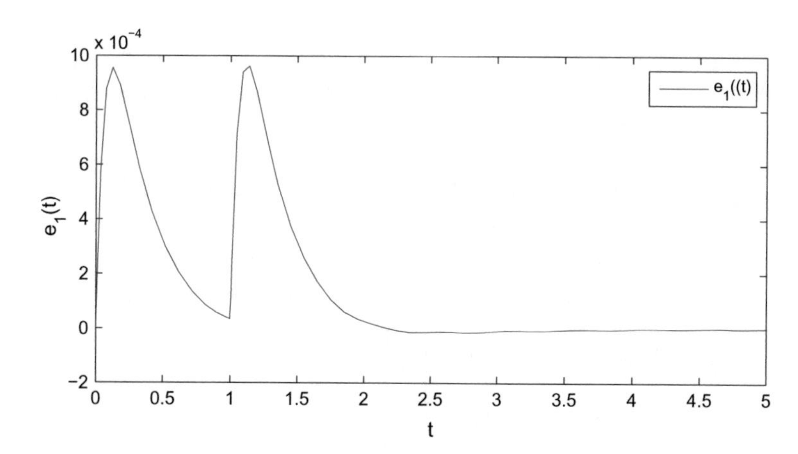

Fig. 12.5 Response of the error $e_1(t)$ for Case 1

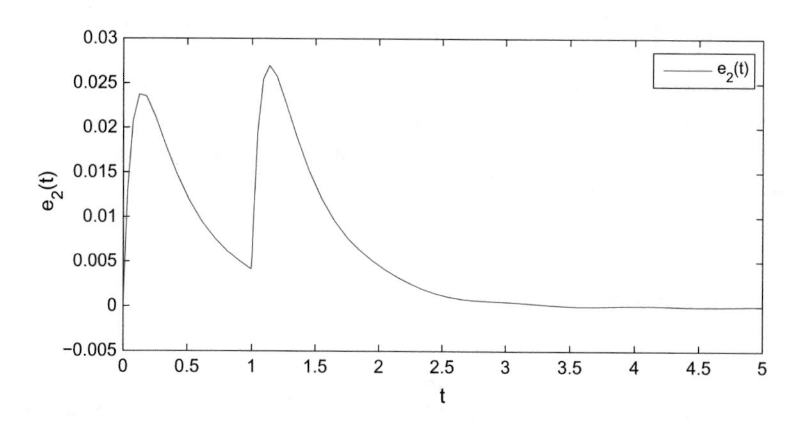

Fig. 12.6 Response of the error $e_2(t)$ for Case 1

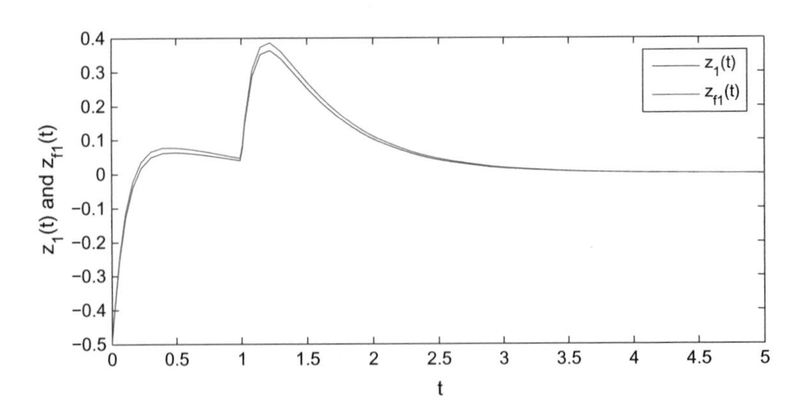

Fig. 12.7 $z_1(t)$ and its estimation $z_{f1}(t)$ for Case 2

(2) Strict dissipative case: Let $Q = -0.9$, $R = 2$, $S = 0.6$, the performance index is 0.3352 by solving LMIs in (12.28)–(12.30), and the corresponding filter parameter matrices can be got as follows:

$$A_{f11} = \begin{bmatrix} -1.8535 & -0.3518 \\ -1.2394 & -10.1253 \end{bmatrix}, A_{f12} = \begin{bmatrix} -1.8233 & -0.2952 \\ -1.2725 & -10.1269 \end{bmatrix},$$
$$A_{f21} = \begin{bmatrix} -1.6079 & -0.0670 \\ -2.5516 & -12.2916 \end{bmatrix}, A_{f22} = \begin{bmatrix} -1.4555 & -0.0196 \\ -2.5606 & -12.2777 \end{bmatrix},$$
$$B_{f11} = \begin{bmatrix} -1.2764 \\ 21.5233 \end{bmatrix}, B_{f12} = \begin{bmatrix} -1.5826 \\ 17.8615 \end{bmatrix},$$
$$B_{f21} = \begin{bmatrix} -0.2572 \\ 25.2683 \end{bmatrix}, B_{f22} = \begin{bmatrix} -0.4368 \\ 22.2754 \end{bmatrix},$$
$$F_{f11} = \begin{bmatrix} 0.1310 & -0.5175 \end{bmatrix}, F_{f12} = \begin{bmatrix} 0.1563 & -0.5394 \end{bmatrix},$$
$$F_{f21} = \begin{bmatrix} 0.0283 & -0.7732 \end{bmatrix}, F_{f22} = \begin{bmatrix} 0.0459 & -0.7736 \end{bmatrix}.$$

For the above gain matrices, $z_i(t)$ and $z_{fi}(t)$ are depicted in Figs. 12.7 and 12.8, and the responses of the error are shown in Figs. 12.9 and 12.10. For these simulation results, it can be shown that the effectiveness of the proposed method for the filter design is proved.

12.5 Conclusion

In this chapter, we have studied the problem of dissipative filtering for nonlinear interconnected systems with interval time-varying delays. Based on the construction of augmented Lyapunov-Krasovskii functional, and applying of the proposed new

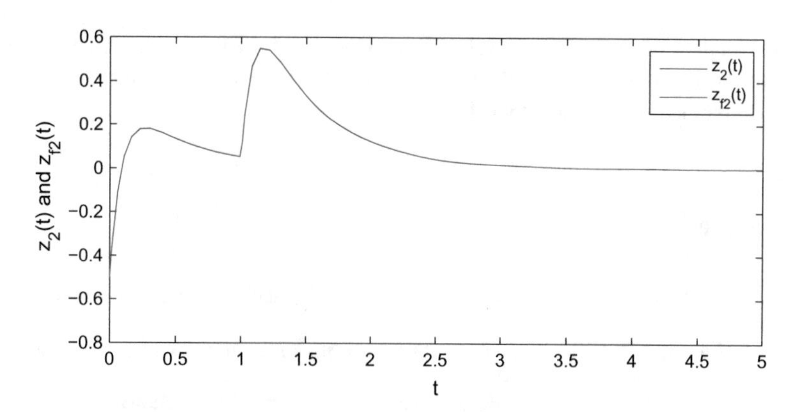

Fig. 12.8 $z_2(t)$ and its estimation $z_{f2}(t)$ for Case 2

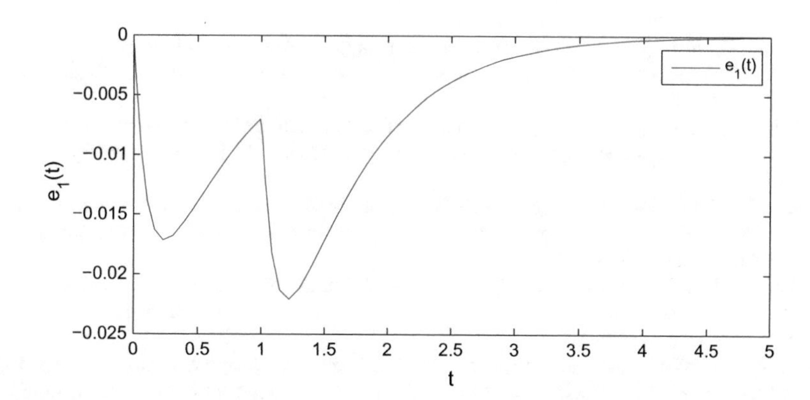

Fig. 12.9 Response of the error $e_1(t)$ for Case 2

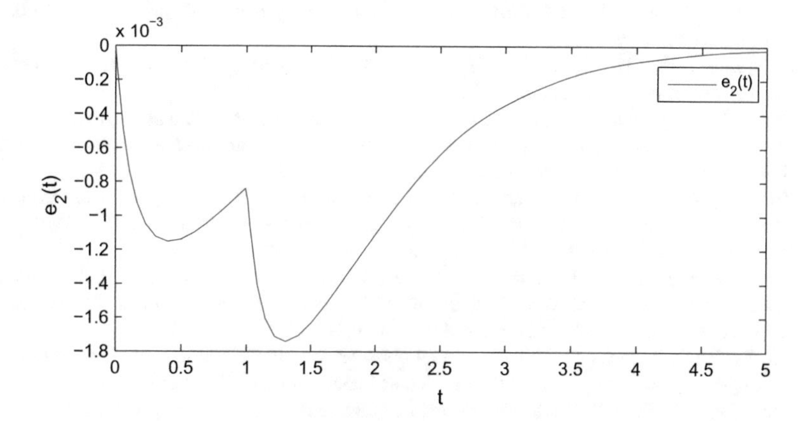

Fig. 12.10 Response of the error $e_2(t)$ for Case 2

integral inequality, some criteria related to time-delay have been established. Finally, an example has been presented to describe the feasibility of the proposed method and improvements of existing works.

References

1. Siljak D (1978) Large-scale dynamic systems: stability and structure. North-Holland, New York
2. Yin X, Liu J (2017) Distributed moving horizon state estimation of two-time-scale nonlinear systems. Automatica 79:152–161
3. Davison E (1974) The decentralized stabilization and control of unknown nonlinear time-varying systems. Automatica 10:309–316
4. Hu Z (1994) Decentralized stabilization of large-scale interconnection systems with delays. IEEE Trans Autom Control 39:180–182
5. Zhang C, Hu J, Qiu J, Chen Q (2017) Reliable output feedback control for T-S fuzzy systems with decentralized event-triggering communication and actuator failures. IEEE Trans Cybern 47(9):2592–2602
6. Wang H, Liu W, Qiu J, Liu P (2017) Adaptive fuzzy decentralized control for a class of strong interconnected nonlinear systems with unmodeled dynamics. IEEE Trans Fuzzy Syst 26:836–846
7. Lee TH, Ji DH, Park JH, Jung HY (2012) Decentralized guaranteed cost dynamic control for synchronization of a complex dynamical network with randomly switching topology. Appl Math Comput 219:996–1010
8. Lee TH, Park JH, Wu ZG, Lee SC, Lee DH (2012) Robust $\mathscr{H}_\infty$ decentralized dynamic control for synchronization of a complex dynamical network with randomly occurring uncertainties. Nonlinear Dyn 70:559–570
9. Park JH (2002) Robust non-fragile decentralized controller design for uncertain large-scale interconnected systems with time-delays. ASME J Dyn Syst Meas Control 124:332–336
10. Zhang X, Han Q (2015) Event-based $\mathscr{H}_\infty$ filtering for sampled-data systems. Automatica 51:55–69
11. Sahereh B, Aliakbar J, Ail K (2017) $\mathscr{H}_\infty$ for descriptor systems with strict LMI conditions. Automatica 80:88–94
12. Li X, Lam J, Gao H, Xiong J (2016) $\mathscr{H}_\infty$ and $\mathscr{H}_2$ filtering for linear systems with uncertain Markov transitions. Automatica 67:252–266
13. Yan H, Yang Q, Zhang H, Yang F, Zhan X (2018) Distributed $\mathscr{H}_\infty$ state estimation for a class of filtering networks with time-varying switching topologies and packet losses. IEEE Trans Syst Man Cybern: Syst 48:2047–2057
14. Zhang H, Zheng X, Yan H, Peng C, Wang Z, Chen Q (2017) Codesign of event-triggered and distributed $\mathscr{H}_\infty$ filtering for active semi-vehicle suspension systems. IEEE/ASME Trans Mechatron 22:1047–1058
15. Yan H, Zhang H, Yang F, Huang C, Chen S (2018) Distributed $\mathscr{H}_\infty$ filtering for switched repeated scalar nonlinear systems with randomly occurred sensor nonlinearities and asynchronous switching. IEEE Trans Syst Man Cybern: Syst 48:2263–2270
16. Xie X, Yue D, Peng C (2017) Multi-instant observer design of discrete-time fuzzy systems: a ranking-based switching approach. IEEE Trans Fuzzy Syst 25:1281–1292
17. Xia J, Sun W, Zhang B (2018) Finite-time adaptive fuzzy control for nonlinear systems with full state constraints. IEEE Trans Syst Man Cybern: Syst. https://doi.org/10.1109/TSMC.2018.2854770
18. Zhang L, Ning Z, Wang Z (2015) Distributed filtering for fuzzy time-delay systems with packet dropouts and redundant channels. IEEE Trans Syst Man Cybern: Syst 46:559–572

19. Liu J, Wu C, Wang Z, Wu L (2017) Reliable filter design for sensor networks using Type-2 fuzzy framework. IEEE Trans Ind Inf 13:1742–1752
20. Li F, Shi P, Lim CC, Wu L (2018) Fault detection filtering for nonhomogenenous Markovian jump systems via a fuzzy approach. IEEE Trans Fuzzy Syst 26:131–141
21. Chang XH, Park JH, Shi P (2017) Fuzzy resilient energy-to peak filtering for continuous-time nonlinear systems. IEEE Trans Fuzzy Syst 25:1576–1588
22. Liu Y, Guo BZ, Park JH, Lee SM (2018) Event-based reliable dissipative filtering for T-S fuzzy systems with asynchronous constraints. IEEE Trans Fuzzy Syst 26:2089–2098
23. Shen H, Su L, Park JH (2017) Reliable mixed $\mathscr{H}_\infty$/passive control for T-S fuzzy delayed systems based on a semi-Markov jump model approach. Fuzzy Sets Syst 314:79–98
24. Shen H, Park JH, Wu ZG (2014) Finite-time reliable $\mathscr{L}_2 - \mathscr{L}_\infty/\mathscr{H}_\infty$ control for T-S fuzzy systems with actuator faults. IET Control Theory Appl 8:688–696
25. Hsiao FH, Hwang JD (2002) Stability analysis of fuzzy large-scale systems. IEEE Trans Syst Man Cybern Part B (Cybern) 32:122–126
26. Wang Y, Karimi HR, Shen H, Fang Z, Liu M (2018) Fuzzy-model-based sliding mode control of nonlinear descriptor systems. IEEE Trans Cybern. https://doi.org/10.1109/TCYB.2018.2842920
27. Zhang H, Li C, Liao X (2006) Stability analysis and $\mathscr{H}_\infty$ controller design of fuzzy large-scale systems based on piecewise Lyapunov functions. IEEE Trans Syst Man Cybern Part B (Cybern) 36:685–698
28. Zhang H, Feng G (2008) Stability analysis and $\mathscr{H}_\infty$ controller design of discrete-time fuzzy large-scale systems based on piecewise Lyapunov functions. IEEE Trans Syst Man Cybern Part B (Cybern) 38:1390–1401
29. Zhang H, Zhong H, Zhang B, Dang C (2012) Delay dependent decentralized $\mathscr{H}_\infty$ filtering for discrete-time interconnected systems with time varying delay based on the T-S fuzzy model. IEEE Trans Fuzzy Syst 20:431–443
30. Zhang H, Dang Y, Zhang J (2010) Decentralized fuzzy $\mathscr{H}_\infty$ filtering for nonlinear interconnected systems with multiple time delays. IEEE Trans Syst Man Cybern Part B (Cybern) 40:1197–1203
31. Zhang H, Yu G, Zhou C, Dang C (2013) Delay-dependent decentralised $\mathscr{H}_\infty$ filtering for fuzzy interconnected systems with time varying delay based on Takagi-Sugeno fuzzy model. IET Control Theory Appl 7:720–729
32. Zhang Z, Lin C, Chen B (2015) New decentralized $\mathscr{H}_\infty$ filter design for nonlinear interconnected systems based on Takagi-Sugeno fuzzy models. IEEE Trans Cybern 45:2914–2924
33. Willems J (1972) Dissipative dynamical systems-part I: general theory. Arch Ration Mech Anal 45:321–351
34. Willems J (1972) Dissipative dynamical systems II: linear systems with quadratic supply rates. Arch Ration Mech Anal 45:352–393
35. Su X, Shi P, Wu L, Basin M (2014) Reliable filtering with strict dissipativity for T-S fuzzy time-delay systems. IEEE Trans Cybern 44:2470–2483
36. Shen H, Zhu Y, Zhang L, Park JH (2017) Extended dissipative state estimation of Markov jump neural networks with unreliable links. IEEE Trans Neural Netw Learn Syst 28:346–358
37. Xia J, Chen G, Sun W (2017) Extended dissipative analysis of generalized Markovian switching neural networks with two delay components. Neurocomputing 260:275–283
38. Liu Y, Guo B, Park JH, Lee S (2018) Eventbased reliable dissipative filtering for T-S fuzzy systems with asynchronous constraints. IEEE Trans Fuzzy Syst 26:2089–2098
39. Park P, Lee W, Lee S (2015) Auxiliary function-based integral inequalities for quadratic functions and their application to time-delay systems. J Frankl Inst 352:1378–1396
40. Liu K, Seuret A, Xia Y (2017) Stability analysis of systems with time-varying delays via the second-order Bassel-Legendre inequality. Automatica 76:138–142

Chapter 13
State Estimation of Genetic Regulatory Networks with Leakage, Constant, and Distributed Time-Delays

13.1 Problem Formulation and Description

13.1.1 System Description

Genetic regulatory networks (GRNs) are a kind of biological networks consisted of the interaction between the transcription of genes and the translation of mRNAs in cells and have attracted much attention from researchers [1–11]. Since there are a large number of genes and proteins with either directly or indirectly interactions, a dynamics of GRNs become more and more complex. So, to establish the exact model of GRNs is an important task and become a powerful tool for studying GRNs. Two types of GRNs models are usually employed for the research, namely, Boolean model [12–15] and ordinary differential equation model [16–19]. Later one is more popular than the former one because of high accuracy for describing the behavior. In ordinary differential equation model, generally GRNs consisted of n mRNAs and n proteins are represented by the following differential equation model:

$$\begin{cases} \dot{m}(t) = -Am(t) + Bk(p(t)) + T, \\ \dot{p}(t) = -Cp(t) + Dm(t), \end{cases} \tag{13.1}$$

where $m(t) = [m_1(t), \ldots, m_n(t)]^T$, $p(t) = [p_1(t), \ldots, p_n(t)]^T$, $k(p(t)) = [k_1(p_1(t)), \ldots, k_n(p_n(t))]^T$, $A = \text{diag}\{a_1, \ldots, a_n\}$, $C = \text{diag}\{c_1, \ldots, c_n\}$, $D = \text{diag}\{d_1, \ldots, d_n\}$, $T = [T_1, \ldots, T_n]^T$, and $m_i(t)$, $p_i(t) \in \mathbb{R}$ are concentrations of mRNA and protein of the ith node, positive real numbers, a_i and c_i, are for the degradation rates of mRNA and protein, respectively, a positive constant, d_i, is the ith translation rate from mRNA to protein, and $k_j(x) = \frac{(x/\beta)^H}{1+(x/\beta)^H}$ contains the information of the feedback regulation, connection, and topology of the proteins, $T_i = \sum_{j \in I_i} \delta_{ij}$, I_i is the set of all the repressor of gene i, and $B = (b_{ij}) \in \mathbb{R}^{n \times n}$ is defined as:

J. H. Park et al., *Dynamic Systems with Time Delays: Stability and Control*,
https://doi.org/10.1007/978-981-13-9254-2_13

$$b_{ij} = \begin{cases} \delta_{ij} & \text{if transcription factor } j \text{ is} \\ & \text{an activator of the gene } i, \\ -\delta_{ij} & \text{if transcription factor } j \text{ is} \\ & \text{an repressor of the gene } i, \\ 0 & \text{if there is no link from gene } j \text{ to } i. \end{cases}$$

Suppose (m^*, p^*) as an equilibrium point of the system (13.1), then the equilibrium points m^* and p^* can be shifted to the origin by a transformation, i.e. $x(t) = m(t) - m^*$ and $y(t) = p(t) - p^*$. Finally, it can be obtained that

$$\begin{cases} \dot{x}(t) = -Ax(t) + Bf(y(t)), \\ \dot{y}(t) = -Cy(t) + Dx(t), \end{cases} \tag{13.2}$$

where $x(t) = [x_1(t),\ldots, x_n(t)]^T$ with $x_i(t) = m_i(t) - m_i^*$, $y(t) = [y_1(t),\ldots, y_n(t)]^T$ with $y_i(t) = p_i(t) - p_i^*$, $f(y(t)) = [f_1(y_1(t)),\ldots, f_n(y_n(t))]^T$ with $f_i(y_i(t)) = k_i(y_i(t) + p^*) - k_i(p^*)$, respectively.

On the other hand, among various kinds of delays, leakage delay is natural in practice but often ignored in the modeling. A leakage delay has a tendency to destroy the stability of the system and is hard to handle compared with other types of delays. Therefore, it is worth to study such a system in the consideration of leakage delay [20–22], furthermore GRNs with leakage delay [19, 23, 24].

For example, let us consider a GRN (13.2) and the following GRN with leakage delay:

$$\begin{cases} \dot{x}(t) = -Ax(t-\rho) + Bf(y(t)), \\ \dot{y}(t) = -Cy(t-\sigma) + Dx(t), \end{cases} \tag{13.3}$$

where ρ and σ are leakage delay for the concentrations of mRNA and protein, respectively.

When we given the following parameters,

$$A = \begin{bmatrix} 9 & 0 \\ 0 & 8 \end{bmatrix}, \quad B = \begin{bmatrix} -1.5 & 0 \\ 1 & 2 \end{bmatrix}, \quad C = \begin{bmatrix} 8 & 0 \\ 0 & 9 \end{bmatrix}, \quad D = \begin{bmatrix} -0.9 & 0 \\ 0 & 1 \end{bmatrix},$$
$$x(0) = \begin{bmatrix} 2 & -2 \end{bmatrix}, \quad y(0) = \begin{bmatrix} 0.5 & -2 \end{bmatrix}, \quad \rho = 0.2, \quad \sigma = 0.2.$$

Then, the simulation results of the GRN (13.2) and (13.3) can be obtained in Figs. 13.1 and 13.2, respectively. By comparing Figs. 13.1 and 13.2, we can notice that a small value of leakage delay destroyed the stability of the GRN.

13.1.2 Problem Statement

In this chapter, we consider the leakage, constant, and distributed time-delays in GRN, then the GRN (13.2) can be reformulated as follows:

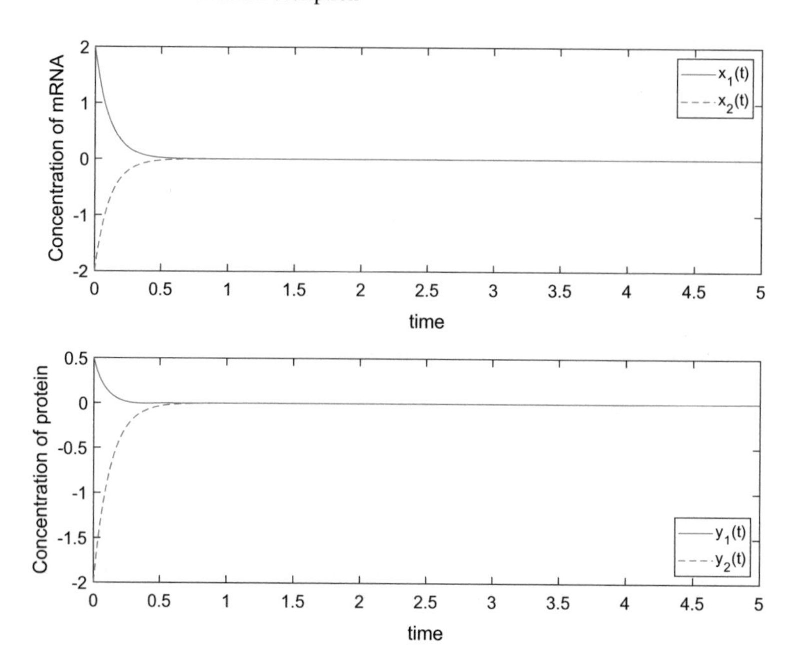

Fig. 13.1 The state trajectories of GRNs (13.2)

$$\begin{cases} \dot{x}(t) = -Ax(t-\rho) + B_1 f(y(t-h)) + B_2 \int_{t-h}^{t} y(s)ds, \\ \dot{y}(t) = -Cy(t-\sigma) + D_1 x(t-\tau) + D_2 \int_{t-\tau}^{t} x(s)ds, \end{cases} . \tag{13.4}$$

On the other hand, it is hard to obtain the information of genes by available data in practice. Therefore, it is important to estimate the genes' information through the available measurements for analysis and applications of GRNs [19, 24–27]. So, this chapter proposes a criterion to design a state estimator for monitoring the concentrations of mRNA and protein of the GRNs (13.4) from the available network output. To this end, we consider that the measurement outputs of the GRN (13.4) are given as

$$\begin{aligned} z_x(t) &= G_x x(t), \\ z_y(t) &= G_y y(t), \end{aligned}$$

where $z_x(t), z_y(t) \in \mathbb{R}^k$ are the measurement outputs, and G_x and G_y are known constant matrices with appropriate dimensions.

And then, we also consider the following state estimator:

$$\begin{cases} \dot{\hat{x}}(t) = -A\hat{x}(t-\rho) + B_1 f(\hat{y}(t-h)) + B_2 \int_{t-h}^{t} \hat{y}(s)ds \\ \qquad\quad +K_1\big(z_x(t) - G_x\hat{x}(t)\big), \\ \dot{\hat{y}}(t) = -C\hat{y}(t-\sigma) + D_1\hat{x}(t-\tau) + D_2 \int_{t-\tau}^{t} \hat{x}(s)ds \\ \qquad\quad +K_2\big(z_y(t) - G_y\hat{y}(t)\big), \end{cases} \tag{13.5}$$

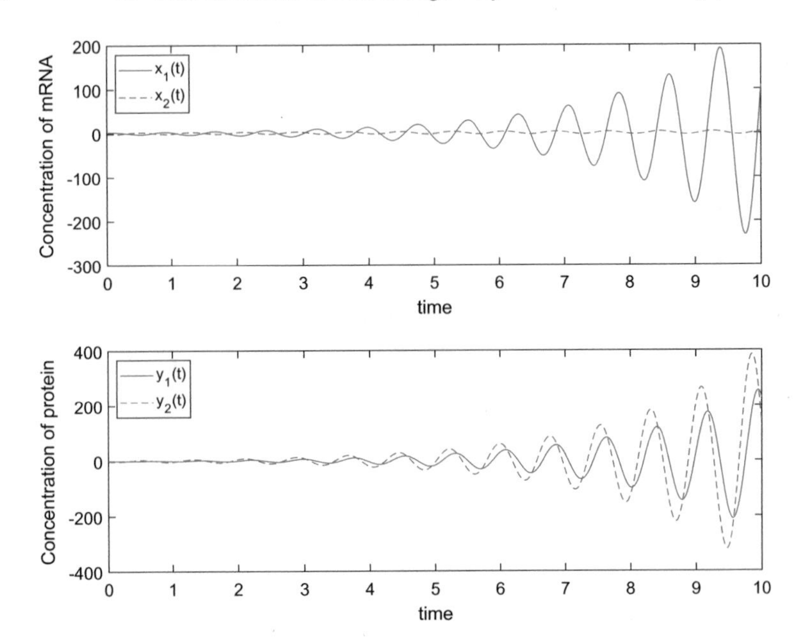

Fig. 13.2 The state trajectories of GRNs (13.3) with leakage delays, $\rho = 0.2$ and $\sigma = 0.2$

where $\hat{x}(t)$ and $\hat{y}(t) \in \mathbb{R}^n$ are the estimation of states $x(t)$ and $y(t)$, respectively, $B_1, B_2, D_1, D_2 \in \mathbb{R}^{n\times n}$ are known constant matrices, and K_1 and K_2 are estimator gain matrices to be designed.

Defining the error vectors as $v(t) = x(t) - \hat{x}(t)$ and $w(t) = y(t) - \hat{y}(t)$, then the error dynamical system is expressed from (13.4) and (13.5):

$$\begin{cases} \dot{v}(t) = -Av(t-\rho) + B_1 g(w(t-h)) + B_2 \int_{t-h}^{t} w(s)ds - K_1 G_x v(t), \\ \dot{w}(t) = -Cw(t-\sigma) + D_1 v(t-\tau) + D_2 \int_{t-\tau}^{t} v(s)ds - K_2 G_y w(t), \end{cases} \tag{13.6}$$

where $g(w(t)) = f(y(t)) - f(\hat{y}(t))$.

From the relationship between $f(\cdot)$ and $g(\cdot)$ and the property of $f(\cdot)$ which is a monotonically increasing function with saturation, the following equation (13.7) can be obtained.

$$g(w(t))(g(w(t)) - L(w(t))) \leq 0, \tag{13.7}$$

where $L = \text{diag}\{l_1, \ldots, l_n\}$ with $l_i > 0$.

13.2 Main Results

In this section, we propose such criteria for designing state estimator for the GRN (13.4) with leakage, constant, and distributed time-delays.

Before beginning the main theorem, we define the block entry matrices as $r_i \in \mathbb{R}^{17n\times n}$ $(i = 1, 2, \ldots, 17)$, for example, $r_2^T = [0,\ I_n,\ \underbrace{0,\ \ldots,\ 0}_{15}]$ and give the following lemma.

Lemma 13.1 ([28]) *Let x be a differentiable function:*$[\alpha,\ \beta] \to \mathbb{R}^n$. *For a positive symmetric matrix* $R \in \mathbb{R}^{n\times n}$, *and any matrices* $N_1, N_2, N_3 \in \mathbb{R}4n \times n$, *the following inequality holds:*

$$-\int_{\alpha}^{\beta} \dot{x}^T(s)R\dot{x}(s)ds \le \varpi_1^T(t)\mathscr{R}\varpi_1(t),$$

where

$$\begin{aligned}\mathscr{R} &= (\beta-\alpha)\Big(N_1R^{-1}N_1^T + \frac{1}{3}N_2R^{-1}N_2^T + \frac{1}{5}N_3R^{-1}N_3^T\Big)\\ &\quad + \mathrm{Sym}\Big\{N_1\varpi_2(t) + N2\varpi_3(t) + N_3\varpi_4(t)\Big\}\\ \varpi_1(t) &= \Big[x^T(\beta),\ x^T(\alpha)\ \tfrac{1}{\beta-\alpha}\textstyle\int_{\alpha}^{\beta} x(s)ds,\ \tfrac{1}{(\beta-\alpha)^2}\int_{\alpha}^{\beta}\int_{\alpha}^{v} x(s)dsdv\Big]^T\\ \varpi_2(t) &= x(\beta) - x(\alpha),\\ \varpi_3(t) &= x(\beta) + x(\alpha) - \frac{2}{\beta-\alpha}\int_{\alpha}^{\beta} x(s)ds,\\ \varpi_4(t) &= x(\beta) - x(\alpha) - \frac{6}{\beta-\alpha}\int_{\alpha}^{\beta} x(s)ds - \frac{6}{(\beta-\alpha)^2}\int_{\alpha}^{\beta}\int_{\alpha}^{v} x(s)dsdv.\end{aligned}$$

Now, here is a main result for the state estimation problem stated above.

Theorem 13.1 *For given positive constants* $\rho,\ \sigma,\ \tau,\ h,\ l_i\ (i = 1, \ldots, n)$, *the error system (13.6) is asymptotically stable, if there exist matrices* $P_1, P_2 \in \mathbb{S}_+^{5n}$, $Q_1, Q_2, R_1, R_2, S_1, S_2, U_1, U_2 \in \mathbb{S}_+^n$, $X \in \mathbb{D}_+^n$, $N_i, M_i \in \mathbb{R}^{4n\times n}$ $(i = 1, \ldots, 6)$, Y_1, Y_2, $J_1, J_2 \in \mathbb{R}^{n\times n}$ *satisfying the following LMI:*

$$\begin{bmatrix} \Upsilon & \sqrt{\rho}\Gamma_5N_1 & \sqrt{\rho}\Gamma_5N_2 & \sqrt{\rho}\Gamma_5N_3 & \sqrt{\sigma}\Gamma_6M_1 & \sqrt{\sigma}\Gamma_6M_2 & \sqrt{\sigma}\Gamma_6M_3 \\ * & -S_1 & 0 & 0 & 0 & 0 & 0 \\ * & * & -3S_1 & 0 & 0 & 0 & 0 \\ * & * & * & -5S_1 & 0 & 0 & 0 \\ * & * & * & * & -S_2 & 0 & 0 \\ * & * & * & * & * & -3S_2 & 0 \\ * & * & * & * & * & * & -5S_2 \\ * & * & * & * & * & * & * \\ * & * & * & * & * & * & * \\ * & * & * & * & * & * & * \\ * & * & * & * & * & * & * \\ * & * & * & * & * & * & * \\ * & * & * & * & * & * & * \end{bmatrix}$$

$$\left.\begin{matrix} \sqrt{\tau}\Gamma_7 N_4 & \sqrt{\tau}\Gamma_7 N_5 & \sqrt{\tau}\Gamma_7 N_6 & \sqrt{h}\Gamma_8 M_4 & \sqrt{h}\Gamma_8 M_5 & \sqrt{h}\Gamma_8 M_6 \\ 0 & 0 & 0 & 0 & 0 & 0 \\ 0 & 0 & 0 & 0 & 0 & 0 \\ 0 & 0 & 0 & 0 & 0 & 0 \\ 0 & 0 & 0 & 0 & 0 & 0 \\ 0 & 0 & 0 & 0 & 0 & 0 \\ 0 & 0 & 0 & 0 & 0 & 0 \\ -U_1 & 0 & 0 & 0 & 0 & 0 \\ * & -3U_1 & 0 & 0 & 0 & 0 \\ * & * & -6U_1 & 0 & 0 & 0 \\ * & * & * & -U_2 & 0 & 0 \\ * & * & * & * & -3U_2 & 0 \\ * & * & * & * & * & -6U_2 \end{matrix}\right] < 0 \tag{13.8}$$

where

$$\begin{aligned} \Upsilon &= \Xi_1 + \Xi_2 + \Xi_3 + \Xi_4 + \Xi_5 + \Xi_6 + \Xi_7 + \Xi_8, \\ \Xi_1 &= \mathrm{Sym}\Big\{\Gamma_1 P_1 \Gamma_2^T + \Gamma_3 P_2 \Gamma_4^T\Big\}, \\ \Xi_2 &= r_1 Q_1 r_1^T - r_2 Q_1 r_2^T + r_8 Q_2 r_8 - r_9 Q_2 r_9^T, \\ \Xi_3 &= r_1 R_1 r_1^T - r_5 R_1 r_5^T + r_8 R_2 r_8 - r_{12} R_2 r_{12}^T, \\ \Xi_4 &= \rho r_{16} S_1 r_{16}^T + \sigma r_{17} S_2 r_{17}^T + \mathrm{Sym}\Big\{\Gamma_5(N_1 \Pi_1^T + N_2 \Pi_2^T + N_3 \Pi_3^T) \\ &\quad + \Gamma_6(M_1 \Pi_4^T + M_2 \Pi_5^T + M_3 \Pi_6^T)\Big\}, \\ \Xi_5 &= \tau r_{16} U_1 r_{16}^T + h r_{17} U_2 r_{17}^T + \mathrm{Sym}\Big\{\Gamma_7(N_4 \Pi_7^T + N_5 \Pi_8^T + N_6 \Pi_9^T) \\ &\quad + \Gamma_8(M_4 \Pi_{10}^T + M_5 \Pi_{11}^T + M_6 \Pi_{12}^T)\Big\}, \end{aligned} \tag{13.9}$$

$$\begin{aligned} \Xi_6 &= \mathrm{Sym}\Big\{r_{15} X (L r_{12}^T - r_{15}^T)\Big\}, \\ \Xi_7 &= \mathrm{Sym}\Big\{\Big(r_1 + r_{16}\Big)\Big(-Y_1 r_{16}^T - Y_1 A r_2^T + Y_1 B_1 r_{15}^T + h Y_1 B_2 r_{13}^T - J_1 G_v r_1^T\Big)\Big\}, \\ \Xi_8 &= \mathrm{Sym}\Big\{\Big(r_8 + r_{17}\Big)\Big(-Y_2 r_{17}^T - Y_2 C r_9^T + Y_2 D_1 r_5^T + \tau Y_2 D_2 r_6^T - J_2 G_w r_8^T\Big)\Big\}, \\ \Gamma_1 &= \big[r_1 - \rho r_3 A^T,\ \rho r_3,\ \rho^2 r_4,\ \tau r_6,\ \tau^2 r_7\big], \\ \Gamma_2 &= \big[r_{16} - r_1 A^T + r_2 A^T,\ r_1 - r_2,\ r_1 - \rho r_3,\ r_1 - r_5,\ r_1 - \tau r_6\big], \\ \Gamma_3 &= \big[r_8 - \sigma r_{10} C^T,\ \sigma r_{10},\ \sigma^2 r_{11},\ h r_{13},\ h^2 r_{14}\big], \\ \Gamma_4 &= \big[r_{17} - r_8 C^T + r_9 C^T,\ r_8 - r_9,\ r_8 - \sigma r_{10},\ r_8 - r_{12},\ r_8 - h r_{13}\big], \\ \Gamma_5 &= \big[r_1,\ r_2,\ r_3,\ r_4\big], \\ \Gamma_6 &= \big[r_8,\ r_9,\ r_{10},\ r_{11}\big], \end{aligned}$$

$$\begin{aligned}
\Gamma_7 &= \left[r_1,\, r_5,\, r_6,\, r_7 \right],\\
\Gamma_8 &= \left[r_8,\, r_{12},\, r_{13},\, r_{14} \right],\\
\Pi_1 &= r_1 - r_2,\\
\Pi_2 &= r_1 + r_2 - 2r_3,\\
\Pi_3 &= r_1 - r_2 - 6r_3 + 6r_4,\\
\Pi_4 &= r_8 - r_9,\\
\Pi_5 &= r_8 + r_9 - 2r_{10},\\
\Pi_6 &= r_8 - r_9 - 6r_{10} + 6r_{11},\\
\Pi_7 &= r_1 - r_5,\\
\Pi_8 &= r_1 + r_5 - 2r_6,\\
\Pi_9 &= r_1 - r_5 - 6r_6 + 6r_7,\\
\Pi_{10} &= r_8 - r_{12},\\
\Pi_{11} &= r_8 + r_{12} - 2r_{13},\\
\Pi_{12} &= r_8 - r_{12} - 6r_{13} + 6r_{14},\\
J_1 &= Y_1 K_1,\\
J_2 &= Y_2 K_2.
\end{aligned}$$

And the state estimator gains can be calculated $K_1 = Y_1^{-1} J_1$ *and* $K_2 = Y_2^{-1} J_2$.

Proof Let us consider the following Lyapunov functional:

$$V(t) = V_1(t) + V_2(t) + V_3(t) + V_4(t) + V_5(t), \tag{13.10}$$

where

$$\begin{aligned}
V_1(t) &= \eta_1^T(t) P_1 \eta_1 + \eta_2^T(t) P_2 \eta_2(t),\\
V_2(t) &= \int_{t-\rho}^{t} v^T(s) Q_1 v(s) dt + \int_{t-\sigma}^{t} w^T(s) Q_2 w(s) dt,\\
V_3(t) &= \int_{t-\tau}^{t} v^T(s) R_1 v(s) dt + \int_{t-h}^{t} w^T(s) Q_2 w(s) dt,\\
V_4(t) &= \int_{t-\rho}^{t} \int_{\theta}^{t} \dot{v}^T(s) S_1 \dot{v}(s) ds d\theta + \int_{t-\sigma}^{t} \int_{\theta}^{t} \dot{w}^T(s) S_2 \dot{w}(s) ds d\theta,\\
V_5(t) &= \int_{t-\tau}^{t} \int_{\theta}^{t} \dot{v}^T(s) U_1 \dot{v}(s) ds d\theta + \int_{t-h}^{t} \int_{\theta}^{t} \dot{w}^T(s) U_2 \dot{w}(s) ds d\theta,
\end{aligned}$$

with

$$\eta_1(t) = \begin{bmatrix} v(t) - A\int_{t-\rho}^{t} v(s)ds \\ \int_{t-\rho}^{t} v(s)ds \\ \int_{t-\rho}^{t}\int_{\theta}^{t} v(s)dsd\theta \\ \int_{t-\tau}^{t} v(s)ds \\ \int_{t-\tau}^{t}\int_{\theta}^{t} v(s)dsd\theta \end{bmatrix},$$

$$\eta_2(t) = \begin{bmatrix} w(t) - C\int_{t-\sigma}^{t} w(s)ds \\ \int_{t-\sigma}^{t} w(s)ds \\ \int_{t-\sigma}^{t}\int_{\theta}^{t} w(s)dsd\theta \\ \int_{t-h}^{t} w(s)ds \\ \int_{t-h}^{t}\int_{\theta}^{t} w(s)dsd\theta \end{bmatrix}.$$

The time derivative along the system trajectories of (13.6) can be calculated as:

$$\dot{V}_1(t) = 2\eta_1^T(t)P_1\eta_3(t) + 2\eta_2^T(t)P_2\eta_4(t), \tag{13.11}$$

$$\begin{aligned} \dot{V}_2(t) =& v^T(t)Q_1v(t) - v^T(t-\rho)Q_1v(t-\rho) \\ &+ w^T(t)Q_2w(t) - w^T(t-\sigma)Q_2w(t-\sigma), \end{aligned} \tag{13.12}$$

$$\begin{aligned} \dot{V}_3(t) =& v^T(t)R_1v(t) - v^T(t-\tau)R_1v(t-\tau) \\ &+ w^T(t)R_2w(t) - w^T(t-h)R_2w(t-h), \end{aligned} \tag{13.13}$$

$$\begin{aligned} \dot{V}_4(t) =& \rho\dot{v}^T(t)S_1\dot{v}(t) - \int_{t-\rho}^{t} \dot{v}^T(s)S_1\dot{v}(s)ds \\ &+ \sigma\dot{w}^T(t)S_2\dot{w}(t) - \int_{t-\sigma}^{t} \dot{w}^T(s)S_2\dot{w}(s)ds, \end{aligned} \tag{13.14}$$

$$\begin{aligned} \dot{V}_5(t) =& \tau\dot{v}^T(t)U_1\dot{v}(t) - \int_{t-\tau}^{t} \dot{v}^T(s)U_1\dot{v}(s)ds \\ &+ h\dot{w}^T(t)U_2\dot{w}(t) - \int_{t-h}^{t} \dot{w}^T(s)U_2\dot{w}(s)ds, \end{aligned} \tag{13.15}$$

where

$$\eta_3(t) = \begin{bmatrix} \dot{v}(t) - Av(t) + Av(t-\rho) \\ v(t) - v(t-\rho) \\ v(t) - \int_{t-\rho}^{t} v^T(s)ds \\ v(t) - v(t-\tau) \\ v(t) - \int_{t-\tau}^{t} v^T(s)ds \end{bmatrix},$$

$$\eta_4(t) = \begin{bmatrix} \dot{w}(t) - Cw(t) + Cw(t-\sigma) \\ w(t) - w(t-\sigma) \\ w(t) - \int_{t-\sigma}^{t} w^T(s)ds \\ w(t) - w(t-h) \\ w(t) - \int_{t-h}^{t} w^T(s)ds \end{bmatrix}.$$

For any matrices $N_1, N_2, N_3, M_1, M_2, M_3 \in \mathbb{R}^{4n\times n}$, applying Lemma 13.1 to the two integral terms in $\dot{V}_4(t)$ leads the following inequalities:

$$
\begin{aligned}
&-\int_{t-\rho}^{t} \dot{v}^T(s)S_1\dot{v}(s)ds \\
&\le \rho\eta_5^T(t)\Big(N_1S_1^{-1}N_1^T + \frac{1}{3}N_2S_1^{-1}N_2^T + \frac{1}{5}N_3S_1^{-1}N_3^T\Big)\eta_5(t) \\
&\quad + 2\eta_5^T(t)\Bigg(N_1\Big(v(t)-v(t-\rho)\Big) + N_2\Big(v(t)+v(t-\rho)-\frac{2}{\rho}\int_{t-\rho}^{t} v(s)ds\Big) \\
&\quad + N_3\Big(v(t)-v(t-\rho)-\frac{6}{\rho}\int_{t-\rho}^{t} v(s)ds + \frac{6}{\rho^2}\int_{t-\rho}^{t}\int_{\theta}^{t} v(s)dsd\theta\Big)\Bigg), \quad (13.16)
\end{aligned}
$$

and

$$
\begin{aligned}
&-\int_{t-\sigma}^{t} \dot{w}^T(s)S_2\dot{w}(s)ds \\
&\le \sigma\eta_6^T(t)\Big(M_1S_2^{-1}M_1^T + \frac{1}{3}M_2S_2^{-1}M_2^T + \frac{1}{5}M_3S_2^{-1}M_3^T\Big)\eta_6(t) \\
&\quad + 2\eta_6^T(t)\Bigg(M_1\Big(w(t)-w(t-\sigma)\Big) + M_2\Big(w(t)+w(t-\sigma)-\frac{2}{\sigma}\int_{t-\sigma}^{t} w(s)ds\Big) \\
&\quad + M_3\Big(w(t)-w(t-\sigma)-\frac{6}{\sigma}\int_{t-\sigma}^{t} w(s)ds + \frac{6}{\sigma^2}\int_{t-\sigma}^{t}\int_{\theta}^{t} w(s)dsd\theta\Big)\Bigg),
\end{aligned}
\quad (13.17)
$$

where

$$
\eta_5(t) = \begin{bmatrix} v(t) \\ v(t-\rho) \\ \frac{1}{\rho}\int_{t-\rho}^{t} v(s)ds \\ \frac{1}{\rho^2}\int_{t-\rho}^{t}\int_{\theta}^{t} v(s)dsd\theta \end{bmatrix},
$$

$$
\eta_6(t) = \begin{bmatrix} w(t) \\ w(t-\sigma) \\ \frac{1}{\sigma}\int_{t-\sigma}^{t} w(s)ds \\ \frac{1}{\sigma^2}\int_{t-\sigma}^{t}\int_{\theta}^{t} w(s)dsd\theta \end{bmatrix}.
$$

As the same procedure to the above, by Lemma 13.1 with any matrices N_4, N_5, N_6, M_4, M_5, $M_6 \in \mathbb{R}^{4n\times n}$, the two integral terms in $\dot{V}_5(t)$ can be:

$$
\begin{aligned}
&-\int_{t-\tau}^{t} \dot{v}^T(s)U_1\dot{v}(s)ds \\
&\le \tau\eta_7^T(t)\Big(N_4U_1^{-1}N_4^T + \frac{1}{3}N_5U_1^{-1}N_5^T + \frac{1}{5}N_6U_1^{-1}N_6^T\Big)\eta_7(t)
\end{aligned}
$$

$$+ 2\eta_7^T(t)\Big(N_4\big(v(t) - v(t-\tau)\big) + N_5\Big(v(t) + v(t-\tau) - \frac{2}{\tau}\int_{t-\tau}^{t} v(s)ds\Big)$$

$$+ N_6\Big(v(t) - v(t-\tau) - \frac{6}{\tau}\int_{t-\tau}^{t} v(s)ds + \frac{6}{\tau^2}\int_{t-\tau}^{t}\int_{\theta}^{t} v(s)dsd\theta\Big)\Big), \quad (13.18)$$

and

$$-\int_{t-h}^{t} \dot{w}^T(s)U_2\dot{w}(s)ds$$

$$\leq h\eta_8^T(t)\Big(M_4U_2^{-1}M_4^T + \frac{1}{3}M_5U_2^{-1}M_5^T + \frac{1}{5}M_6U_2^{-1}M_6^T\Big)\eta_8(t)$$

$$+ 2\eta_8^T(t)\Big(M_4\big(w(t) - w(t-h)\big) + M_5\Big(w(t) + w(t-h) - \frac{2}{h}\int_{t-h}^{t} w(s)ds\Big)$$

$$+ M_6\Big(w(t) - w(t-h) - \frac{6}{h}\int_{t-h}^{t} w(s)ds + \frac{6}{h^2}\int_{t-h}^{t}\int_{\theta}^{t} w(s)dsd\theta\Big)\Big), \quad (13.19)$$

where

$$\eta_8(t) = \begin{bmatrix} w(t) \\ w(t-h) \\ \frac{1}{h}\int_{t-h}^{t} w(s)ds \\ \frac{1}{h^2}\int_{t-h}^{t}\int_{\theta}^{t} w(s)dsd\theta \end{bmatrix},$$

$$\eta_7(t) = \begin{bmatrix} v(t) \\ v(t-\tau) \\ \frac{1}{\tau}\int_{t-\tau}^{t} v(s)ds \\ \frac{1}{\tau^2}\int_{t-\tau}^{t}\int_{\theta}^{t} v(s)dsd\theta \end{bmatrix}.$$

In addition, for matrices $X \in \mathbb{D}_+^n$, $Y_1, Y_2 \in \mathbb{R}^{n\times n}$, the following is true according to (13.7) and (13.6):

$$0 \leq 2g^T(w(t-h))X(Lw(t-h) - g(w(t-h))), \quad (13.20)$$

$$0 = 2\big(v^T(t) + \dot{v}^T(t)\big)Y_1\Big[-\dot{v}(t) - Av(t-\rho) + B_1g(w(t-h))$$

$$+ B_2\int_{t-h}^{t} w(s)ds - K_1G_xv(t)\Big], \quad (13.21)$$

$$0 = 2\big(w^T(t) + \dot{w}^T(t)\big)Y_2\Big[-\dot{w}(t) - Cw(t-\sigma) + D_1v(t-\tau)$$

$$+ D_2\int_{t-\tau}^{t} v(s)ds - K_2G_yw(t)\Big]. \quad (13.22)$$

Define the following vector:

$$\begin{aligned}\zeta(t) =\Big[& v^T(t),\ v^T(t-\rho),\ \tfrac{1}{\rho}\int_{t-\rho}^{t} v^T(s)ds,\ \tfrac{1}{\rho^2}\int_{t-\rho}^{t}\int_{\theta}^{t} v^T(s)dsd\theta, \\ & v^T(t-\tau),\ \tfrac{1}{\tau}\int_{t-\tau}^{t} v^T(s)ds,\ \tfrac{1}{\tau^2}\int_{t-\tau}^{t}\int_{\theta}^{t} v^T(s)dsd\theta,\ w^T(t), \\ & w^T(t-\sigma),\ \tfrac{1}{\sigma}\int_{t-\sigma}^{t} w^T(s)ds,\ \tfrac{1}{\sigma^2}\int_{t-h}^{t}\int_{\theta}^{t} w^T(s)dsd\theta,\ w^T(t-h), \\ & \tfrac{1}{h}\int_{t-h}^{t} w^T(s)ds,\ \tfrac{1}{h^2}\int_{t-h}^{t}\int_{\theta}^{t} w^T(s)dsd\theta,\ f^T(t),\ \dot{v}^T(t),\ \dot{w}^T(t)\Big]^T,\end{aligned}$$

then using Eqs. (13.11)–(13.22), we can express the time derivative of Lyapunov functional (13.10), respectively, as follows:

$$\begin{aligned}\dot{V}_1(t) &= \zeta^T(t)\Xi_1\zeta(t), \\ \dot{V}_2(t) &= \zeta^T(t)\Xi_2\zeta(t), \\ \dot{V}_3(t) &= \zeta^T(t)\Xi_3\zeta(t), \\ \dot{V}_4(t) &= \zeta^T(t)\Big(\Xi_4 + \rho\Psi_1 + \sigma\Psi_2\Big)\zeta(t), \\ \dot{V}_5(t) &= \zeta^T(t)\Big(\Xi_5 + \tau\Psi_3 + h\Psi_4\Big)\zeta(t),\end{aligned}$$

where

$$\begin{aligned}\Psi_1 &= \Gamma_5\Big(N_1S_1^{-1}N_1^T + \frac{1}{3}N_2S_1^{-1}N_2^T + \frac{1}{5}N_3S_1^{-1}N_3^T\Big)\Gamma_5^T, \\ \Psi_2 &= \Gamma_6\Big(M_1S_2^{-1}M_1^T + \frac{1}{3}M_2S_2^{-1}M_2^T + \frac{1}{5}M_3S_2^{-1}M_3^T\Big)\Gamma_6^T, \\ \Psi_3 &= \Gamma_7\Big(N_4U_1^{-1}N_4^T + \frac{1}{3}N_5U_1^{-1}N_5^T + \frac{1}{5}N_6U_1^{-1}N_6^T\Big)\Gamma_7^T, \\ \Psi_4 &= \Gamma_8\Big(M_4U_2^{-1}M_4^T + \frac{1}{3}M_5U_2^{-1}M_5^T + \frac{1}{5}M_6U_2^{-1}M_6^T\Big)\Gamma_8^T.\end{aligned}$$

and the other notations defined in Theorem 13.1.

This implies that

$$\dot{V}(t) \le \zeta^T(t)\Big(\Upsilon + \rho\Psi_1 + \sigma\Psi_2 + \tau\Psi_3 + h\Psi_4\Big)\zeta(t), \tag{13.23}$$

where Υ is defined in Theorem 13.1. Then, by Schur complement it is clear that LMI (13.8) is equivalent to $\dot{V}(t) < 0$ which means the error system is asymptotically stable by Lyapunov stability theory. This completes the proof. ■

When we consider no leakage delay case, the GRN (13.4) and estimator (13.5) can be expressed as:

$$\begin{cases} \dot{x}(t) = -Ax(t) + B_1 f(y(t-h)) + B_2 \int_{t-h}^{t} w(s)ds, \\ \dot{y}(t) = -Cy(t) + D_1 x(t-\tau) + D_2 \int_{t-\tau}^{t} v(s)ds, \end{cases} \tag{13.24}$$

$$\begin{cases} \dot{\hat{x}}(t) = -A\hat{x}(t) + B_1 f(\hat{y}(t-h)) + B_2 \int_{t-h}^{t} w(s)ds + K_1\big(z_x(t) - G_x\hat{x}(t)\big), \\ \dot{\hat{y}}(t) = -C\hat{y}(t) + D_1\hat{x}(t-\tau) + D_2 \int_{t-\tau}^{t} v(s)ds + K_2\big(z_y(t) - G_y\hat{y}(t)\big), \end{cases} \tag{13.25}$$

then, the error dynamics between the GRN (13.24) and estimator (13.25) can be obtained as:

$$\begin{cases} \dot{v}(t) = -Av(t) + B_1 g(w(t-h)) + B_2 \int_{t-h}^{t} w(s)ds - K_1 G_x v(t), \\ \dot{w}(t) = -Cw(t) + D_1 v(t-\tau) + D_2 \int_{t-\tau}^{t} v(s)ds - K_2 G_y w(t). \end{cases} \tag{13.26}$$

For deriving a criterion of designing state estimator, define the following vector:

$$\begin{aligned} \bar{\zeta}(t) = \Big[& v^T(t),\ v^T(t-\tau),\ \tfrac{1}{\tau}\int_{t-\tau}^{t} v^T(s)ds,\ \tfrac{1}{\tau^2}\int_{t-\tau}^{t}\int_{\theta}^{t} v^T(s)dsd\theta, \\ & w^T(t),\ w^T(t-h),\ \tfrac{1}{h}\int_{t-h}^{t} w^T(s)ds,\ \tfrac{1}{h^2}\int_{t-h}^{t}\int_{\theta}^{t} w^T(s)dsd\theta, \\ & f^T(t),\ \dot{v}^T(t),\ \dot{w}^T(t)\Big]^T, \end{aligned}$$

and the block entry matrices as $\bar{r}_i \in \mathbb{R}^{11n\times n}$ $(i = 1, 2, \ldots, 11)$, for example, $\bar{r}_2^T = [0,\ I_n,\ \underbrace{0,\ \ldots,\ 0}_{9}]$. Then, finally we can get the following corollary.

Corollary 13.1 *For given positive constants τ, h, l_i $(i = 1, \ldots, n)$, the error system (13.6) is asymptotically stable, if there exist matrices $P_1, P_2 \in \mathbb{S}_+^{3n}$, $R_1, R_2, U_1, U_2 \in \mathbb{S}_+^n$, $X \in \mathbb{D}_+^n$, $N_i, M_i \in \mathbb{R}^{4n\times n}$ $(i = 1, \ldots, 3)$, $Y_1, Y_2, J_1, J_2 \in \mathbb{R}^{n\times n}$ satisfying the following LMI:*

$$\begin{bmatrix} \bar{\Upsilon} & \sqrt{\tau}\bar{\Gamma}_5 N_1 & \sqrt{\tau}\bar{\Gamma}_5 N_2 & \sqrt{\tau}\bar{\Gamma}_5 N_3 & \sqrt{h}\bar{\Gamma}_6 M_1 & \sqrt{h}\bar{\Gamma}_6 M_2 & \sqrt{h}\bar{\Gamma}_6 M_3 \\ * & -U_1 & 0 & 0 & 0 & 0 & 0 \\ * & * & -3U_1 & 0 & 0 & 0 & 0 \\ * & * & * & -5U_1 & 0 & 0 & 0 \\ * & * & * & * & -U_2 & 0 & 0 \\ * & * & * & * & * & -3U_2 & 0 \\ * & * & * & * & * & * & -5U_2 \end{bmatrix} < 0 \tag{13.27}$$

where

$$\bar{\Upsilon} = \bar{\Xi}_1 + \bar{\Xi}_2 + \bar{\Xi}_3 + \bar{\Xi}_4 + \bar{\Xi}_5 + \bar{\Xi}_6,$$

$$\bar{\Xi}_1 = \mathrm{Sym}\Big\{\bar{\Gamma}_1 P_1 \bar{\Gamma}_2^T + \bar{\Gamma}_3 P_2 \bar{\Gamma}_4^T\Big\},$$

$$\bar{\Xi}_2 = \bar{r}_1 R_1 \bar{r}_1^T - \bar{r}_2 R_1 \bar{r}_2^T + \bar{r}_5 R_2 \bar{r}_5 - \bar{r}_6 R_2 \bar{r}_6^T,$$
$$\bar{\Xi}_3 = \tau \bar{r}_{10} U_1 \bar{r}_{10}^T + h \bar{r}_{11} U_2 \bar{r}_{11}^T + \mathrm{Sym}\Big\{\bar{\Gamma}_5(N_1 \bar{\Pi}_1^T + N_2 \bar{\Pi}_2^T + N_3 \bar{\Pi}_3^T)$$
$$+ \bar{\Gamma}_6(M_1 \bar{\Pi}_4^T + M_2 \bar{\Pi}_5^T + M_3 \bar{\Pi}_6^T)\Big\},$$
$$\bar{\Xi}_4 = \mathrm{Sym}\Big\{\bar{r}_9 X(L\bar{r}_6^T - \bar{r}_9^T)\Big\},$$
$$\bar{\Xi}_5 = \mathrm{Sym}\Big\{\Big(\bar{r}_1 + \bar{r}_{10}\Big)\Big(-Y_1\bar{r}_{10}^T - Y_1 A \bar{r}_1^T + Y_1 B_1 \bar{r}_9^T + h Y_1 B_2 \bar{r}_7^T - J_1 G_v \bar{r}_1^T\Big)\Big\},$$
$$\bar{\Xi}_6 = \mathrm{Sym}\Big\{\Big(\bar{r}_5 + \bar{r}_{11}\Big)\Big(-Y_2\bar{r}_{11}^T - Y_2 C \bar{r}_5^T + Y_2 D_1 \bar{r}_2^T + \tau Y_2 D_2 \bar{r}_3^T - J_2 G_w \bar{r}_5^T\Big)\Big\},$$
$$\bar{\Gamma}_1 = \big[\bar{r}_1,\ \tau\bar{r}_3,\ \tau^2\bar{r}_4\big],$$
$$\bar{\Gamma}_2 = \big[\bar{r}_{10},\ \bar{r}_1 - \bar{r}_2,\ \bar{r}_1 - \tau\bar{r}_3\big],$$
$$\bar{\Gamma}_3 = \big[\bar{r}_5,\ h\bar{r}_7,\ h^2\bar{r}_8\big],$$
$$\bar{\Gamma}_4 = \big[\bar{r}_{11},\ \bar{r}_5 - \bar{r}_6,\ \bar{r}_5 - h\bar{r}_7\big],$$
$$\bar{\Gamma}_5 = \big[\bar{r}_1,\ \bar{r}_2,\ \bar{r}_3,\ \bar{r}_4\big],$$
$$\bar{\Gamma}_6 = \big[\bar{r}_5,\ \bar{r}_6,\ \bar{r}_7,\ \bar{r}_8\big],$$
$$\bar{\Pi}_1 = \bar{r}_1 - \bar{r}_2,$$
$$\bar{\Pi}_2 = \bar{r}_1 + \bar{r}_2 - 2\bar{r}_3,$$
$$\bar{\Pi}_3 = \bar{r}_1 - \bar{r}_2 - 6\bar{r}_3 + 6\bar{r}_4,$$
$$\bar{\Pi}_4 = \bar{r}_5 - \bar{r}_6,$$
$$\bar{\Pi}_5 = \bar{r}_5 + \bar{r}_6 - 2\bar{r}_7,$$
$$\bar{\Pi}_6 = \bar{r}_5 - \bar{r}_6 - 6\bar{r}_7 + 6\bar{r}_8,$$
$$J_1 = Y_1 K_1,$$
$$J_2 = Y_2 K_2.$$

And the state estimator gains can be calculated $K_1 = Y_1^{-1} J_1$ *and* $K_2 = Y_2^{-1} J_2$.

Proof Consider Lyapunov functional as:

$$V(t) = V_1(t) + V_2(t) + V_3(t), \tag{13.28}$$

where

$$V_1(t) = \bar{\eta}_1^T(t) P_1 \bar{\eta}_1 + \bar{\eta}_2^T(t) P_2 \bar{\eta}_2(t),$$
$$V_2(t) = \int_{t-\tau}^{t} v^T(s) R_1 v(s) dt + \int_{t-h}^{t} w^T(s) Q_2 w(s) dt,$$
$$V_3(t) = \int_{t-\tau}^{t}\int_{\theta}^{t} \dot{v}^T(s) U_1 \dot{v}(s) ds d\theta + \int_{t-h}^{t}\int_{\theta}^{t} \dot{w}^T(s) U_2 \dot{w}(s) ds d\theta,$$

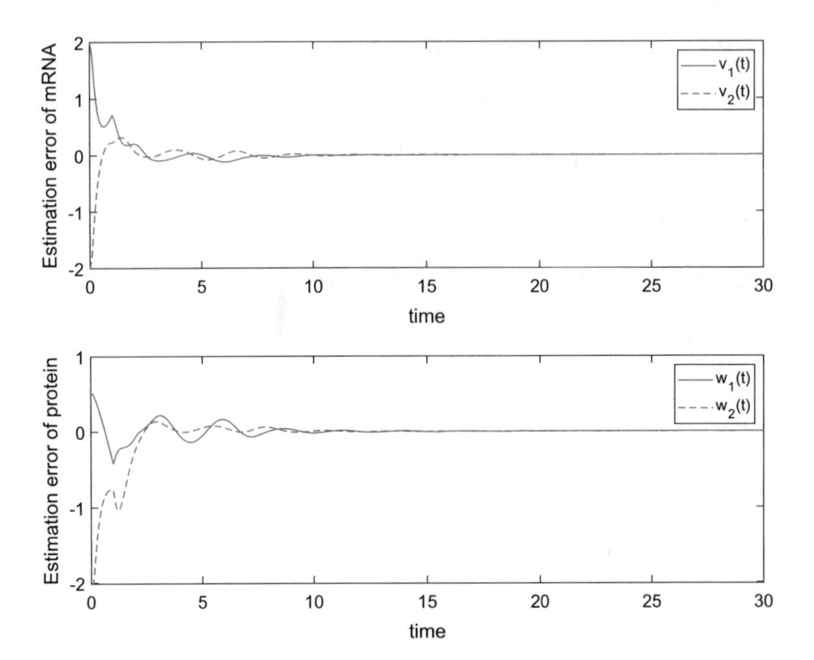

Fig. 13.3 The estimation error between the GRN (13.4) and estimator (13.5)

with

$$\bar{\eta}_1(t) = \left[v^T(t), \int_{t-\tau}^{t} v^T(s)ds, \int_{t-\tau}^{t} \int_{\theta}^{t} v^T(s)dsd\theta \right],$$
$$\bar{\eta}_2(t) = \left[w^T(t), \int_{t-h}^{t} w^T(s)ds, \int_{t-h}^{t} \int_{\theta}^{t} w^T(s)dsd\theta \right].$$

Then, as the same procedure with Theorem 13.1, Corollary 13.1 can be easily obtained. This completes the proof. ■

13.3 Numerical Examples

Two numerical examples are given in this section for checking the validity of the proposed theorems.

Example 1 For the first example, we consider a GRN (13.4) and state estimator (13.5) with the following parameters:

$$A = \begin{bmatrix} 2 & 0 \\ 0 & 2 \end{bmatrix}, \quad B_1 = \begin{bmatrix} 1 & -2 \\ 0.8 & 0 \end{bmatrix}, \quad B_2 = \begin{bmatrix} 1 & -2 \\ 0.8 & 0 \end{bmatrix},$$

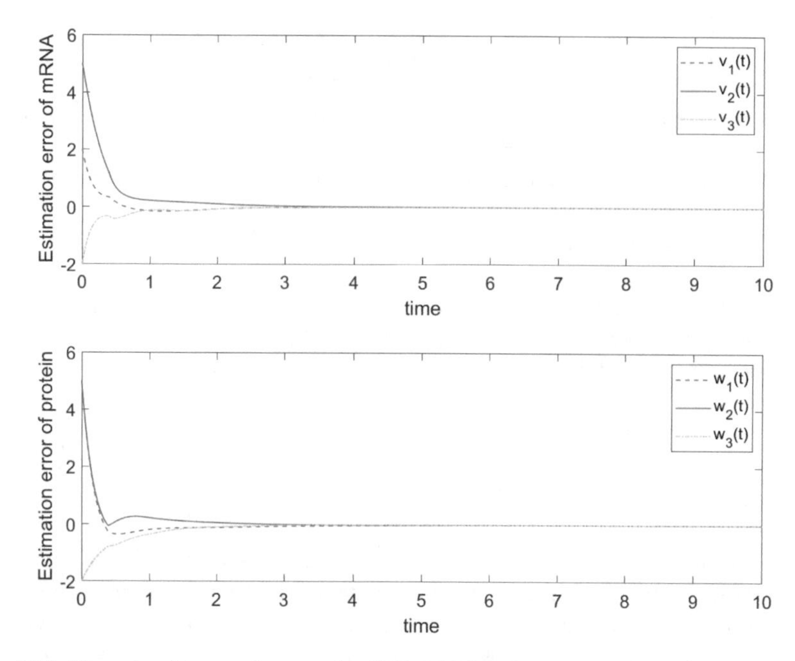

Fig. 13.4 The estimation error between the GRN (13.24) and estimator (13.25)

$$C = \begin{bmatrix} 2 & 0 \\ 0 & 2 \end{bmatrix}, \quad D_1 = \begin{bmatrix} 2 & 0 \\ 0 & 2 \end{bmatrix}, \quad D_2 = \begin{bmatrix} 2 & 0 \\ 0 & 2 \end{bmatrix},$$

$$G_v = \begin{bmatrix} 1 & 1 \end{bmatrix}, \quad G_w = \begin{bmatrix} -1 & 1 \end{bmatrix}, \quad f(a) = \frac{a^2}{1+a^2},$$

then, it is clear that $l_i = 0.65$.

For the simulation, let us define $\rho = 0.1$, $\sigma = 0.1$, $\tau = 1$, $h = 1$, and $x(0) = (2, -2)$, $y(0) = (0.5, -2)$, $\hat{x}(0) = (0, 0)$, $\hat{y}(t) = (0, 0)$, then the state estimator gains can be obtained by Theorem 13.1 as follows:

$$K_1 = \begin{bmatrix} 2.7186 & 1.2349 \end{bmatrix}^T, \quad K_2 = \begin{bmatrix} -0.9639 & 0.9317 \end{bmatrix}^T.$$

With the above gains, the estimation errors, i.e. $v(t)$ and $w(t)$, are given in Fig. 13.3 which proves the designed state estimator well follows the target system.

Example 2 In this example, we consider no leakage delay case, i.e. the GRN (13.24) and state estimator (13.25), with the following parameters:

$$A = \text{diag}\{3, 3.1, 2.9\}, \quad B_1 = \begin{bmatrix} 0 & 0 & -2.5 \\ -2.5 & 0 & 0 \\ 0 & -2.5 & 0 \end{bmatrix}, \quad B_2 = \begin{bmatrix} 0 & 0 & -2.6 \\ -2.6 & 0 & 0 \\ 0 & -2.6 & 0 \end{bmatrix},$$

$$C = \text{diag}\{2.6, 2.5, 2.4\}, \quad D_1 = \text{diag}\{0.8, 0.8, 0.8\}, \quad D_2 = \text{diag}\{1, 1, 1\},$$

$$G_v = \left[1\ 1\ 0\right], \quad G_w = \left[-1\ 0\ 1\right], \quad f(a) = \frac{a^2}{1+a^2}, \quad \tau = 0.4, \quad h = 0.4,$$

$$x(0) = (2, 5, -2), \quad y(0) = (5, 5, -2), \quad \hat{x}(0) = (0, 0, 0), \quad \hat{y}(t) = (0, 0, 0).$$

Then, the state estimator gains can be calculated by Corollary 13.1 as follows:

$$K_1 = \left[2.6905\ 2.6433\ -0.0464\right], \quad K_2 = \left[-0.9281\ 0.0869\ 1.2188\right].$$

Finally, the error trajectories of $v(t)$ and $w(t)$, are displayed in Fig. 13.4 which shows the error signals converse to zero which implies the states of designed estimator converse to the state of the GRN.

13.4 Conclusion

The state estimation problem of a GRN with leakage, constant, and distributed delays was investigated in this chapter. Based on augmented Lyapunov functional approach and a recent integral inequality, such criteria were proposed for designing state estimator. Two numerical examples were given to illustrate the validity of the proposed theorems.

References

1. Chaouiya C, Jong HD, Thieffry D (2006) Dynamical modeling of biological regulatory networks. BioSystems 84:77–80
2. Keller AD (1995) Model genetic circuits encoding autoregulatory transcription factors. J Theor Biol 172:169–185
3. Wang Z, Gao H, Cao J, Liu X (2008) On delayed genetic regulatory networks with polytopic uncertainties: robust stability analysis. IEEE Trans Nanobiosci 7:154–163
4. Noor A, Serpedin E, Nounou M, Nounou HN (2012) Inferring gene regulatory networks via nonlinear state-space models and exploiting sparsity. IEEE/ACM Trans Comput Biol Bioinform 9:1203–1211
5. Julius A, Zavlanos M, Boyd S, Pappas GJ (2009) Genetic network identification using convex programming. IET Syst Biol 3:155–166
6. Wang H, Qian J, Dougherty E (2010) Inference of gene regulatory networks using S-system: a unified approach. IET Syst Biol 4:145–156
7. Elowitz MB, Leibler S (2000) A synthetic oscillatory network of transcriptional regulators. Nature 403:335–338
8. Lee TH, Park MJ, Kwon OM, Park JH, Lee SM (2013) State estimation for genetic regulatory networks with time-varying delay using stochastic sampled-data. In: Proceedings of the 2013 9th Asian control conference
9. Sakthivel R, Mathiyalagan K, Lakshmanan S, Park JH (2013) Robust state estimation for discrete-time genetic regulatory networks with randomly occurring uncertainties. Nonlinear Dyn 74:1297–1315

10. Lakshmanan S, Park JH, Jung HY, Balasubramaniam P, Lee SM (2013) Design of state estimator for genetic regulatory networks with time-varying delays and randomly occurring uncertainties. Biosystems 111:51–70
11. Koo JH, Ji DH, Won SC, Park JH (2012) An improved robust delay-dependent stability criterion for genetic regulatory networks with interval time delays. Commun Nonlinear Sci Numer Simul 17:3399–3405
12. Jong H (2002) Modeling and simulation of genetic regulatory systems: a literature review. J Comput Biol 9:67–103
13. Han M, Liu Y, Tu YS (2014) Controllability of Boolean control networks with time delays both in states and inputs. Neurocomputing 129:467–475
14. Zheng Y, Li H, Ding X, Liu Y (2017) Stabilization and set stabilization of delayed boolean control networks based on trajectory stabilization. J Frankl Inst 354:7812–7827
15. Meng M, Lam J, Feng J, Cheung KC (2018) Stability and stabilization of boolean networks with stochastic delays. IEEE Trans Autom Control. https://doi.org/10.1109/TAC.2018.2835366
16. Yue D, Guan ZH, Li J, Liu F, Xiao JW, Ling G (2019) Stability and bifurcation of delay-coupled genetic regulatory networks with hub structure. J Frankl Inst 356:2847–2869
17. Li L, Yang Y (2015) On sampled-data control for stabilization of genetic regulatory networks with leakage delays. Neurocomputing 149:1225–1231
18. Lakshmanan S, Rihan FA, Rakkiyappan R, Park JH (2014) Stability analysis of the differential genetic regulatory networks model with time-varying delays and Markovian jumping parameters. Nonlinear Anal: Hybrid Syst 14:1–15
19. Lee TH, Lakshmanan S, Park JH, Balasubramaniam P (2013) State estimation for genetic regulatory networks with mode-dependent leakage and time-varying delays and Markovian jumping parameters. IEEE Trans Nanobiosci 12:363–375
20. Samidurai R, Rajavel S, Sriraman R, Cao J, Alsaedi A, Alsaadi FE (2017) Novel results on stability analysis of neutral-type neural networks with additive time-varying delay components and leakage delay. Int J Control, Autom Syst 15:1888–1900
21. Liu J, Xu R (2017) Global dissipativity analysis for memristor-based uncertain neural networks with time delay in the leakage term. Int J Control, Autom Syst 15:2406–2415
22. Xu C, Li P (2018) Periodic dynamics for memristor-based bidirectional associative memory neural networks with leakage delays and time-varying delays. Int J Control, Autom Syst 16:535–549
23. Ratnavelu K, Kalpana M, Balasubramaniam P (2016) Asymptotic stability of Markovian switching genetic regulatory networks with leakage and mode-dependent time delays. J Frankl Inst 353:1615–1638
24. Pandiselvi S, Raja R, Zhu Q, Rajchakit G (2018) A state estimation $\mathscr{H}_\infty$ issue for discrete-time stochastic impulsive genetic regulatory networks in the presence of leakage, multiple delays and Markovian jumping parameters. J Frankl Inst 355:2735–2761
25. Huang Z, Xia J, Wang J, Wei Y, Wang Z, Wang J (2019) Mixed $\mathscr{H}_\infty / \updownarrow_2 - \updownarrow_\infty$ state estimation for switched genetic regulatory networks subject to packet dropouts: A persistent dwell-time switching mechanism. Appl Math Comput 355:198–212
26. Li Q, Shen B, Liu Y, Alsaadi FE (2016) Event-triggered $\mathscr{H}_\infty$ state estimation for discrete-time stochastic genetic regulatory networks with Markovian jumping parameters and time-varying delays. Neurocomputing 174:912–920
27. Vembarasan V, Nagamani G, Balasubramaniam P, Park JH (2013) State estimation for delayed genetic regulatory networks based on passivity theory. Math Biosci 244:165–175
28. Zeng HB, He Y, Wu M, She J (2015) New results on stability analysis for systems with discrete distributed delay. Automatica 60:189–192

Chapter 14
Secure Communication Based on Synchronization of Uncertain Chaotic Systems with Propagation Delays

14.1 Introduction

Dynamic systems carry large amounts of signals and information. In particular, the development of IT technology makes modern dynamic systems naturally connected to communication networks. Recently, a communication channel is strongly demanded to have high security for the protection of privacy and personal information. To develop effective encryption and decryption technology have naturally arisen for various communication circumstances [1–12].

On the other hand, chaos is a natural nonlinear phenomenon and has abundant properties such as the sensitive dependence on initial conditions and system parameters, no periodicity and topological transitivity, and pseudo random property. Among some research topics in chaotic systems, a hot issue in recent years is the chaotic synchronization. As an application of utilization of the chaotic synchronization, one of the useful techniques to improve security in the communication channel is secure communication [13–22]. Chaotic synchronization was firstly reported by Pecora and Carroll for two identical chaotic systems [23]. After the pioneering work chaotic synchronization widely investigated in various fields [23–27]. Therefore, secure communication based on chaotic synchronization utilizes the basic feature of chaotic systems that is sensitivity to initial conditions and parameters. Since the chaotic systems would have totally different dynamic behaviors according to the tiny difference of the initial values and parameters of the system, we cannot recover original signals in secure communication based on chaotic synchronization, if we don't know exact initial values and parameters. Therefore, secure communication based on chaotic synchronization is a hot topic in the field of security systems. In [10], $\mathscr{H}_\infty$ observer-based periodic event-triggered controller for the control of a class of cyber-physical systems was designed under consideration of DoS attacks where the designed controller maximized the frequency and duration of the DoS attacks. In [11], distributed event-triggered $\mathscr{H}_\infty$ filters were designed on sensor networks in the presence of sensor saturations and randomly occurring cyber-attacks. In [12], the consensus problem of nonlinear multi-agent systems was studied by using

J. H. Park et al., *Dynamic Systems with Time Delays: Stability and Control*,
https://doi.org/10.1007/978-981-13-9254-2_14

sampled agents' states and considering cyber-attacks on the connectivity of the network topology but is recoverable.

Transmission delay or propagation delay is a natural issue in exchange of signals and this cause poor performance and instability of the system, and fail of recovering the message. Therefore, it is necessary to consider propagation delays in signal transmission for the reliability of security systems. There are two types of propagation delays, i.e. constant and time-varying propagation delay. In this chapter, we study both cases of propagation delays. Firstly, a constant propagation delay is dealt by a Wirtinger-based multiple integral inequality, and then, a new Lyapunov functional constructed based on matrix-refined-function for time-varying propagation delay case.

14.2 Problem Formulation

For the investigation of secure communication based on chaotic synchronization, we first consider the following chaotic secure communication system with constant propagation delay in the signal transmission:

$$Transmitter: \begin{cases} \dot{x}(t) = (A + A_{\Delta}(t))x(t) + (B + B_{\Delta}(t))f(x(t)) - Km_e(t) \\ v_x(t) = Cx(t-h) + m_e(t) \end{cases} \tag{14.1}$$

$$Receiver: \begin{cases} \dot{y}(t) = (A + A_{\Delta}(t))y(t) + (B + B_{\Delta}(t))f(y(t)) + u(t) \\ v_y(t) = Cy(t-h) \end{cases} \tag{14.2}$$

$$Controller: u(t) = K(v_y(t) - v_x(t)) \tag{14.3}$$

where $x(t) \in \mathbb{R}^n$, and $y(t) \in \mathbb{R}^n$ are the state vectors, $u(t) \in \mathbb{R}^n$ is control input, $v_x(t) \in \mathbb{R}$, and $v_y(t) \in \mathbb{R}$ are the output of transmitter and receiver, respectively, $m_e(t) \in \mathbb{R}$ is the encrypted message signal. $A, B \in \mathbb{R}^{n \times n}$ and $C \in \mathbb{R}^{1 \times n}$ are known system matrices, $K \in \mathbb{R}^n$ is the controller gain to be determined, h is a constant delay satisfying $h > 0$, $f : \mathbb{R}^n \to \mathbb{R}^n$ is a nonlinear function which satisfies the following global Lipschitz condition:

$$f(a) - f(b) = l\|a - b\|, \quad \forall a, b \in \mathbb{R}^n, \tag{14.4}$$

for a positive scalar l. And, $A_{\Delta}(t)$ and $B_{\Delta}(t)$ are the uncertainties of system matrices of the form

$$[A_{\Delta}(t) \;\; B_{\Delta}(t)] = DG(t)[E_a \;\; E_b], \tag{14.5}$$

where D, E_a, E_b are known constant matrices and the time-varying nonlinear function $G(t)$ satisfies

$$G^T(t)G(t) \leq I, \quad \forall t \geq 0. \tag{14.6}$$

Assume the original message signal $m_o(t)$ encrypted by N-shift cipher [17] and the encrypted message signal $m_e(t)$ is transmitted to receiver. So, we define encryption key signal as $k_e(t) = \sum_{i=1}^{n} a_i x_i(t)$ and decryption key signal as $k_d(t) = \sum_{i=1}^{n} a_i y_i(t)$ in which $a_i (i = 1, \ldots, n)$ are constants and

$$m_e(t) = \underbrace{En(\cdots(En(En}_{N}(m_0(t), k_e(t)), k_e(t)), \ldots), k_e(t)), \tag{14.7}$$

where the nonlinear function, $En(\cdot, \cdot)$, is defined as

$$En(a, b) = \begin{cases} (a+b) + 2\tau, & -2\tau \leq (a+b) \leq -\tau \\ (a+b), & -\tau < (a+b) < \tau \\ (a+b) - 2\tau, & \tau \leq (a+b) \leq 2\tau \end{cases} \tag{14.8}$$

with τ is a positive constant satisfying $-\tau < m_o(t) < \tau$ and $-\tau < k_e(t) < \tau$.

It is clear that the synchronization between the chaotic systems in transmitter and receiver ensures $k_e(t) \approx k_d(t)$. Then, we can calculate the recovered message signal $m_r(t)$ by using received message signal $m_e(t)$ and decryption key signal $k_d(t)$ by the following decryption function

$$m_r(t) = En(\cdots(En(En(m_e(t), -k_d(t)), -k_d(t)), \cdots), -k_d(t)). \tag{14.9}$$

Therefore, the main aim of this chapter is designing a control gain K such that the state of a chaotic system in the receiver, $y(t)$, synchronized up to the state of the chaotic system in the transmitter, $x(t)$.

Now, let us define the synchronization error $e(t)$ as $e(t) = y(t) - x(t)$, then, the error dynamics can be

$$\dot{e}(t) = (A + A_\Delta(t))e(t) + (B + B_\Delta(t))\bar{f}(e(t)) + KCe(t-h), \tag{14.10}$$

where $\bar{f}(e(t)) = f(y(t)) - f(x(t))$.

Here, we can rewrite the system (14.10) as follows:

$$\begin{cases} \dot{e}(t) = Ae(t) + KCe(t-h) + B\bar{f}(e(t)) + Dp(t), \\ p(t) = G(t)q(t), \\ q(t) = E_a e(t) + E_b \bar{f}(e(t)). \end{cases} \tag{14.11}$$

14.3 Design of Controller for Secure Communication with Propagation Delay

This section consists of two subsections, the first subsection considers a constant propagation delay and the second one deal with time-varying propagation delay. In each subsection, we design the feedback controller for secure communication which guarantees robust synchronization between uncertain chaotic systems in the transmitter (14.1) and the receiver (14.2).

The following lemmas are necessary to derive theorems.

Lemma 14.1 (Wirtinger-based multiple integral inequality [28–30]) *For given a constant l, a matrix $W \in \mathbb{S}_+^n$, and all continuous function x in $[a, b] \to \mathbb{R}^n$ the following inequality holds:*

$$G_l(x, a, b, W) \geq \frac{(l+1)!}{(b-a)^{l+1}} g_l(x, a, b) W g_l(x, a, b) + \frac{(l!)(l+3)}{(b-a)^{l+1}} \Upsilon_l(x, a, b)^T W \Upsilon_l(x, a, b),$$

where

$$G_l(x, a, b, W) = \int_a^b \int_{v_1}^b \cdots \int_{v_l}^b x^T(s) W x(s) ds dv_l \cdots dv_1,$$

$$G_0(x, a, b, W) = \int_a^b x^T(s) W x(s) ds,$$

$$g_l(x, a, b) = \int_a^b \int_{v_1}^b \cdots \int_{v_l}^b x(s) ds dv_l \cdots dv_1,$$

$$g_0(x, a, b) = \int_a^b x(s) ds,$$

$$\Upsilon_l(x, a, b) = g_l(x, a, b) - \frac{(l+2)}{b-a} g_{l+1}(x, a, b).$$

Lemma 14.2 (Lower bound lemma for reciprocal convexity [31]) *Let $f_1, f_2, \ldots, f_N$: $\mathbf{R}^m \mapsto \mathbf{R}$ have positive values in an open subset $\mathbf{D}$ of $\mathbf{R}^m$. Then, the reciprocally convex combination of f_i over $\mathbf{D}$ satisfies*

$$\min_{\alpha_i | \alpha_i > 0, \sum_{i=1}^N \alpha_i = 1} \sum_{i=1}^N \frac{1}{\alpha_i} f_i(t) = \sum_{i=1}^N f_i(t) + \max_{d_{ij}(t)} \sum_{i \neq j}^N d_{ij}(t)$$

subject to

$$\left\{ d_{ij} : \mathbf{R}^m \mapsto \mathbf{R}, d_{ji}(t) \triangleq d_{ij}(t), \begin{bmatrix} f_i(t) & d_{ij}(t) \\ d_{ij}(t) & f_j(t) \end{bmatrix} \geq 0 \right\}.$$

14.3.1 *Synchronization of Uncertain Chaotic System with Constant Propagation Delay*

For the system (14.11), we have the following theorem.

Theorem 14.1 *For given positive constants h, l, ρ, known matrices E_a, E_b, $D \in \mathbb{R}^{n\times n}$, the feedback controller (14.3) guarantees robust synchronization between chaotic systems in transmitter (14.1) and receiver (14.2), if there exist positive scalars α, β, matrices $P \in \mathbb{S}_+^{3n}$, $Q_1, Q_2, R_1, R_2 \in \mathbb{S}_+^n$, $H \in \mathbb{R}^{n\times n}$, $J \in \mathbb{R}^{n\times 1}$ satisfying the following LMI:*

$$\begin{bmatrix} \Omega_1 & \Omega_2^T \\ * & -\beta I \end{bmatrix} < 0, \tag{14.12}$$

where

$$\Omega_1 = \begin{bmatrix} \Pi_1 & JC - P_2 - 2R_1 & \frac{6}{h}R_1 - P_3 + P_4^T + hP_5^T & P_5 + hP_6^T + 3R_2 & P_1 - H - \rho A^T H^T & HB & HD \\ * & \frac{6}{h}R_1 - Q_1 - 4R_1 & -P_4^T & -P_5 & \rho C^T J^T & 0 & 0 \\ * & * & \Pi_2 & \frac{6}{h}(Q_2 + R_2) - P_6^T & R_2^T & 0 & 0 \\ * & * & * & -\frac{12}{h^2}Q_2 - \frac{18}{h^2}R_2 & P_3^T & 0 & 0 \\ * & * & * & * & h^2R_1 + \frac{h^4}{4}R_2 - \rho(H + H^T) & \rho HB & \rho HD \\ * & * & * & * & * & -\alpha I & 0 \\ * & * & * & * & * & * & -\beta I \end{bmatrix},$$

$$\begin{aligned}
\Omega_2 &= \begin{bmatrix} \beta E_a \; 0 \; 0 \; 0 \; 0 \; \beta E_b \; 0 \end{bmatrix}, \\
\Pi_1 &= P_2 + P_2^T + h(P_3 + P_3^T) + Q_1 + h^2 Q_2 - 4R_1 - \frac{3h^2}{2}R_2 + \alpha L, \\
&\quad + HA + A^T H^T, \\
\Pi_2 &= -P_5 - P_5^T - 4Q_2 - \frac{12}{h^2}R_1 - 3R_2, \\
L &= l^2 I_n.
\end{aligned}$$

Also, the desired control gain matrix is given by $K = H^{-1}J$.

Proof Let us define the following Lyapunv functional candidate:

$$V(t) = V_1(t) + V_2(t) + V_3(t), \tag{14.13}$$

where

$$\begin{aligned} V_1(t) &= \eta_1^T(t) P \eta_1(t), \\ V_2(t) &= \int_{t-h}^{t} e^T(s) Q_1 e(s) ds + h \int_{t-h}^{t} \int_{v}^{t} e^T(s) Q_2 e(s) ds dv, \\ V_3(t) &= h \int_{t-h}^{t} \int_{v}^{t} \dot{e}^T(s) R_1 \dot{e}(s) ds dv + \frac{h^2}{2} \int_{t-h}^{t} \int_{r}^{t} \int_{v}^{t} \dot{e}^T(s) R_2 \dot{e}(s) ds dv dr, \end{aligned}$$

with

$$\eta_1^T(t) = \left[e^T(t),\ \int_{t-h}^{t} e^T(s) ds,\ \int_{t-h}^{t} \int_{v}^{t} e^T(s) ds dv \right],$$

$$P = \begin{bmatrix} P_1 & P_2 & P_3 \\ * & P_4 & P_5 \\ * & * & P_6 \end{bmatrix}.$$

Time derivative of Lyapunov functional (14.13) can be

$$\dot{V}_1(t) = 2\eta_1^T(t) P \eta_2(t), \tag{14.14}$$

$$\begin{aligned} \dot{V}_2(t) &= e^T(t) Q_1 e(t) - e^T(t-h) Q_1 e(t-h) + h^2 e^T(t) Q_2 e(t) \\ &\quad - h \int_{t-h}^{t} e^T(s) Q_2 e(s) ds, \end{aligned} \tag{14.15}$$

$$\begin{aligned} \dot{V}_3(t) &= h^2 \dot{e}^T(t) R_1 \dot{e}(t) - h \int_{t-h}^{t} \dot{e}^T(s) R_1 \dot{e}(s) ds + \frac{h^4}{4} \dot{e}^T(t) R_2 \dot{e}(t) \\ &\quad - \frac{h^2}{2} \int_{t-h}^{t} \int_{v}^{t} \dot{e}^T(s) R_2 \dot{e}(s) ds dv, \end{aligned} \tag{14.16}$$

where

$$\eta_2^T(t) = \left[\dot{e}^T(t),\ e^T(t) - e^T(t-h),\ h e^T(t) - \int_{t-h}^{t} e^T(s) ds \right].$$

By Lemma 14.1, integral terms in Eqs. (14.15) and (14.16) can be estimated as

$$\begin{aligned} &- h \int_{t-h}^{t} e^T(s) Q_2 e(s) ds \\ &\le - \begin{bmatrix} \int_{t-h}^{t} e(s) ds \\ \int_{t-h}^{t} e(s) ds - \frac{2}{h} \int_{t-h}^{t} \int_{v}^{t} e(s) ds dv \end{bmatrix}^T \begin{bmatrix} Q_2 & 0 \\ 0 & 3Q_2 \end{bmatrix} \end{aligned}$$

$$\times \begin{bmatrix} \int_{t-h}^{t} e(s)ds \\ \int_{t-h}^{t} e(s)ds - \frac{2}{h}\int_{t-h}^{t}\int_{v}^{t} e(s)dsdv \end{bmatrix}, \tag{14.17}$$

$$\begin{aligned} &-h\int_{t-h}^{t} \dot{e}^T(s)R_1\dot{e}(s)ds \\ &\leq -\begin{bmatrix} \int_{t-h}^{t} \dot{e}(s)ds \\ \int_{t-h}^{t} \dot{e}(s)ds - \frac{2}{h}\int_{t-h}^{t}\int_{v}^{t} \dot{e}(s)dsdv \end{bmatrix}^T \begin{bmatrix} R_1 & 0 \\ 0 & 3R_1 \end{bmatrix} \\ &\quad\times \begin{bmatrix} \int_{t-h}^{t} \dot{e}(s)ds \\ \int_{t-h}^{t} \dot{e}(s)ds - \frac{2}{h}\int_{t-h}^{t}\int_{v}^{t} \dot{e}(s)dsdv \end{bmatrix} \\ &= -\begin{bmatrix} e(t) - e(t-h) \\ e(t) + e(t-h) - \frac{2}{h}\int_{t-h}^{t} e(s)ds \end{bmatrix}^T \begin{bmatrix} R_1 & 0 \\ 0 & 3R_1 \end{bmatrix} \\ &\quad\times \begin{bmatrix} e(t) - e(t-h) \\ e(t) + e(t-h) - \frac{2}{h}\int_{t-h}^{t} e(s)ds \end{bmatrix}, \end{aligned} \tag{14.18}$$

$$\begin{aligned} &-\frac{h^2}{2}\int_{t-h}^{t}\int_{v}^{t} \dot{e}^T(s)R_2\dot{e}(s)dsdv \\ &\leq -\begin{bmatrix} \int_{t-h}^{t}\int_{v}^{t} \dot{e}(s)dsdv \\ \int_{t-h}^{t}\int_{v}^{t} \dot{e}(s)dsdv - \frac{3}{h}\int_{t-h}^{t}\int_{r}^{t}\int_{v}^{t} \dot{e}(s)dsdvdr \end{bmatrix}^T \begin{bmatrix} R_2 & 0 \\ 0 & 2R_2 \end{bmatrix} \\ &\quad\times \begin{bmatrix} \int_{t-h}^{t}\int_{v}^{t} \dot{e}(s)dsdv \\ \int_{t-h}^{t}\int_{v}^{t} \dot{e}(s)dsdv - \frac{3}{h}\int_{t-h}^{t}\int_{r}^{t}\int_{v}^{t} \dot{e}(s)dsdvdr \end{bmatrix} \\ &= -\begin{bmatrix} he(t) - \int_{t-h}^{t} e(s)ds \\ \frac{h}{2}e(t) + \int_{t-h}^{t} e(s)ds - \frac{3}{h}\int_{t-h}^{t}\int_{v}^{t} e(s)dsdv \end{bmatrix}^T \begin{bmatrix} R_2 & 0 \\ 0 & 2R_2 \end{bmatrix} \\ &\quad\times \begin{bmatrix} he(t) - \int_{t-h}^{t} e(s)ds \\ \frac{h}{2}e(t) + \int_{t-h}^{t} e(s)ds - \frac{3}{h}\int_{t-h}^{t}\int_{v}^{t} e(s)dsdv \end{bmatrix}. \end{aligned} \tag{14.19}$$

From Eqs. (14.4)–(14.6) and (14.11), for given ρ, positive constants α, β, and any matrix $H \in \mathbb{R}^{n\times n}$, the followings are obtained

$$0 \leq \alpha(e^T(t)Le(t) - f^T(e(t))f(e(t))), \tag{14.20}$$

$$0 \leq \beta(E_a e(t) + E_b f(e(t)))^T(E_a e(t) + E_b f(e(t))) - \beta p^T(t)p(t), \tag{14.21}$$

$$0 = 2[e^T(t) + \rho\dot{e}^T(t)]H[-\dot{e}(t) + Ae(t) + Bf(e(t)) + KCe(t-h) + Dp(t)]. \tag{14.22}$$

By using Eqs. (14.17)–(14.19) and adding Eqs. (14.20)–(14.22) to $\dot{V}(t)$, we can obtain

$$\dot{V}(t) \leq \zeta^T(t)(\Omega_1 + \beta\Omega_2^T\Omega_2)\zeta(t), \tag{14.23}$$

where Ω_1 and Ω_2 are defined in Theorem 14.1 and

$$\zeta^T(t) = \Big[e^T(t),\ e^T(t-h),\ \int_{t-h}^{t} e^T(s)ds,\ \int_{t-h}^{t}\int_{v}^{s} e^T(s)dsdv,\ \dot{e}^T(t), \\ f^T(e(t)),\ p^T(t) \Big].$$

Then, by Schur complement, $\Omega_1 + \beta\Omega_2^T\Omega_2 < 0$ is equivalent to the LMI (14.12). Therefore, if LMI (14.12) holds, then, the error system is asymptotically stable which implies that the synchronization between chaotic systems in the transmitter (14.1) and the receiver (14.2) is achieved by the controller. This completes the proof. ■

Without uncertainties, the transmitter and receiver can be rewritten as

$$Transmitter: \begin{cases} \dot{x}(t) = Ax(t) + Bf(x(t)) - Km_e(t) \\ v_x(t) = Cx(t-h) + m_e(t) \end{cases} \tag{14.24}$$

$$Receiver: \begin{cases} \dot{y}(t) = Ay(t) + Bf(y(t)) + u(t) \\ v_y(t) = Cy(t-h) \end{cases} \tag{14.25}$$

then we propose the following theorem.

Corollary 14.1 *For given positive constants h, l, ρ, the feedback controller (14.3) guarantees robust synchronization between chaotic systems in transmitter (14.24) and receiver (14.25), if there exist a positive constant α, matrices $P \in \mathbb{S}_{+}^{5n}$, Q_1, Q_2, R_1, $R_2 \in \mathbb{S}_{+}^{n}$, $H \in \mathbb{R}^{n\times n}$, $J \in \mathbb{R}^{n\times 1}$ satisfying the following LMI:*

$$\begin{bmatrix} \Pi_1 & JC - P_2 - 2R_1 & \frac{6}{h}R_1 - P_3 + P_4^T + hP_5^T & P_5 + hP_6^T + 3R_2 & P_1 - H - \rho A^T H^T & HB \\ * & \frac{6}{h}R_1 - Q_1 - 4R_1 & -P_4^T & -P_5 & \rho C^T J^T & 0 \\ * & * & \Pi_2 & \frac{6}{h}(Q_2 + R_2) - P_6^T & R_2^T & 0 \\ * & * & * & -\frac{12}{h^2}Q_2 - \frac{18}{h^2}R_2 & P_3^T & 0 \\ * & * & * & * & h^2R_1 + \frac{h^4}{4}R_2 - \rho(H + H^T) & \rho HB \\ * & * & * & * & * & -\alpha I \end{bmatrix} < 0, \tag{14.26}$$

where notations are the same as Theorem 14.1. Also, the desired control gain matrix is given by $K = H^{-1}J$.

Proof It can be easily proven by following the same procedure of Theorem 14.1 without (14.21), so it is omitted here. ■

14.3.2 Synchronization of Uncertain Chaotic Systems with Time-Varying Propagation Delay

In this subsection, we consider the time-varying propagation delay, $h(t)$, which satisfying $0 \le h(t) \le h$ and $\dot{h}(t) \le \mu \le 1$. Then, the transmitter and receiver can be rewritten as

$$Transmitter : \begin{cases} \dot{x}(t) = (A + A_{\Delta}(t))x(t) + (B + B_{\Delta}(t))f(x(t)) - Km_e(t) \\ v_x(t) = Cx(t - h(t)) + m_e(t) \end{cases} \tag{14.27}$$

$$Receiver : \begin{cases} \dot{y}(t) = (A + A_{\Delta}(t))y(t) + (B + B_{\Delta}(t))f(y(t)) + u(t) \\ v_y(t) = Cy(t - h(t)) \end{cases} \tag{14.28}$$

and error dynamics can be

$$\begin{cases} \dot{e}(t) = \; Ae(t) + KCe(t - h(t)) + B\bar{f}(e(t)) + Dp(t), \\ p(t) = G(t)q(t), \\ q(t) = \; E_a e(t) + E_b \bar{f}(e(t)). \end{cases} \tag{14.29}$$

Theorem 14.2 *For given positive constants h, μ, l, ρ, known matrices E_a, E_b, $D \in \mathbb{R}^{n\times n}$, the feedback controller (14.3) guarantees robust synchronization between chaotic systems in transmitter (14.27) and receiver (14.28), if there exist positive scalars α, β, matrices $P \in \mathbb{S}_{+}^{5n}$, $Q_1, Q_2 \in \mathbb{S}_{+}^{2n}$, R, W_1, $W_2 \in \mathbb{S}_{+}^{n}$, $S \in \mathbb{R}^{2n\times 2n}$, $H \in \mathbb{R}^{n\times n}$, $J \in \mathbb{R}^{n\times 1}$ satisfying the following LMIs: for $h(t) \in \{0, h\}$ and $\dot{h}(t) \in \{-\mu, \mu\}$*

$$\begin{bmatrix} \Upsilon_{1,[\dot{h}(t)]} & \Upsilon_2^T \\ * & -\beta I \end{bmatrix} < 0, \tag{14.30}$$

$$\begin{bmatrix} \mathrm{diag}\{R, 3R\} & S \\ * & \mathrm{diag}\{R, 3R\} \end{bmatrix} > 0, \tag{14.31}$$

where

$$\Upsilon_{1,[h(t),\dot{h}(t)]} = \begin{bmatrix} \Gamma_{1,[h(t)]} & \Gamma_{2,[\dot{h}(t)]} & \Gamma_3 & h(t)P_{44}^T + 6R & h(t)P_{45} + 2(S_2 + S_4) \\ * & \Gamma_{5,[\dot{h}(t)]} & \Gamma_{6,[\dot{h}(t)]} & \Gamma_{7,[h(t),\dot{h}(t)]} & \Gamma_{8,[h(t),\dot{h}(t)]} \\ * & * & \Gamma_{10} & -h(t)P_{45}^T - 2S_3^T + 2S_4^T & 6R - (h - h(t))P_{55}^T \\ * & * & * & -12R & -4S_4 \\ * & * & * & * & -12R \\ * & * & * & * & * \\ * & * & * & * & * \\ * & * & * & * & * \\ * & * & * & * & * \\ * & * & * & * & * \end{bmatrix}$$

$$
\left.\begin{array}{ccccc}
\Gamma_4 & h_D(t)(P_{12}+W_1) & P_{13} & HB & HD \\
P_{12}^T+W_1+\rho C^T J^T & \Gamma_9 & P_{23}+W_2 & 0 & 0 \\
P_{13}^T & h_D(t)(P_{23}^T+W_2) & P_{33}-Q_{22}-W_2 & 0 & 0 \\
P_{14}^T & h_D(t)P_{24}^T & P_{34}^T & 0 & 0 \\
P_{15}^T & h_D(t)P_{25}^T & P_{35}^T & 0 & 0 \\
\Gamma_{11} & 0 & 0 & \rho HB & \rho HD \\
* & h_D(t)(-Q_{13}-hW_1+hW_2) & 0 & 0 & 0 \\
* & * & -Q_{23}-hW_2 & 0 & 0 \\
* & * & * & -\alpha I & 0 \\
* & * & * & * & -\beta I
\end{array}\right],
$$

$$
\begin{aligned}
\Upsilon_2 &= \left[\beta E_a\ 0\ 0\ 0\ 0\ 0\ 0\ 0\ \beta E_b\ 0\right], \\
\Gamma_{1,[h(t)]} &= P_{14}+P_{14}^T+Q_{11}+Q_{21}-4R+\alpha L+HA+A^T H^T, \\
\Gamma_{2,[\dot{h}(t)]} &= h_D(t)(-P_{14}+P_{15})+P_{24}^T-2R-S_1-S_2-S_3-S_4+JC, \\
\Gamma_3 &= -P_{15}+P_{34}^T+S_1-S_2+S_3-S_4, \\
\Gamma_4 &= P_{11}+Q_{12}+Q_{22}-W_1-H+\rho A^T H^T, \\
\Gamma_{5,[\dot{h}(t)]} &= h_D(t)(P_{25}+P_{25}^T-P_{24}-P_{24}^T-Q_{11})-8R+S_1+S_2-S_3-S_4 \\
&\qquad +S_1^T+S_2^T-S_3^T-S_4^T, \\
\Gamma_{6,[\dot{h}(t)]} &= -P_{25}+h_D(t)(P_{35}^T-P_{34}^T)-S_1+S_2+S_3-S_4-2R, \\
\Gamma_{7,[h(t),\dot{h}(t)]} &= h(t)h_D(t)(P_{45}^T-P_{44}^T)+6R+2(S_3^T+S_4^T), \\
\Gamma_{8,[h(t),\dot{h}(t)]} &= h_D(t)((h-h(t))P_{55}^T-h(t)P_{45}^T)+2(S_4-S_2)+6R, \\
\Gamma_9 &= h_D(t)(P_{22}-Q_{12}-W_1-W_2), \\
\Gamma_{10} &= -P_{35}-P_{35}^T-Q_{21}-4R, \\
\Gamma_{11} &= Q_{13}+Q_{23}+h^2 R+hW_1-\rho(H+H^T), \\
h_D(t) &= 1-\dot{h}(t).
\end{aligned}
$$

Also, the desired control gain matrix is given by $K = H^{-1}J$.

Proof Consider the following Lyapunv functional candidate:

$$
\bar{V}(t) = \bar{V}_1(t)+\bar{V}_2(t)+\bar{V}_3(t)+\bar{V}_4(t), \tag{14.32}
$$

where

$$
\begin{aligned}
\bar{V}_1(t) &= \bar{\eta}_1^T(t)P\bar{\eta}_1(t), \\
\bar{V}_2(t) &= \int_{t-h(t)}^{t} \bar{\eta}_2^T(s)Q_1\bar{\eta}_2(s)ds + \int_{t-h}^{t-h(t)} \bar{\eta}_2^T(s)Q_2\bar{\eta}_2(s)ds, \\
\bar{V}_3(t) &= h\int_{t-h}^{t}\int_{v}^{t} \dot{e}^T(s)R\dot{e}(s)dsdv,
\end{aligned}
$$

$$\bar{V}_4(t) = h\left(\int_{t-h(t)}^{t} \dot{e}^T(s)W_1\dot{e}(s)ds + \int_{t-h}^{t-h(t)} \dot{e}^T(s)W_2\dot{e}(s)ds\right)$$
$$- (e(t) - e(t-h(t)))^T W_1(e(t) - e(t-h(t)))$$
$$- (e(t-h(t)) - e(t-h))^T W_2(e(t-h(t)) - e(t-h)),$$

with

$$\bar{\eta}_1^T(t) = \left[e^T(t),\ e^T(t-h(t)),\ e^T(t-h),\ \textstyle\int_{t-h(t)}^{t} e^T(s)ds,\ \int_{t-h}^{t-h(t)} e^T(s)ds\right],$$
$$\bar{\eta}_2^T(t) = \left[e^T(t)\ \dot{e}^T(t)\right],$$
$$P = \begin{bmatrix} P_{11} & P_{12} & P_{13} & P_{14} & P_{15} \\ * & P_{22} & P_{23} & P_{24} & P_{25} \\ * & * & P_{33} & P_{34} & P_{35} \\ * & * & * & P_{44} & P_{45} \\ * & * & * & * & P_{55} \end{bmatrix}.$$

We firstly show the positiveness of Lyapunov functional $V_4(t)$ because it contains negative terms.

By Jensen's inequality [32] and similar way to [33], it is clear that

$$\bar{V}_4(t) \geq h(t)\int_{t-h(t)}^{t} \dot{e}^T(s)W_1\dot{e}(s)ds + (h - h(t))\int_{t-h}^{t-h(t)} \dot{e}^T(s)W_2\dot{e}(s)ds$$
$$- (e(t) - e(t-h(t)))^T W_1(e(t) - e(t-h(t)))$$
$$- (e(t-h(t)) - e(t-h))^T W_2(e(t-h(t)) - e(t-h))$$
$$\geq \int_{t-h(t)}^{t} \dot{e}^T(s)ds W_1 \int_{t-h(t)}^{t} \dot{e}(s)ds + \int_{t-h}^{t-h(t)} \dot{e}^T(s)ds W_2 \int_{t-h}^{t-h(t)} \dot{e}(s)ds$$
$$- (e(t) - e(t-h(t)))^T W_1(e(t) - e(t-h(t)))$$
$$- (e(t-h(t)) - e(t-h))^T W_2(e(t-h(t)) - e(t-h))$$
$$= 0.$$

Therefore, the considered Lyapunov functional is positive function.

Time derivative of Lyapunov functional (14.32) can be

$$\dot{\bar{V}}_1(t) = 2\bar{\eta}_1^T(t)P\bar{\eta}_3(t), \tag{14.33}$$

$$\dot{\bar{V}}_2(t) = \bar{\eta}_2^T(t)(Q_1 + Q_2)\bar{\eta}_2(t) - h_D(t)\bar{\eta}_2^T(t-h(t))Q_1\bar{\eta}_2(t-h(t))$$
$$- \bar{\eta}_2^T(t-h)Q_2\bar{\eta}_2(t-h), \tag{14.34}$$

$$\dot{\bar{V}}_3(t) = h^2\dot{e}^T(t)R\dot{e}(t) - h\int_{t-h}^{t} \dot{e}^T(s)R\dot{e}(s)ds, \tag{14.35}$$

$$\dot{\bar{V}}_4(t) = h\dot{e}^T(t)W_1\dot{e}(t) - h_D(t)h\dot{e}^T(t-h(t))W_1\dot{e}(t-h(t))$$

$$
\begin{aligned}
&- 2(e(t) - e(t-h(t)))^T W_1(\dot{e}(t) - h_D(t)\dot{e}(t-h(t))) \\
&+ h_D(t)h\dot{e}^T(t-h(t))W_2\dot{e}(t-h(t)) - h\dot{e}^T(t-h)W_2\dot{e}(t-h) \\
&- 2(e(t-h(t)) - e(t-h))^T W_2(h_D(t)\dot{e}(t-h(t)) - \dot{e}(t-h)),
\end{aligned} \tag{14.36}
$$

where

$$
\begin{aligned}
\bar{\eta}_3^T(t) = \Big[&\dot{e}^T(t),\ h_D(t)\dot{e}^T(t-h(t)),\ \dot{e}^T(t-h),\ e(t) - h_D(t)e^T(t-h(t)), \\
&h_D(t)e^T(t-h(t)) - e^T(t-h) \Big].
\end{aligned}
$$

By combining Lemmas 14.1 and 14.2, if LMI (14.31) holds, the integral term in Eq. (14.35) can be estimated as

$$
\begin{aligned}
&-h\int_{t-h}^{t} \dot{e}^T(s)R\dot{e}(s)ds \\
&= -h\left(\int_{t-h-h(t)}^{t} \dot{e}^T(s)R\dot{e}(s)ds + \int_{t-h}^{t-h(t)} \dot{e}^T(s)R\dot{e}(s)ds\right) \\
&= -\frac{h}{h(t)}\begin{bmatrix} e(t) - e(t-h(t)) \\ e(t) + e(t-h(t)) - \frac{2}{h(t)}\int_{t-h(t)}^{t} e(s)ds \end{bmatrix}^T \\
&\quad \times \mathrm{diag}\{R, 3R\}\begin{bmatrix} e(t) - e(t-h(t)) \\ e(t) + e(t-h(t)) - \frac{2}{h(t)}\int_{t-h(t)}^{t} e(s)ds \end{bmatrix} \\
&\quad - \frac{h}{h-h(t)}\begin{bmatrix} e(t-h(t)) - e(t-h) \\ e(t-h(t)) + e(t-h) - \frac{2}{h-h(t)}\int_{t-h}^{t-h(t)} e(s)ds \end{bmatrix}^T \\
&\quad \times \mathrm{diag}\{R, 3R\}\begin{bmatrix} e(t-h(t)) - e(t-h) \\ e(t-h(t)) + e(t-h) - \frac{2}{h-h(t)}\int_{t-h}^{t-h(t)} e(s)ds \end{bmatrix} \\
&\le -\begin{bmatrix} e(t) - e(t-h(t)) \\ e(t) + e(t-h(t)) - \frac{2}{h(t)}\int_{t-h(t)}^{t} e(s)ds \\ e(t-h(t)) - e(t-h) \\ e(t-h(t)) + e(t-h) - \frac{2}{h-h(t)}\int_{t-h}^{t-h(t)} e(s)ds \end{bmatrix}^T \left[\begin{array}{c|c} \begin{bmatrix} R & 0 \\ 0 & 3R \end{bmatrix} & S \\ \hline * & \begin{bmatrix} R & 0 \\ 0 & 3R \end{bmatrix} \end{array}\right] \\
&\quad \times \begin{bmatrix} e(t) - e(t-h(t)) \\ e(t) + e(t-h(t)) - \frac{2}{h(t)}\int_{t-h(t)}^{t} e(s)ds \\ e(t-h(t)) - e(t-h) \\ e(t-h(t)) + e(t-h) - \frac{2}{h-h(t)}\int_{t-h}^{t-h(t)} e(s)ds \end{bmatrix}.
\end{aligned} \tag{14.37}
$$

Using Eq. (14.37) and adding Eqs. (14.20)–(14.22) to $\dot{\bar{V}}(t)$ leads the new upper bound of $\dot{\bar{V}}(t)$ as follows

$$
\dot{\bar{V}}(t) \le \bar{\zeta}^T(t)(\Upsilon_{1,[h(t),\dot{h}(t)]} + \beta\Upsilon_2^T\Upsilon_2)\bar{\zeta}(t), \tag{14.38}
$$

where $\Upsilon_{1,[h(t),\dot{h}(t)]}$ and Υ_2 are defined in Theorem 14.2 and

$$\bar{\zeta}^T(t) = \Big[e^T(t),\ e^T(t-h(t)),\ e^T(t-h),\ \tfrac{1}{h(t)}\int_{t-h(t)}^{t} e^T(s)ds,$$
$$\tfrac{1}{h-h(t)}\int_{t-h}^{t-h(t)} e^T(s)ds,\ \dot{e}^T(t),\ \dot{e}^T(t-h(t)),\ \dot{e}^T(t-h),\ f^T(e(t)),\ p^T(t) \Big].$$

By the same procedure to Theorem 14.1, it is clear that $\Upsilon_{1,[h(t),\dot{h}(t)]} + \beta \Upsilon_2^T \Upsilon_2 < 0$ is equivalent to the LMI (14.30). In addition, it is true LMI (14.30) is affinely dependent on $h(t)$ and $\dot{h}(t)$ independently, so Theorem 14.2 guarantee the synchronization of transmitter (14.27) and receiver (14.28). This completes the proof. ∎

When we consider no uncertainties case, the transmitter and receiver can be

$$Transmitter: \begin{cases} \dot{x}(t) = Ax(t) + Bf(x(t)) - Km_e(t) \\ v_x(t) = Cx(t-h(t)) + m_e(t) \end{cases} \tag{14.39}$$

$$Receiver: \begin{cases} \dot{y}(t) = Ay(t) + Bf(y(t)) + u(t) \\ v_y(t) = Cy(t-h(t)) \end{cases} \tag{14.40}$$

then we derive the following theorem.

Corollary 14.2 *For given positive constants h, μ, l, ρ, the feedback controller (14.3) guarantees robust synchronization between chaotic systems in transmitter (14.39) and receiver (14.40), if there exist a positive constant α, matrices $P \in \mathbb{S}_+^{5n}$, $Q_1, Q_2 \in \mathbb{S}_+^{2n}$, $R, W_1, W_2 \in \mathbb{S}_+^{n}$, $S \in \mathbb{R}^{2n\times 2n}$, $H \in \mathbb{R}^{n\times n}$, $J \in \mathbb{R}^{n\times 1}$ satisfying LMI (14.31) and the following LMI: for $h(t) \in \{0, h\}$ and $\dot{h}(t) \in \{-\mu, \mu\}$*

$$\begin{bmatrix} \Gamma_{1,[h(t)]} & \Gamma_{2,[\dot{h}(t)]} & \Gamma_3 & P_{44}^T + 6R & P_{45} + 2(S_2 + S_4) & \Gamma_4 & h_D(t)(P_{12} + W_1) & P_{13} & HB \\ * & \Gamma_{5,[\dot{h}(t)]} & \Gamma_{6,[\dot{h}(t)]} & \Gamma_{7,[h(t),\dot{h}(t)]} & \Gamma_{8,[h(t),\dot{h}(t)]} & P_{12}^T + W_1 + \rho C^T J^T & \Gamma_9 & P_{23} + W_2 & 0 \\ * & * & \Gamma_{10} & -P_{45}^T - 2S_3^T + 2S_4^T & 6R - P_{55}^T & P_{13}^T & h_D(t)(P_{23}^T + W_2) & P_{33} - Q_{22} - W_2 & 0 \\ * & * & * & -12R & -4S_4 & P_{14}^T & h_D(t)P_{24}^T & P_{34}^T & 0 \\ * & * & * & * & -12R & P_{15}^T & h_D(t)P_{25}^T & P_{35}^T & 0 \\ * & * & * & * & * & \Gamma_{11} & 0 & 0 & \rho HB \\ * & * & * & * & * & * & D(t)(-Q_{13} - hW_1 + hW_2) & 0 & 0 \\ * & * & * & * & * & * & * & -Q_{23} - hW_2 & 0 \\ * & * & * & * & * & * & * & * & -\alpha I \end{bmatrix} < 0, \tag{14.41}$$

where the notations are the same with Theorem 14.2. Also, the desired control gain matrix is given by $K = H^{-1}J$.

Proof Corollary 14.2 can be easily proven by following the same procedure of Theorem 14.2 without (14.21), so it is omitted here. ■

14.4 Numerical Example with Simulation

To show the effectiveness of the proposed method, Chua's circuit [34] are considered which is described by following parameters:

$$A = \begin{bmatrix} -am_1 & a & 0 \\ 1 & -1 & 1 \\ 0 & -b & 0 \end{bmatrix}, \quad B = \begin{bmatrix} -a(m_0 - m_1) & 0 & 0 \\ 0 & 0 & 0 \\ 0 & 0 & 0 \end{bmatrix},$$

$$C = \begin{bmatrix} 1 & 0 & 9 \end{bmatrix}, \quad f(a) = \frac{1}{2}(|a + c| - |a - c|),$$

with the parameters are $a = 9, b = 14.28, c = 1, m_0 = -1/7, m_1 = 2/7$, and the nonlinear function $f(\cdot)$ satisfies the Lipschitz condition with $l = 0.5$.

The parameters associated with system uncertainties are given $D = 0.1I_n$, $E_a = 0.3I_n$, $E_b = 0.4I_n$, $G(t) = 0.4 + 0.2\sin t$, and initial conditions are chosen as $x(0) = [-0.1\ \ -0.5\ \ -0.7]$, $y(0) = [-0.1\ \ -0.4\ 0.3]$.

For secure communication, we consider the constant and time-varying propagation delay as $h = 0.1$ and $h(t) = 0.05\sin(5t) + 0.05$, respectively, where it is clear $0 \leq h(t) \leq h = 0.1$ and $|\dot{h}(t)| \leq \mu = 0.3$. Then the control gains can be calculated by Theorems 14.1 and 14.2, we can obtain the following control gains

$$\text{Theorem 14.1}: K = \begin{bmatrix} 0.8455 & 0.1590 & -0.9377 \end{bmatrix}^T,$$

$$\text{Theorem 14.2}: K = \begin{bmatrix} 0.9780 & 0.1828 & -1.1069 \end{bmatrix}^T.$$

Consider the original message as $m_o(t) = 0.15\cos(\pi t)$, parameters for encryption/decryption key signal as $a_1 = 0.05$, $a_2 = 0.02$, $a_3 = 0.01$, then we can choose $\tau = 0.2$ to meet $|m_o(t)| < \tau$ and $|k_e(t)| < \tau$. With above parameters, the state trajectories of the chaotic system in transmitter, the original message signal $m_o(t)$, the encryption key signal $k_e(t)$, and the encrypted message signal are shown in Figs. 14.1, 14.2, 14.3 and 14.4, respectively, in which we can confirm both conditions, $|m_o(t)| < \tau$ and $|k_e(t)| < \tau$ by Figs. 14.2 and 14.3.

Figures 14.5 and 14.7 display the state trajectories of the error signals by Theorem 14.1 and 14.2, respectively. As seen in Figs. 14.5 and 14.7, the error signal converse to zero as time goes to infinity which means the synchronization between the chaotic systems in transmitter and receiver is achieved by the designed controller. Finally, the recovered message signal $m_r(t)$ by decryption function (14.9) and error between

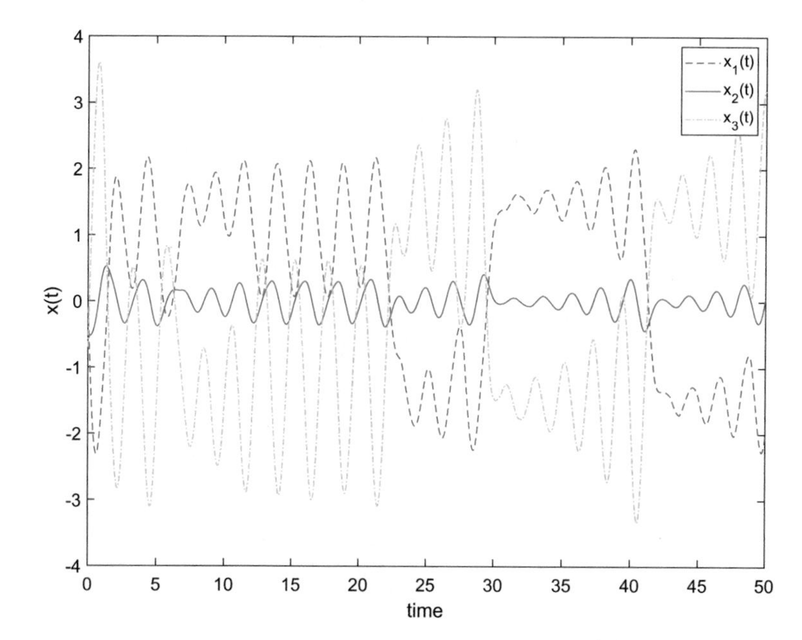

Fig. 14.1 The state trajectories of the transmitter

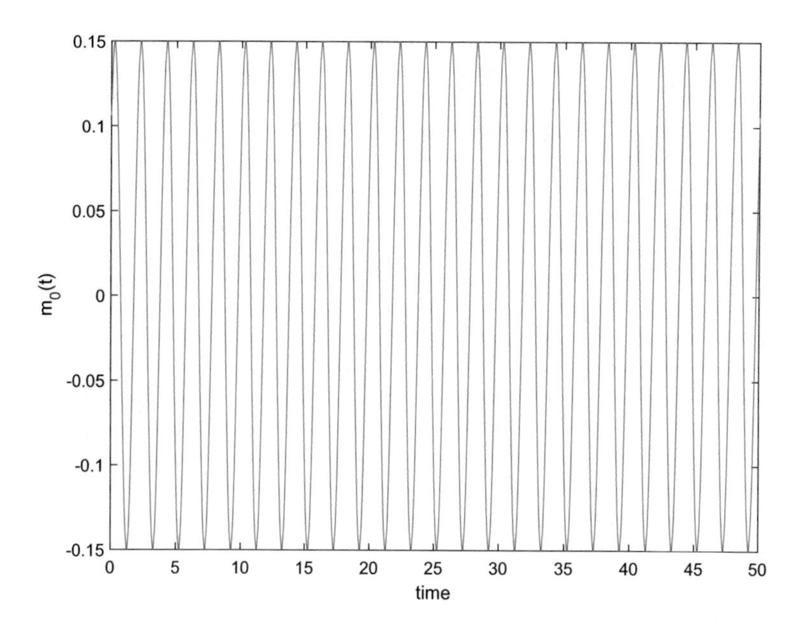

Fig. 14.2 The original message signal

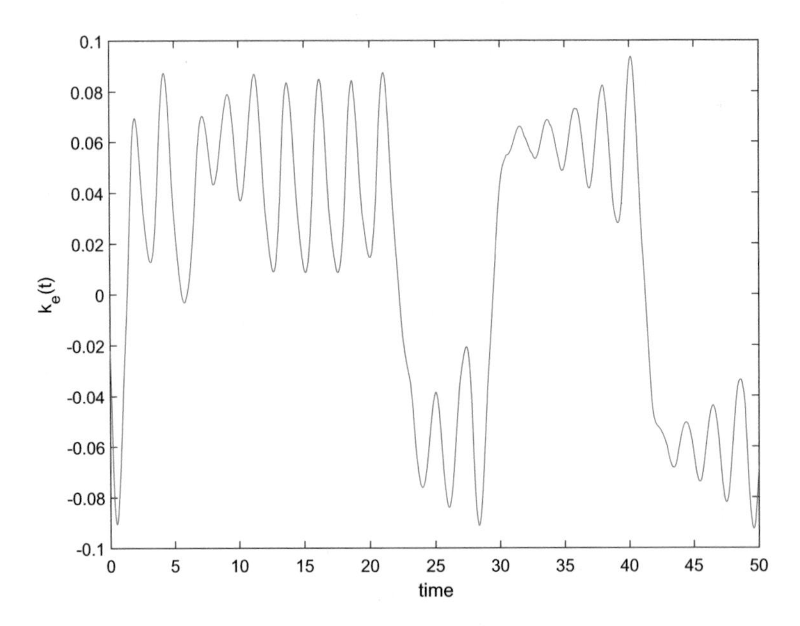

Fig. 14.3 The encryption key signal

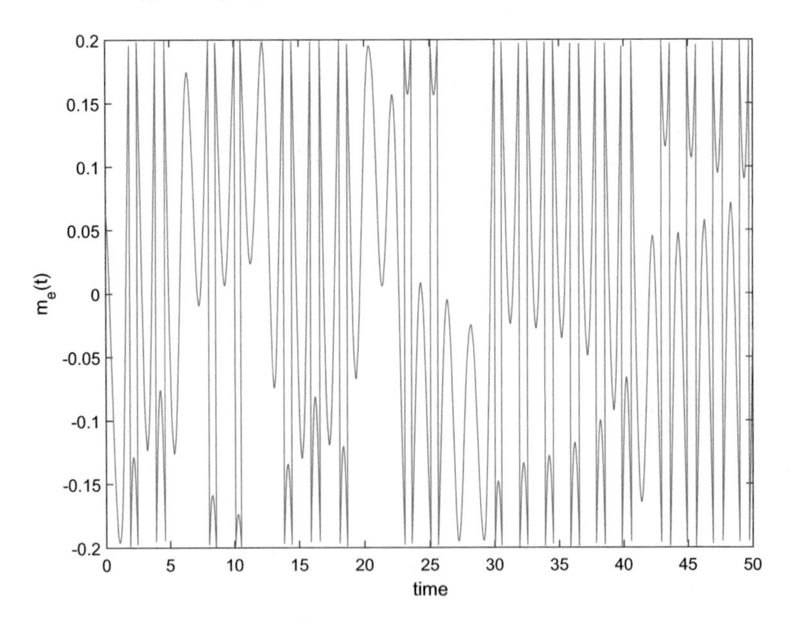

Fig. 14.4 The encrypted message signal

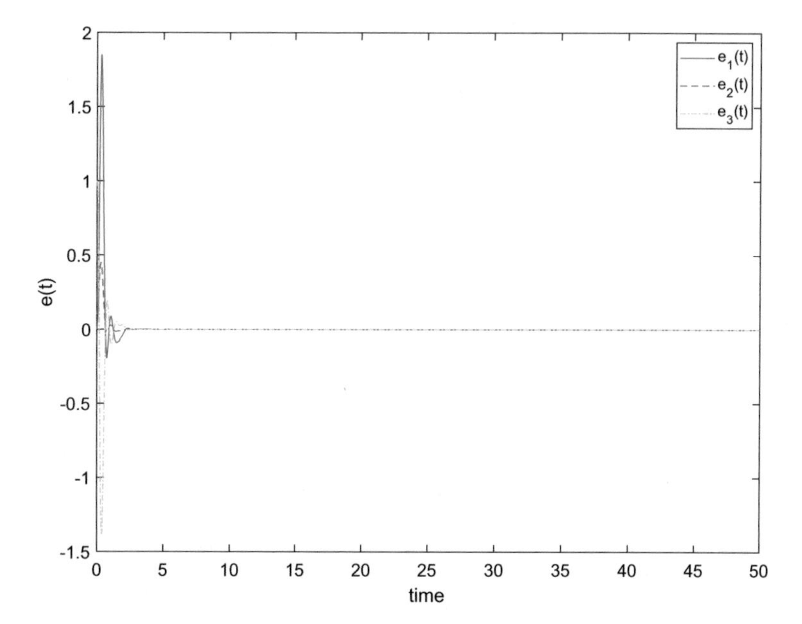

Fig. 14.5 The error signals by Theorem 14.1

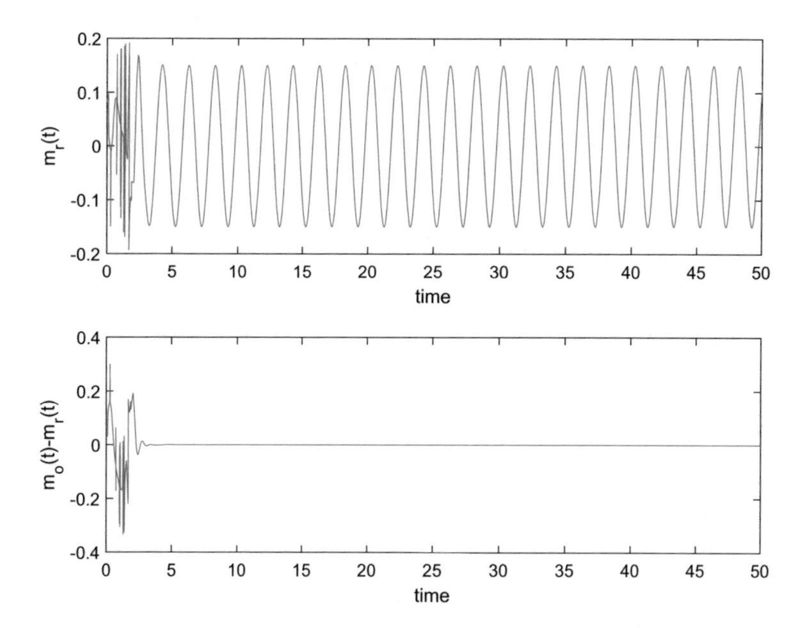

Fig. 14.6 The recovered message signal and the error between original and recovered message signals by Theorem 14.1

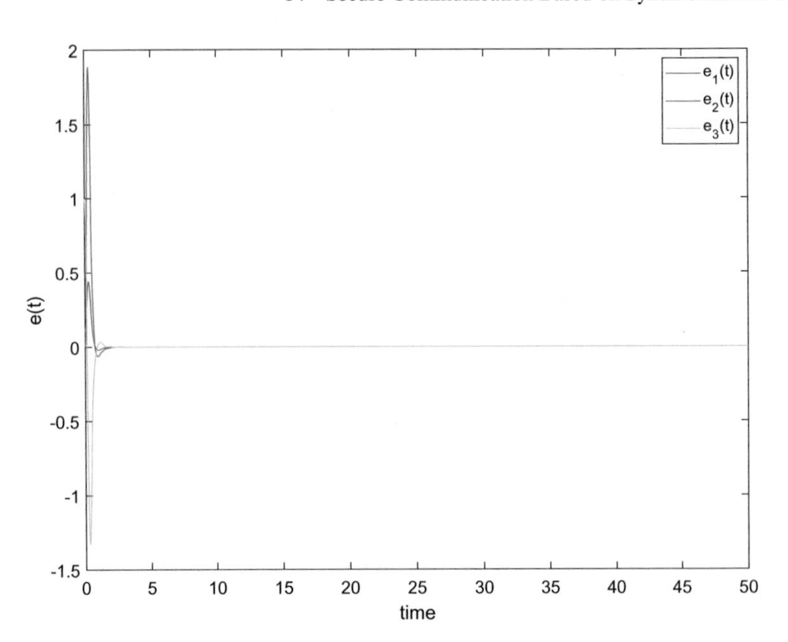

Fig. 14.7 The error signals by Theorem 14.2

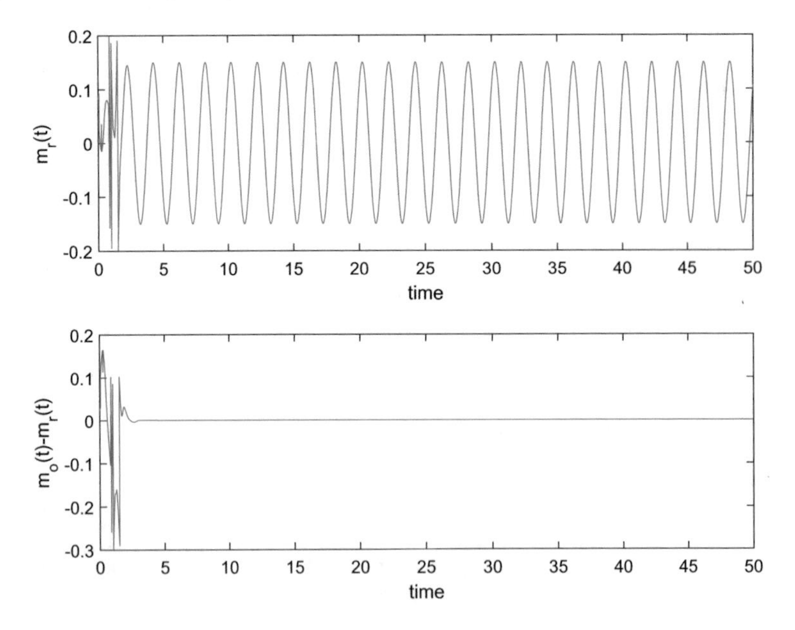

Fig. 14.8 The recovered message signal and the error between original and recovered message signals by Theorem 14.2

original and recovered message signal $m_o(t) - m_r(t)$ by Theorems 14.1 and 14.2 are depicted in Figs. 14.6 and 14.8, respectively, which implies that we can recover the original message successfully.

14.5 Conclusion

This chapter was concerned with designing a state feedback controller for secure communication systems. Secure communication was constructed based on chaotic synchronization and both constant and time-varying propagation delay were considered in the signal transmission. For higher security of the communication system, the original message was encrypted by N-shift cipher scheme. The Wirtinger-based multiple integral inequality was employed for the constant propagation delay case and the theorems for time-varying propagation delay case utilized a new Lyapunov functional based on matrix-refined function. Furthermore, parameter uncertainties were taken into account for the practice. Finally, designing conditions for the controller were derived in terms of LMIs. Through a numerical example, we have shown the aim of this chapter successfully accomplished.

References

1. Pasqualetti F, Dörfler F, Bullo F (2013) Attack detection and identification in cyber-physical systems. IEEE Trans Autom Control 58:2715–2729
2. Fawzi H, Tabuada P, Diggavi S (2014) Secure estimation and control for cyber-physical systems under adversarial attacks. IEEE Trans Autom Control 59:1454–1467
3. Hug G, Giampapa JA (2012) Vulnerability assessment of AC state estimation with respect to false data injection cyber-attacks. IEEE Trans Smart Grid 3:1362–1370
4. Ye N, Zhang Y, Borror CM (2004) Robustness of the Markov-chain model for cyber-attack detection. IEEE Trans Reliab 53:116–123
5. Zhang H, Cheng P, Shi L, Chen J (2015) Optimal denial-of-service attack scheduling with energy constraint. IEEE Trans Autom Control 60:3023–30288
6. Wang Z, Wang D, Shen B, Alsaadi FE (2018) Centralized security-guaranteed filtering in multirate-sensor fusion under deception attacks. J Frankl Inst 355:406–420
7. Teixeira A, Shames I, Sandberg H, Johansson KH (2015) A secure control framework for resource-limited adversaries. Automatica 51:135–148
8. Sandberg H, Amin S, Johansson KH (2015) Cyberphysical security in networked control systems: an introduction to the issue. IEEE Control Syst Mag 35:20–23
9. Yuan Y, Zhang P, Guo L, Yang HJ (2017) Towards quantifying the impact of randomly occurred attacks on a class of networked control systems. J Frankl Inst 354:4966–4988
10. Sun YC, Yang GH (2018) Periodic event-triggered resilient control for cyber-physical systems under denial-of-service attacks. J Frankl Inst 355:5613–5631
11. Liu J, Gu Y, Cao J, Fei S (2018) Distributed event-triggered $\mathscr{H}_\infty$ filtering over sensor networks with sensor saturations and cyber-attacks. ISA Trans 81:63–75
12. Zhang W, Wang Z, Liu Y, Ding D, Alsaadi FE (2018) Sampled-data consensus of nonlinear multiagent systems subject to cyber attacks. Int J Robust Nonlinear Control 28:53–67
13. Liao TL, Tsai SH (2000) Adaptive synchronization of chaotic systems and its application to secure communications. Chaos, Solitons Fractals 11:1387–1396

14. Wu CW, Chua LO (1993) A simple way to synchronize chaotic systems with application to secure communication systems. Int J Bifurc Chaos 3:1619–1627
15. Kwon OM, Park JH, Lee SM (2011) Secure communication based on chaotic synchronization via interval time-varying delay feedback control. Nonlinear Dyn 63:239–252
16. Lakshmanan S, Prakash M, Lim CP, Rakkiyappan R, Balasubramaniam P, Nahavandi S (2018) Synchronization of an inertial neural network with time-varying delays and its application to secure communication. IEEE Trans Neural Netw Learn Syst 29:195–207
17. Yang T, Wu CW, Chua LO (1997) Cryptography based on chaotic systems. IEEE Trans Circuits Syst I: Fundam Theory Appl 44:469–472
18. Lee TH, Lim CP, Nahavandi S, Park JH (2018) Network-based synchronization of T-S Fuzzy chaotic systems with asynchronous samplings. J Frankl Inst 355:5736–5758
19. Lee TH, Park JH (2017) Improved sampled-data control for synchronization of chaotic Lur'e systems using two new approaches. Nonlinear Anal: Hybrid Syst 24:132–145
20. Ji DH, Jeong SC, Park JH, Won SC (2012) Robust adaptive backstepping synchronization for a class of uncertain chaotic systems using fuzzy disturbance observer. Nonlinear Dyn 69:1125–1136
21. Park JH (2009) Further results on functional projective synchronization of Genesio-Tesi chaotic system. Mod Phys Lett B 24:1889–1895
22. Park JH, Lee SM, Kwon OM (2007) Adaptive synchronization of Genesio-Tesi chaotic system via a novel feedback control. Phys Lett A 371:263–270
23. Pecora LM, Carroll TL (1990) Synchronization in chaotic systems. Phys Rev Lett 64:821–825
24. Boccaletti S, Kurths J, Osipov G, Valladares DL, Zhou SC (2002) The synchronization of chaotic systems. Phys Rep 366:1–101
25. Lee TH, Park JH, Lee SM, Kwon OM (2013) Robust synchronisation of chaotic systems with randomly occurring uncertainties via stochastic sampled-data control. Int J Control 86:107–119
26. Rulkov NF, Sushchik MM, Tsimring LS, Abarbanel HDI (1995) Generalized synchronization of chaos in directionally coupled chaotic systems. Phys Rev E 51:980–994
27. Rosenblum MG, Pikovsky AS, Kurths J (1996) Phase synchronization of chaotic oscillators. Phys Rev Lett 76:1804–1807
28. Seuret A, Gouaisbaut F (2013) Wirtinger-based integral inequality: application to time-delay systems. Automatica 49:2860–2866
29. Park MJ, Kwon OM, Park JH, Lee SM, Cha EJ (2015) Stability of time-delay systems via Wirtinger-based double integral inequality. Automatica 55:204–208
30. Lee TH, Park MJ, Park JH, Kwon OM, Jung HY (2015) On stability criteria for neural networks with time-varying delay using Wirtinger-based multiple integral inequality. J Frankl Inst 352:5627–5645
31. Park PG, Ko JW, Jeong C (2011) Reciprocally convex approach to stability of systems with time-varying delays. Automatica 47:235–238
32. Gu K, Kharitonov VL, Chen J (2003) Stability of time-delay systems. Birkhauser, Basel
33. Lee TH, Park JH (2018) Improved stability conditions of time-varying delay systems based on new Lyapunov functionals. J Frankl Inst 355:1176–1191
34. Chua LO, Komuro M, Matsumoto T (2000) The double scroll family. IEEE Trans Circuits Syst I 33:1072–1118

Index

J. H. Park et al., *Dynamic Systems with Time Delays: Stability and Control*,
https://doi.org/10.1007/978-981-13-9254-2

Printed in the United States
By Bookmasters